Metacyclic Groups and the D(2) Problem

Metacyclic Groups and the D(2) Problem

Francis E A Johnson
University College London, UK

NEW JERSEY · LONDON · SINGAPORE · BEIJING · SHANGHAI · HONG KONG · TAIPEI · CHENNAI · TOKYO

Published by

World Scientific Publishing Co. Pte. Ltd.
5 Toh Tuck Link, Singapore 596224
USA office: 27 Warren Street, Suite 401-402, Hackensack, NJ 07601
UK office: 57 Shelton Street, Covent Garden, London WC2H 9HE

Library of Congress Control Number: 2020053073

British Library Cataloguing-in-Publication Data
A catalogue record for this book is available from the British Library.

METACYCLIC GROUPS AND THE D(2) PROBLEM

ISBN 978-981-122-275-7 (hardcover)
ISBN 978-981-122-276-4 (ebook for institutions)
ISBN 978-981-122-277-1 (ebook for individuals)

For any available supplementary material, please visit
https://www.worldscientific.com/worldscibooks/10.1142/11897#t=suppl

Desk Editor: Soh Jing Wen

Typeset by Stallion Press
Email: enquiries@stallionpress.com

Printed in Singapore

Preface

Perhaps the place to start this book is with an explanation of the title. The $D(2)$ problem, the second named item, is a fundamental problem in low dimensional topology. It is stated formally in the Introduction and we will not anticipate that here except to make the general point that its precise formulation depends on having made a specific choice of group G which plays the role of fundamental group throughout. On the basis of accumulated evidence, its difficulty varies on a case-by-case basis with the nature of G; the more complicated the group, the more difficult the problem.

This book is a continuation of the author's work in this area ([30], [31], [32], [33], [35]). In that sense the $D(2)$ problem is its motivating principle. However, that may not be immediately apparent; rather like Fortinbras in another context, it receives only a brief mention at the beginning and is otherwise absent until it makes its appearance at the very end. Whether also to survey the resulting carnage we leave for the reader to decide.

The rest of the book is principally concerned with the first item in the title, namely finite metacyclic groups. These are the semi-direct products $G(p,q) = C_p \rtimes C_q$ where p is an odd prime, q is a positive integral divisor of $p-1$ and C_q acts via the canonical imbedding $C_q \hookrightarrow \mathrm{Aut}(C_p) \cong C_{p-1}$. In its original conception this book was envisaged simply as an *ad hoc* verification that the $D(2)$-property is satisfied for what was then a comparatively small collection of finite metacyclic groups. However, as the collection grew and complications multiplied, it became clear that a more natural setting for such an investigation was a systematic application of homological algebra to the integral representation theory of metacyclic groups. And, for the most part, that is what this book is about.

I have endeavoured to write throughout at a level suitable for a committed graduate student The first four chapters are preliminary; thereafter the material is progressively specialized. In treating the preliminary material I have taken a number of prerequisites for granted. As regards the theory of rings and modules I have assumed a standard knowledge of the most elementary aspects; other slightly more advanced topics for which there are excellent published accounts, Morita's Theorem and Nakayama's Lemma for example, are treated summarily where they occur. There are other some other implicit assumptions; Galois theory; familiarity with categories and functors; an acquaintance with the basics of Algebraic K-theory.

Chapter One is a summary of topics which are too important to omit entirely but which considerations of space require us to treat only briefly. For all these there are excellent published accounts; the book of Fröhlich and Taylor [21] for Steinitz' Theorem; Pierce [50] or Bourbaki [7] for the theory of Wedderburn-Maschke; Milnor's own account [46] for the fibre square description of projective modules. A problem throughout is knowing when modules can be cancelled from direct sums. For this the theorem of Swan-Jacobinski [28], [62] is absolutely fundamental. Our summary of it is inevitably cursory and is no substitute for Swan's comprehensive account.

Chapter Two covers the relevant aspects of homological algebra from a conventional standpoint, in particular, the cohomological interpretation of Baer's theory of module extensions. Chapter Three covers the same general area from a less familiar standpoint, namely the theory of the derived module category. This is a distinctive feature of the subsequent development; almost all the arguments are expressed in these terms. For a parallel treatment in a simpler setting the reader may wish to compare Carlson's book [9] wherein the theory of modular representations is developed from this standpoint.

As mentioned, we have assumed familiarity with the standard representation theory of Wedderburn and Maschke over fields of characteristic zero. However, the theory of integral representations, within which much of this book is situated, is altogether more complicated. Chapter Four is a survey of some rudimentary aspects in this more general context. We would also refer the reader to the definitive treatise of Curtis and Reiner [14].

The author wishes to record his thanks to a number of former students whose various theses and dissertations provided worked examples on which to base more general considerations; in historical order, Jonathan Remez [53], Laura Guthrie [23], Jamil Nadim [48], Gemma Eastlund [16], John Evans [18], Jason Vittis [65].

Finally, the author is especially grateful to his colleague Dr. R.M. Hill for clarifying the status of a certain number theoretic hypothesis, $\mathrm{Inj}(p, q)$, which appears in Chapter Fourteen. Dr Hill's analysis is explained briefly in Chapter Fourteen and set out in detail in Appendix B.

I am indebited to the staff of World Scientific; in particular to Dr Lim Swee Cheng whose initial interest was the catalyst for converting hitherto unstructured notes into definite form. Especially I am grateful to my copy editor Ms Soh Jing Wen, whose diligence was of enormous help in proof reading. Needless to say, for what remains the author takes sole responsibility.

F.E.A. Johnson

University College London
October 23, 2020

Contents

Introduction

Let G be a finitely presented group and X a finite connected 3-complex with $\pi_1(X) = G$ and universal cover $\widetilde{X}$ such that for all local coefficient systems $\mathcal{B}$ on X

$$H^3(X, \mathcal{B}) = H_3(\widetilde{X}, \mathbf{Z}) = 0.$$

The $D(2)$-problem for G (cf [66], also [31], [33]) asks the following question:

D(2) problem: Does there exist a finite 2-complex K such that $X \simeq K$?

We say that G has the $D(2)$-property when the answer to (*) is always 'yes'.

In this book we consider the metacyclic groups $G(p, q)$ where p is an odd prime, q is a positive integral divisor of $p - 1$ and C_q acts via the canonical imbedding $C_q \hookrightarrow \mathrm{Aut}(C_p) \cong C_{p-1}$. We introduce two numerical conditions which guarantee that the $D(2)$-property holds for $G(p, q)$. The first is a straightforward restriction on the projective class group $\widetilde{K}_0(\mathbb{Z}[C_q])$, namely :

(*) $\widetilde{K}_0(\mathbb{Z}[C_q]) = 0$.

The second is more subtle; let $\zeta_p = \exp(2\pi i/p)$ and let $A = \mathbb{Z}(\zeta_p)^{C_q}$ be the fixed point ring under the Galois action of C_q on the ring of integers of the cyclotomic field $\mathbb{Q}(\zeta_p)$. The correspondence $\mathfrak{a} \mapsto \mathfrak{a} \cdot \mathbb{Z}(\zeta_p)$ defines a homomorphism of ideal class groups $\nu : Cl(A) \to Cl(\mathbb{Z}(\zeta_p))$. Our second condition is:

(**) The natural homomorphism $\nu : Cl(A) \to Cl(\mathbb{Z}(\zeta_p))$ is injective.

We shall prove:

Main Theorem: If conditions (*) and (**) hold then $G(p, q)$ has the $D(2)$-property.

The groups $G(p, q)$

Throughout $\mathbb{F}_p$ will denote the field with p elements where p is an odd prime. For most purposes it will suffice to describe the group $G(p, q)$ by means of the following naive presentation with two generators and three relations:

$$G(p, q) = \langle x, y | yxy^{-1} = x^a, x^p = 1, y^q = 1 \rangle;$$

here p is an odd prime, q is an integral divisor of $p-1$ and $a \in \{2, \ldots, p-1\}$ is an integer whose residue class mod p has order q in $\mathbb{F}_p^*$. However we will eventually need to describe $G(p, q)$ using only two relations. We accomplish this by choosing an integer m such that $(a-1) \cdot m \equiv 1 (\mathrm{mod}\ p)$. Then as we shall see in Chapter Fourteen, a special case of the theorem of Wamsley [67] shows that the following presentation $\mathcal{G}(p, q; m)$ also defines $G(p, q)$:

$$\mathcal{G}(p, q; m) = \langle X, Y | Y X^m Y^{-1} = X^{m+1}, X^p = Y^q \rangle.$$

Clearly $G(p, q)$ occurs in an exact sequence $1 \to C_p \to G(p, q) \to C_q \to 1$ which in turn gives a cohomology spectral sequence ([42] p. 351)

$$E_2^{r,s} = H^r(C_q, H^s(C_p, \mathbb{Z})) \Longrightarrow H^{r+s}(G, \mathbb{Z}).$$

It is a straightforward deduction from the spectral sequence that $G(p, q)$ has cohomological period $2q$. Writing $\Lambda = \mathbb{Z}[G(p, q)]$ this statement is realized in the form of an exact sequence

(0.1) $$\mathcal{P} = (0 \to \mathbb{Z} \to P_{2q-1} \to P_{2q-2} \to \cdots \to P_1 \to P_0 \to \mathbb{Z} \to 0)$$

where each P_i is a finitely generated projective Λ-module and $\mathbb{Z}$ denotes the Λ-module whose underlying abelian group is the integers on which Λ acts trivially. We note that, beyond the above general insight, spectral sequences play no further part in our considerations as, in Chapter Nine, we shall construct such sequences directly. In fact, a stronger statement than (0.1) is true as the groups $G(p, q)$ also have free period $2q$; that is there is an exact sequence of Λ homomorphisms

(0.2) $$\mathcal{F} = (0 \to \mathbb{Z} \to F_{2q-1} \to F_{2q-2} \to \cdots \to F_1 \to F_0 \to \mathbb{Z} \to 0)$$

in which each F_i is a finitely generated and free over Λ. This is by no means obvious; there are examples where the free period exceeds the cohomological period ([5], [47]). When q is odd it is a consequence of a topological theorem of Petrie [49], proved by showing the vanishing of certain surgery obstruction. However, a celebrated theorem of Milnor [44] presents a fundamental obstruction to Petrie's method when q is even. By contrast, the proof given in Chapter Nine, which simplifies that in [38], is purely algebraic and holds for all integral divisors of $p-1$, odd or even.

The derived module category

Turning now to the methods employed in this book the reader will find that they are largely ring and module theoretic. Typically the rings encountered are subrings and quotient rings of group rings $R[G]$ where R is a commutative Noetherian ring and G is finite. Such rings are necessarily Noetherian and it will cause us no difficulty if we agree from the outset to impose this condition on all the rings that we consider. Thus the symbol Λ, whenever encountered, will be assumed to be a Noetherian ring.

We shall interpret homological algebra over Λ via its *derived module category*; that is, the quotient $\mathcal{D}\mathrm{er}(\Lambda)$ of the category of right Λ-modules by the relation '*projective module* $\equiv 0$'. This is entirely natural given that projective modules are generally invisible to the standard constructions of homological algebra; for example, when P is projective and $n \geq 1$ then $\mathrm{Ext}^n(P, -) \equiv 0$. In this connection we introduce a relation '$\approx$' on Λ-modules M_1, M_2 as follows:

$$M_1 \approx M_2 \iff M_1 \oplus P_1 \cong_\Lambda M_2 \oplus P_2$$

for some finitely generated projective Λ-modules P_1, P_2. As we shall see, the criterion for finitely generated Λ-modules M_1, M_2 to become isomorphic in $\mathcal{D}\mathrm{er}(\Lambda)$ is :

(0.3) $$M_1 \cong_{\mathcal{D}\mathrm{er}(\Lambda)} M_2 \iff M_1 \approx M_2.$$

We write $\langle M \rangle$ for the equivalence class of M under '$\approx$'. Given an exact sequence

$$0 \to J \to P \to M \to 0$$

where P is finitely generated projective, then Schanuel's Lemma (see below, p. 20) implies that the equivalence class of J under '$\approx$' depends only upon M and we put

$$D_1(M) = \langle J \rangle.$$

$D_1(M)$ is the *first generalized syzygy* of M. It is straightforward to see that

(0.4) $$M \approx M' \Longrightarrow D_1(M) = D_1(M').$$

Thus we may iterate the construction and define the nth *generalized syzygy* of M by

$$D_n(M) = D_1(D_{n-1}(M)).$$

For the modules M, N we encounter we shall see the that following relation holds:

$$\mathrm{Ext}^n(M, N) \cong \mathrm{Hom}_{\mathcal{D}\mathrm{er}}(D_n(M), N)$$

that is, $\mathrm{Ext}^n(M, -)$ is corepresentable in the derived module category by $D_n(M)$.

Stable modules and syzygies

There is an analogous equivalence relation to '$\approx$' and with which it is frequently confused, namely the stability relation '$\sim$' defined on Λ modules M_1, M_2 by

$$M_1 \sim M_2 \Longleftrightarrow M_1 \oplus \Lambda^{n_1} \cong M_2 \oplus \Lambda^{n_2}$$

for some integers $n_1, n_2 \geq 0$. For any Λ-module M, we denote by $[M]$ the corresponding *stable module*; that is, the set of isomorphism classes of modules N such that $N \sim M$. Clearly $M_1 \sim M_2 \Longrightarrow M_1 \approx M_2$ but, as we shall see, the converse is false in general. We note that:

(0.5) M is finitely generated if and only if each $N \in [M]$ is finitely generated.

When M is a finitely generated Λ-module, the stable module $[M]$ can be represented graphically in the manner introduced by Dyer and Sieradski [15] as follows; take the vertices of $[M]$ to be the isomorphism classes of modules $N \in [M]$ and draw an edge $N_1 \to N_2$ whenever $N_2 \cong N_1 \oplus \Lambda$. As each module $N \in [M]$ has a unique arrow which exits the vertex represented by N, namely the arrow $N \to N \oplus \Lambda$, it follows that the only way of having a nontrivial loop in $[M]$ would be if $N \cong N \oplus \Lambda^a$ for some $a > 0$, which is clearly impossible when Λ is Noetherian; thus we have:

(0.6) If M is a finitely generated Λ-module $[M]$ is an infinite (directed) tree.

The trees which arise this way are far from arbitrary. Thus if M is a finitely generated Λ-module we say that $M_0 \in [M]$ is a *minimal module* for $[M]$ when given $N \in [M]$ there exist integers a, b such that $0 \leq a \leq b$ such that

$$N \oplus \Lambda^a \cong M_0 \oplus \Lambda^b.$$

It is a straightforward deduction from the Noetherian condition that when M is finitely generated over Λ the stable class $[M]$ contains a minimal module. That is, $[M]$ is a tree with roots which *do not extend infinitely downwards.* As examples, consider the following diagrams; **(A)** below represents a tree with a single root and no branching above level two; **(B)** represents a tree with three roots but with no branching above level one.

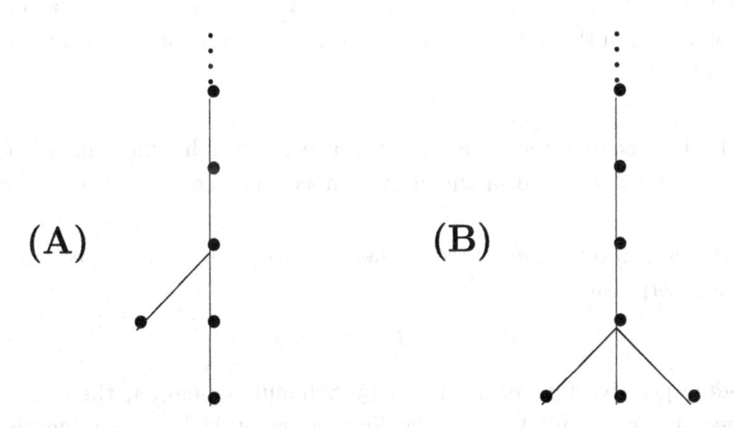

Both of these examples actually arise as stable modules; if $Q(4n)$ denotes the quaternion group of order $4n$ then, from the calculations of Swan [63], **(A)** represents the stable class of 0 over the integral group ring $\mathbb{Z}[Q(32)]$ whilst **(B)** represents the stable class of the augmentation ideal over $\mathbb{Z}[Q(24)]$. Of greater subsequent interest to us is the case where a stable module $[M]$ is *straight*; that is, when it has a single root and no branching as in **(C)** below:

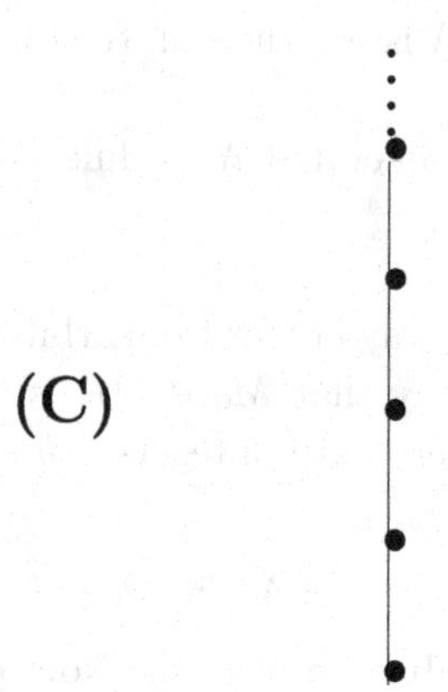

Explicitly, $[M]$ is *straight* when it has a single minimal module M_0 so that

$$[M] = \{M_0 \oplus \Lambda^n | n \in \mathbb{N}\}.$$

We will say that a module M is straight when its stable class $[M]$ is straight. As we have observed, $M_1 \sim M_2 \Longrightarrow M_1 \approx M_2$. In Chapter Thirteen we shall give an explicit example where the converse is false. However, we note the following, in the light of which it is useful to think of $[M]$ as a '*polarized state*' of $\langle M \rangle$:

(0.7) $\langle M \rangle$ also has the structure of a directed graph which, in general, is disconnected and in which $[M]$ imbeds as a connected component.

If M is finitely generated, there exists an exact sequence of Λ-homomorphisms

(0.8) $$0 \to J \to \Lambda^a \to M \to 0$$

for some positive integer a. Again by Schanuel's Lemma, there is a well defined stable module $\Omega_1(M)$, the *first syzygy* of M(1), determined by the rule that

$$\Omega_1(M) = [J].$$

As Λ is Noetherian then J is also finitely generated. By (0.5) above it follows that every module in $\Omega_1(M)$ is finitely generated. More generally,

(1) The notation Ω was introduced by Heller ([25]) by analogy with the loopspace operator in homotopy theory.

when M is finitely generated, then for each $n \geq 0$ there exists an exact sequence

(0.9) $$0 \to J \to F_{n-1} \to \cdots \to F_1 \to F_0 \to M \to 0$$

in which each F_r is a free Λ module of finite rank, and in this case J is again finitely generated. The stable class $[J]$ then defines the nth-syzygy $\Omega_n(M)$. If $J \in \Omega_n(M)$ then $J \in D_n(M)$ but again the converse is false in general. In what follows we maintain a clear distinction between $\Omega_n(M)$ and $D_n(M)$.

The Swan homomorphism

Any Λ-module J has a surjective ring homomorphism $[\ ] : \mathrm{End}_\Lambda(J) \twoheadrightarrow \mathrm{End}_{\mathcal{D}\mathrm{er}}(J)$. In calculations, $\mathrm{End}_{\mathcal{D}\mathrm{er}}(J)$ is usually easier to compute than $\mathrm{End}_\Lambda(J)$ but comes with the complication that the induced homomorphism of unit groups

$$[\ ] : \mathrm{Aut}_\Lambda(J) \to \mathrm{Aut}_{\mathcal{D}\mathrm{er}}(J)$$

in general fails to be surjective. This leads to a significant construction, the *Swan homomorphism* (cf. Chapter Five). Thus if the Λ-module J occurs in an exact sequence

$$0 \to J \xrightarrow{i} \Lambda^m \xrightarrow{p} M \to 0$$

for some positive integer m then we may form the pushout extension

$$\begin{array}{ccccccc} 0 \to & J & \xrightarrow{i} & \Lambda^m & \xrightarrow{p} & M \to & 0 \\ & \downarrow \alpha & & \downarrow \nu & & \downarrow \mathrm{Id} & \\ 0 \to & J & \xrightarrow{i'} & \varinjlim(\alpha, i) & \xrightarrow{p'} & M \to & 0. \end{array}$$

Subject to the condition $\mathrm{Ext}^1(M, \Lambda) \equiv 0$, the further significance of which will become clear at the appropriate point, we make the following observation:

(0.10) $$\varinjlim(\alpha, i) \text{ is projective} \iff [\alpha] \in \mathrm{Aut}_{\mathcal{D}\mathrm{er}}(J)\ .$$

With this condition, and assuming J is finitely generated (which is the case when Λ is Noetherian) we obtain the *Swan mapping*

$$\begin{array}{rcl} S_J : \mathrm{Aut}_{\mathcal{D}\mathrm{er}}(J) & \to & \widetilde{K}_0(\Lambda) \\ S_J([\alpha]) & = & [\varinjlim(\alpha, i)] \end{array}$$

taking values in the reduced projective class group $\widetilde{K}_0(\Lambda)$. In Chapter Five we shall show that S_J is a homomorphism. If $\alpha : J \to J$ is an Λ-isomorphism then the natural mapping $\nu : \Lambda^m \to \varinjlim(\alpha, i)$ is an isomorphism and hence $[\varinjlim(\alpha, i)] = 0$. Thus $\mathrm{Im}([\ \ ]) \subset \mathrm{Ker}(S_J)$. We say that J is *full* when

(0.11) $$\mathrm{Ker}(S_J) = \mathrm{Im}([\,])$$

Modules over metacyclic groups

Without further qualification the theory as indicated so far works for finitely generated modules over Noetherian rings. In particular, it is directly applicable to modules over group algebras of the form $R[G]$ where R is a commutative Noetherian ring and G is a finite group. However, as we are primarily concerned with the finite metacyclic groups $G(p, q)$ we give a more detailed consideration of the theory for modules over an integral group ring $\Lambda = \mathbb{Z}[G(p, q)]$.

We denote by $\Omega_k(\mathbb{Z})$ the kth syzygy of the trivial module $\mathbb{Z}$; that is, the stable class of any module J which occurs in an exact sequence

$$0 \to J \to F_{k-1} \to F_{k-2} \to \cdots \to F_1 \to F_0 \to \mathbb{Z} \to 0$$

where each F_i is free of finite rank over Λ. To proceed further we must analyze the structure of Λ in detail. Thus put $C_p = \langle x, | x^p = 1\rangle$, and let $\zeta_p = \exp(2\pi i/p)$ and let $\pi : \mathbb{Z}[C_p] \to \mathbb{Z}(\zeta_p)$ be the ring homomorphism to the ring of cyclotomic integers $\mathbb{Z}(\zeta_p)$ given by $\pi(x) = \zeta_p$. We have an *extension of scalars* functor $i_* : \mathcal{M}\mathrm{od}_{\mathbb{Z}[C_p]} \to \mathcal{M}\mathrm{od}_\Lambda$ given by

$$i_*(M) = M \otimes_{\mathbb{Z}[C_p]} \Lambda.$$

Noting that $i_*(\mathbb{Z}[C_p]) = \Lambda$, application of i_* to π gives a surjective ring homomorphism

$$\pi_+ = i_*(\pi) : \Lambda \twoheadrightarrow i_*(\mathbb{Z}(\zeta_p)).$$

In addition, the natural projection $G(p, q) \twoheadrightarrow C_q$ gives a surjective ring homomorphism $\pi_- : \Lambda \twoheadrightarrow \mathbb{Z}[C_q]$ and the two homomorphisms combine to give a fibre product decomposition

$$(0.12)\qquad \begin{array}{ccc} \Lambda & \stackrel{\pi_+}{\to} & i_*(\mathbb{Z}(\zeta_p)) \\ \pi_- \downarrow & & \downarrow \\ \mathbb{Z}[C_q] & \to & \mathbb{F}_p[C_q] \end{array}$$

where $\mathbb{F}_p$ is the field with p elements. In this form, the above fibre product decomposition is relatively well known ([14], volume 1, p. 748). What appears to be less well known is the related decomposition arising from a different description of the ring $i_*(\mathbb{Z}(\zeta_p))$. Recall that $\mathrm{Gal}(\mathbb{Q}(\zeta_p)/\mathbb{Q}) \cong C_{p-1}$. We denote by A the fixed ring

$$A = \mathbb{Z}(\zeta_p)^{C_q}$$

under the Galois action of C_q. Then A is a Dedekind domain in which the prime p ramifies completely. Denote by $\mathfrak{p}$ the unique prime in A over p and define $\mathcal{T}_q(A, \mathfrak{p})$ to be the following subring of *quasi-triangular* matrices in the ring $M_q(A)$ of $q \times q$ matrices over A:

$$\mathcal{T}_q(A, \mathfrak{p}) = \{(\alpha_{ij}) \in M_q(A) \mid \alpha_{ij} \in (\mathfrak{p}) \text{ for } i > j\}.$$

Then $\mathcal{T}_q(A, \mathfrak{p})$ is a subring of $M_q(A)$. In Chapter Seven we will show that:

$$(0.13)\qquad \mathcal{T}_q(A, \mathfrak{p}) \cong i_*(\mathbb{Z}(\zeta_p)).$$

This isomorphism is fundamental in what follows; firstly it gives an alternative description of Λ as a fibre product

$$(0.14)\qquad \begin{array}{ccc} \Lambda & \stackrel{\pi_+}{\to} & \mathcal{T}_q(A, \mathfrak{p}) \\ \pi_- \downarrow & & \downarrow \\ \mathbb{Z}[C_q] & \to & \mathbb{F}_p[C_q]. \end{array}$$

Furthermore, it enables us to decompose $\mathcal{T}_q(A, \mathfrak{p})$ into right submodules thus:

$$\mathcal{T}_q(A, \mathfrak{p}) = \bigoplus_{k=1}^q R(k)$$

where $R(k)$ is the kth-row submodule of $\mathcal{T}_q(A, \mathfrak{p})$; that is:

$$R(k) = \{(\alpha_{ij}) \in \mathcal{T}_q(A, \mathfrak{p}) \mid \alpha_{ij} = 0 \text{ for } i \neq k\}.$$

Some structural theorems

Now let $\mathbf{a} = (a_1, \ldots, a_N)$ be a sequence of integers such that

$$1 \leq a_1 < a_2 < \cdots < a_N \leq q$$

and let $\mathbf{e} = (e_1, \ldots, e_N)$ be a sequence of positive integers. We define

$$R(\mathbf{a}, \mathbf{e}) = \bigoplus_{i=1}^n R(a_i)^{e_i}.$$

In the special case where each $e_i = 1$ we write simply $R(\mathbf{a}) = \bigoplus_{i=1}^N R(a_i)$. Let $\eta : \mathbb{Z}[C_q] \to \mathbb{Z}$ be the augmentation homomorphism and put $I_q = \mathrm{Ker}(\eta)$. By a *quasi-Swan module* X we mean one defined by an extension of the form

$$0 \to R(\mathbf{a}, \mathbf{e}) \to X \to Q \to 0.$$

where $Q \cong \mathbb{Z}[C_q]$ or $Q \cong I_q$. We say that such a module is *minimal* when each $e_i = 1$; that is when X is defined by an extension

$$0 \to R(\mathbf{a}) \to X \to Q \to 0.$$

Such an extension is classified up to congruence by a sequence $\mathbf{c} = (c_i)$ where $c_i \in \mathrm{Ext}^1(Q, R(a_i))$. We note that $\mathrm{Ext}^1(Q, R(k)) \cong \mathbb{F}_p$ except when $Q = I_q$ and $k = 1$, in which case we have $\mathrm{Ext}^1(I_q, R(1)) = 0$. A quasi-Swan module X is said to be *nondegenerate* when it is minimal and each $c_i \neq 0$. Otherwise X is *degenerate.*

We are now in a position to state our main general structural theorems: firstly, by a generalization of the arguments of [37], in Chapters Ten and Eleven we shall prove:

Theorem I: *If X is a degenerate quasi-Swan module then X is straight.*

Denote by $[y - 1)$ the right ideal of $\mathbb{Z}[G(p, q)]$ generated by $y - 1$ where $y \in G(p, q)$ is the generator of the subgroup C_q.

We note that $[y - 1)$ is a quasi-Swan module and hence $R(k) \oplus [y - 1)$ is a degenerate quasi-Swan module; as a consequence of Theorem I we have:

Corollary II: The stability class of $R(k) \oplus [y - 1)$ is straight for any k.

In Chapter Twelve we shall also prove

Theorem III: $R(k) \oplus [y-1)$ *is full for each* k.

Identifying syzygies via diagonal resolutions

An underlying theme of this book is the construction of diagonal resolutions. As this procedure is not entirely standard, it seems in order to give a brief account of its salient features.

The method, as a system, began with the author's belated realization that the proof of the $D(2)$-property for the dihedral groups D_{8n+2} given in [30] could be both simplified and extended to all D_{4n+2} by constructing diagonal resolutions in preference to the original naive approach requiring 2×2 matrices over the group ring. Such a diagonal resolution for D_{4n+2} was published in [36] and is given here in Appendix A below. The question of whether the diagonal method can be extended from dihedral groups to groups of higher period is therefore a natural one.

The obvious place to start is with the groups $G(p,q)$ of cohomological period $2q$; in particular, the group $G(7,3)$ of period six, which was assigned as a thesis topic to Jonathan Remez in Autumn 2009 ([53], [54]). Here, when q is prime, the integral representation theory of the groups $G(p,q)$ appears, superficially at least, to be satisfyingly complete, summarized by the footnoted remark below.(2)

In fact, this remark conceals a considerable gap in practice. The actual situation is that a complete list of isomorphism classes of lattices over the p-adic completion $\widehat{\Lambda}_{(p)} = \widehat{\mathbb{Z}}_{(p)}[G(p,q)]$ is known and was given by Pu, a student of Reiner, in [51]. Pu's thesis is summarized between pages 747 to 752 in volume I of Curtis and Reiner's masterwork [14]. However, whilst the proviso in the footnoted remark implicitly concedes the fact that the canonical mapping

$$\{\Lambda\text{-lattices}\} \longrightarrow \{\widehat{\Lambda}_{(p)}\text{-lattices}\}$$

fails to be injective it rather dismisses the task of analyzing its fibres as a minor detail.

Indeed, the nature of Pu's list is in some sense a liability as it makes no distinction concerning relative importance. In consequence, it is not well

(2) 'We shall find all indecomposable $\mathbb{Z}[G]$-lattices apart from questions involving units in rings of algebraic integers' [14], volume 1, p. 747.

adapted to the task of identifying which modules are actually syzygies nor of establishing the sequence in which they occur within a resolution. In practice, for groups of small order, it is simpler to do the calculations *ab initio*, using Pu's list only as a check.

Once the case $G(7,3)$ was settled the next computationally feasible candidate was $G(5,4)$. This was duly assigned as a thesis topic to Jamil Nadim in 2011 (cf [48]). In some sense this is easier than $G(7,3)$ as it is slightly smaller and the appropriate quasi-triangular ring $\mathcal{T}_4(\mathbb{Z},5)$ is defined over the integers rather than, as in $G(7,3)$, over a ring of algebraic integers. In all other respects it is more difficult; the cohomological period being eight, there are more syzygies to consider and crucially, 4 not being prime, Pu's classification is no longer applicable for comparison. In consequence a significant amount of computation is unavoidable, in this case taking the form of explicit checking of cocycles to demonstrate that certain modules are isomorphic; a much abridged summary is given in Appendix A.

These examples form a firm platform on which to theorize. They make it clear that the necessary starting ingredients for a successful application of the diagonal method are firstly, a decomposition of the augmentation ideal $\mathcal{I}_G \cong R(1) \oplus [y-1)$ and secondly, the existence of an exact sequence, the *basic sequence*

$$0 \to R(1) \to \Lambda \longrightarrow \Lambda \to R(q) \to 0.$$

In fact, the former statement was known to Gruenberg and Roggenkamp (cf [24]) although it was never fully exploited by them. By contrast, the existence of the latter, proved in Chapter Nine, is a considerable advance as, after successive p-adic tensoring with a primitive q^{th} root of unity, it generates a collection of derived sequences.

$$0 \to \widehat{R}(k+1) \to \widehat{\Lambda}_{(p)} \longrightarrow \widehat{\Lambda}_{(p)} \to \widehat{R}(k) \to 0$$

and thereby establishes that $R(k+1)$ is at least a generalized $2k^{th}$ syzygy of $R(1)$. Unfortunately, this construction lies outside the theory of integral representations for the obvious reason that primitive q^{th} roots of unity are not integral when $q \geq 3$. Nevertheless, by judicious manipulation of exact sequences, the basic sequence does give rise to a collection of derived sequences

$$0 \to R(k+1) \to P(k) \longrightarrow \Lambda \to R(k) \to 0$$

where $P(k)$ is projective of rank 1. In Chapter Nine we will use these sequences to construct an exact sequence of the periodic form

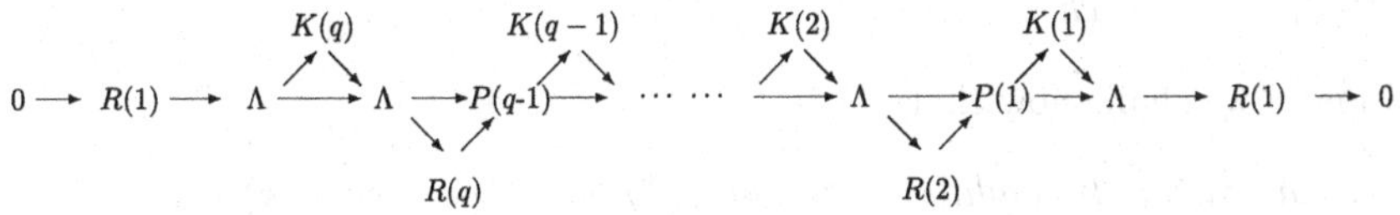

$$\begin{array}{ccccccccccccccccc}
 & & & K(q) & & & K(q-1) & & & K(2) & & & K(1) & & & & \\
 & & & \nearrow\searrow & & & \nearrow\searrow & & & \nearrow\searrow & & & \nearrow\searrow & & & & \\
0 \to R(1) \to & \Lambda & \longrightarrow & & \Lambda & \longrightarrow P(q-1) & \longrightarrow & \cdots\cdots & \longrightarrow & & \Lambda & \longrightarrow P(1) & \longrightarrow & \Lambda & \longrightarrow R(1) & \to 0 \\
 & & & & \searrow\nearrow & & & & & & \searrow\nearrow & & & & & & \\
 & & & & R(q) & & & & & & R(2) & & & & & &
\end{array}$$

where each $P(i)$ is projective of rank 1 over Λ. This is the essential content of the joint paper with Remez [38]. As mentioned above, it follows that:

(0.15) $R(k+1)$ is a minimal element of $D_{2k}(R(1))$ for each k.

Moreover it is also true that $\bigoplus_{k=1}^{q-1} P(k) \cong \Lambda^{(q-1)}$, thereby showing, as previously mentioned, that $G(p,q)$ has free period $2q$ rather than merely cohomological period $2q$. Furthermore, the standard period two resolution of $\mathbb{Z}$ over $\mathbb{Z}[C_q]$ shows that $(y-1)$ is a minimal element of $D_{2k}((y-1))$. By extending scalars from $\mathbb{Z}[C_q]$ to $\Lambda = \mathbb{Z}[G(p,q)]$ we see that:

(0.16) $[y-1)$ is a minimal element of $D_{2k}([y-1))$ for each k.

As the augmentation ideal $\mathcal{I}_G$ splits as a direct sum $\mathcal{I}_G \cong R(1) \oplus [y-1)$ and $\mathcal{I}_G$ represents $D_1(\mathbb{Z})$ we see that:

(0.17) $R(k+1) \oplus [y-1)$ is a minimal element of $D_{2k+1}(\mathbb{Z})$ for each k.

We must now confront the distinction between derivative modules $D_{2k+1}(\mathbb{Z})$ and genuine syzygies $\Omega_{2k+1}(\mathbb{Z})$. In place of **(0.17)** we would wish instead to assert that $R(k+1) \oplus [y-1) \in \Omega_{2k+1}(\mathbb{Z})$. To proceed, note that, in the fibre product (**0.14**), the projections π_-, π_+ together define a homomorphism

$$\pi_* : \widetilde{K}_0(\Lambda) \to \widetilde{K}_0(\mathbb{Z}[C_q]) \oplus \widetilde{K}_0(\mathcal{T}_q(A,\pi))$$

and we define $\mathcal{LF}(\Lambda) = \mathrm{Ker}(\pi_*)$. Consider the following condition on Λ:

$\mathcal{P}(k)$: *There exists an exact sequence* $0 \to R(k+1) \to P(k) \to \Lambda \to R(k) \to 0$ *of* Λ*-modules where* $P(k)$ *is a projective module such that* $[P(k)] \in \mathcal{LF}(\Lambda)$.

We shall prove:

Theorem IV: *If Λ satisfies each condition $\mathcal{P}(k)$ then $R(k+1) \oplus [y-1) \in \Omega_{2k+1}(\mathbb{Z})$ for each k.*

Finally we shall show that:

Theorem V: *If conditions (*) and (**) hold then each $\mathcal{P}(k)$ holds.*

However, as is shown in Chapter Thirteen, condition $\mathcal{P}(1)$ fails in the case of $G(13, 12)$. One might expect that this failure is typical when $\widetilde{K}_0(\mathbb{Z}[C_q]) \neq 0$.

The $D(2)$-property

In relation to the $D(2)$-property we shall prove:

Theorem VI: *If $R(2) \oplus [y-1) \in \Omega_3(\mathbb{Z})$ then $G(p, q)$ has the $D(2)$-property.*

To sketch the proof of Theorem VI, recall that a module $J \in \Omega_3(\mathbb{Z})$ is *geometrically realizable* when there is a finite 2-complex X with $\pi_1(X) \cong G$ and $\pi_2(X) \cong J$. We note the following sufficient condition for the $D(2)$-property to hold for the finite group G (cf [33], (60.3), p. 232).

(0.18) The finite group G satisfies the $D(2)$-property provided all modules at the minimal level of the syzygy $\Omega_3(\mathbb{Z})$ are both full and geometrically realizable.

For the sake of completeness we give a proof of (0.18) in Appendix C.

As previously noted, a theorem of Wamsley [67] shows that $G(p, q)$ has a balanced presentation $\mathcal{G}(p, q)$. Put $J = \pi_2(X_{\mathcal{G}})$ where $X_{\mathcal{G}}$ is the Cayley complex of $\mathcal{G}(p, q)$. As $\mathcal{G}(p, q)$ is balanced then J lies at the minimal level of $\Omega_3(\mathbb{Z})$ and is geometrically realizable. By Corollary II, $R(2) \oplus [y-1)$ is the *unique* minimal representative of its stability class. Thus if $R(2) \oplus [y-1) \in \Omega_3(\mathbb{Z})$ then, by uniqueness, $R(2) \oplus [y-1) \cong J$ and so is geometrically realizable. However, $R(2) \oplus [y-1)$ is full, by Theorem III. It follows now from (0.18) that the $D(2)$-property holds for $G(p, q)$, so proving Theorem VI. This, together with a weaker version of Theorem IV, now shows:

Theorem VII: *If $\mathcal{P}(1)$ is satisfied then the $D(2)$-property holds for $G(p, q)$.*

Theorem VII now implies the following, which is the Main Theorem stated above:

Theorem VIII: *If conditions (*) and (**) hold then $G(p, q)$ has the $D(2)$-property.*

We will consider in detail the practical effect of the restrictions imposed by conditions (*) and (**) in the final chapter.

Chapter One

Projective modules and class groups

A module over a ring Λ is *free* when it possesses a generating set which is linearly independent over Λ. More generally a Λ-module is *projective* when it is isomorphic to a direct summand of a free module. Thus a finitely generated projective Λ-module P takes the form $P \oplus Q \cong \Lambda^n$ for some module Q and some integer $n \geq 1$. Evidently Q is then also projective.

As we shall see, projective modules arise inevitably once cohomology is introduced into the theory of modules. In that context, there is a naive tendency to regard them as being little different from free modules, distinguished from them only by a formal difference in definition. However, as matters develop, projective modules acquire a more individual character and appear as obstructions, potential or actual, to constructions we might wish to make. The literature on projective modules is immense and we will not here attempt any sort of comprehensive account. Rather the present chapter is intended as a brief introduction to those aspects which play a significant role in the rest of the book.

§1: Exact sequences and projective modules:

Let Λ be a ring. We consider sequences of Λ-homomorphisms

$$\mathcal{S} = (\cdots \to A_{n+1} \overset{f_{n+1}}{\to} A_n \overset{f_n}{\to} A_{n-1} \to \cdots)$$

which may be of arbitrary length, finite or infinite. Such a sequence $\mathcal{S}$ is *exact at* n when $\mathrm{Ker}(f_n) = \mathrm{Im}(f_{n+1})$ and is said to be *exact* when it is exact for each possible value of n. The following 'Five Lemma' is fundamental:

(1.1) Suppose given a commutative diagram of Λ-homomorphisms as follows:

$$\begin{array}{ccccccccc} A_4 & \to & A_3 & \to & A_2 & \to & A_1 & \to & A_0 \\ f_4 \downarrow & & f_3 \downarrow & & f_2 \downarrow & & f_1 \downarrow & & f_0 \downarrow \\ B_4 & \to & B_3 & \to & B_2 & \to & B_1 & \to & B_0 . \end{array}$$

in which both rows are exact; if f_0, f_1, f_3 and f_4 are isomorphisms then so also is f_2 an isomorphism.

Given Λ-modules A, C there are canonical Λ-homomorphisms $i_A : A \to A \oplus C$ and $\pi_C : A \oplus C \to C$ allowing the construction of the *trivial exact sequence*

$$\mathcal{T} = (0 \to \ A \ \xrightarrow{i_A} \ A \oplus C \ \xrightarrow{\pi_C} \ C \ \to 0).$$

An exact sequence $\mathcal{E} = (0 \to C \xrightarrow{i} B \xrightarrow{p} A \to 0)$ is said to *split* when it is isomorphic to the trivial exact sequence by means of a commutative diagram as follows:

$$\begin{array}{cccccccc} 0 \to & A & \xrightarrow{i} & B & \xrightarrow{p} & C & \to 0 \\ & \downarrow \mathrm{Id}_A & & \downarrow \psi & & \downarrow \mathrm{Id}_C & \\ 0 \to & A & \xrightarrow{i_A} & A \oplus C & \xrightarrow{\pi_C} & C & \to 0. \end{array}$$

It follows from the Five Lemma that such a splitting ψ is necessarily an isomorphism. We say that $\mathcal{E}$ *splits on the left* when there exists a morphism $r : B \to A$ such that $r \circ i = \mathrm{Id}_A$. Finally we say that $\mathcal{E}$ *splits on the right* when there exists a morphism $s : C \to B$ such that $p \circ s = \mathrm{Id}_C$. If ψ is a splitting of $\mathcal{E}$ then $r = \pi_A \circ \psi$ is a left splitting of $\mathcal{E}$. Conversely if $r : B \to A$ is a left splitting of $\mathcal{E}$ then $\psi = \begin{pmatrix} r \\ p \end{pmatrix} : B \to A \oplus C$ is a splitting. If ψ is a splitting of $\mathcal{E}$ then $s = \psi^{-1} \circ i_C : C \to B$ is a right splitting. Likewise if s is a right splitting then by the Five Lemma, $(i, s) : A \oplus C \to B$ is necessarily an isomorphism and $\psi = (i, s)^{-1}$ is then a splitting. To summarize:

(1.2) $\mathcal{E}$ splits $\Longleftrightarrow \mathcal{E}$ splits on the left $\Longleftrightarrow \mathcal{E}$ splits on the right.

Given a set X we construct a free Λ module F_X with basis X as follows; formally take F_X to be the set of functions $\alpha : X \to \Lambda$ with finite support; then F_X acquires the structure of a right Λ-module by means of the pairing

$$F_X \times \Lambda \to F_X \quad ; \quad (\alpha, \lambda) \mapsto \alpha \cdot \lambda$$

where $\alpha \cdot \lambda(x) = \alpha(x)\lambda$. There is a canonical mapping $\iota_X : X \to F_X$ given by

$$\iota_X(x)(y) = \begin{cases} 1 & x = y \\ 0 & x \neq y \end{cases}$$

and the image of i_X forms a Λ-basis for F_X. The construction possesses the universal property that, given a mapping $h : X \to M$ to a Λ-module

M there is unique Λ-homomorphism $\widehat{h} : F_X \to M$ making the following diagram commute

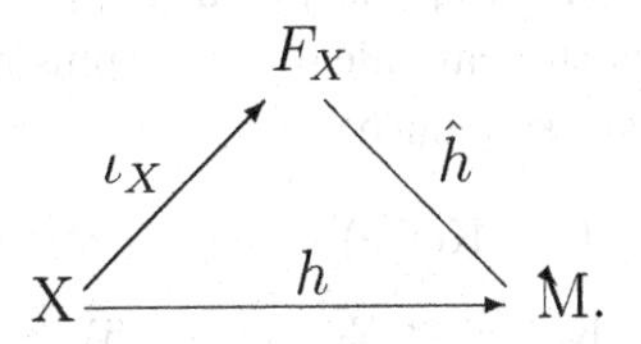

More general than the notion of free module is the notion of *projective module*, defined by the following universal property; we say that the right Λ-module P is projective when, given a surjective Λ-homomorphism $\epsilon : M' \to M$ and a Λ-homomorphism $f : P \to M$ there is a Λ-homomorphism $\widetilde{F} : P \to M'$ making the following commute.

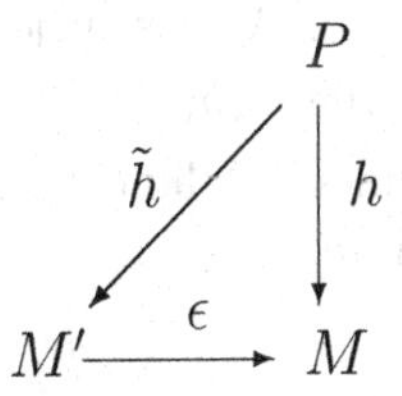

Proposition 1.3: *The following conditions on the Λ-module P are equivalent:*

i) *P is projective;*
ii) *every exact sequence of the form $0 \to Q \to M \to P \to 0$ splits;*
iii) *P is isomorphic to a direct summand of a free Λ-module.*

Proof. (i) $\Longrightarrow$ (ii) Let $\mathcal{E} = (0 \to Q \xrightarrow{i} M \xrightarrow{p} P \to 0)$ be an exact sequence in which P is projective. As p is surjective the universal property of P guarantees the existence of a homomorphism $s : P \to M$ making the following diagram commute.

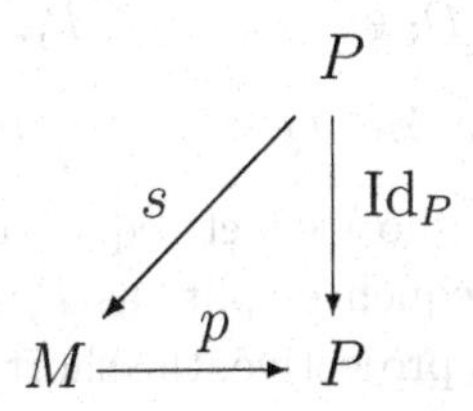

Clearly s splits the exact sequence $\mathcal{E}$ on the right. QED (i) $\Longrightarrow$ (ii)

(ii) $\implies$ (iii). Suppose that each short exact sequence $0 \to Q \to M \to P \to 0$ splits. Let $(e_x)_{x\in X}$ be a set which generates P over Λ, and let $e : F_X \to P$ be the canonical epimorphism $e(\alpha) = \sum_{x\in X} e_x \cdot \alpha(x)$ where $\alpha = \sum_{x\in X} \iota(x) \cdot \alpha(x)$ is the canonical epimorphism from the free module on Λ. We obtain an exact sequence

$$0 \to \mathrm{Ker}(e) \to F_X \overset{e}{\to} P \to 0$$

which splits to give $P \oplus \mathrm{Ker}(e) \cong F_X$. QED (ii) $\implies$ (iii).

(iii) $\implies$ (i). We suppose that P is isomorphic to a direct summand of the free Λ-module F_X for some set X and make the identification $P \oplus N = F_X$ where N is a Λ-submodule of F_X. Let $h : P \to M$ be a homomorphism and $\epsilon : M' \to M$ a surjective homomorphism. We must construct a homomorphism $\widetilde{h} : P \to M'$ such that $\epsilon \circ \widetilde{h} = h$. Extend h to a homomorphism $h : F_X \to M$ by setting $h_{|N} \equiv 0$. As ϵ is surjective then for each $x \in X$ we may choose $\widehat{h}(x) \in M'$ such that $\epsilon(\widehat{h}(x)) = h(\iota_X(x))$ for all $x \in X$. Then the collection $\{\widehat{h}(x)\}_{x\in X}$ defines a Λ-homomorphism $\widehat{h} : F_X \to M'$ making the following diagram commute.

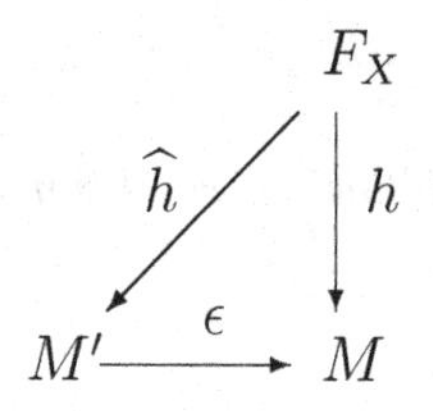

The required statement follows on setting $\widetilde{h} = \widehat{h}_{|P}$ so completing the proof. □

Proposition 1.4: (*Schanuel's Lemma*) *Let* $(0 \to D_r \overset{i_r}{\to} P_r \overset{f_r}{\to} M \to 0)$ *be short exact sequences* $(r = 1, 2)$ *of* Λ*-modules in which* P_1 *and* P_2 *are projective; then*

$$D_1 \oplus P_2 \cong D_2 \oplus P_1.$$

Proof. Form the *fibre product* $Q = P_1 \underset{f_1,f_2}{\times} P_2 = \{(x, y) \in P_1 \times P_2 : f_1(x) = f_2(y)\}$. There is a short exact sequence $0 \to D_2 \to Q \overset{\pi_1}{\to} P_1 \to 0$ where $\pi_1(x, y) = x$. The sequence splits as P_1 is projective, so that $Q \cong D_2 \oplus P_1$. Likewise, as P_2 is projective, the short exact sequence $0 \to D_1 \to Q \overset{\pi_2}{\to} P_2 \to 0$ with $\pi_2(x, y) = y$ also splits, and $Q \cong D_1 \oplus P_2$. Now $D_1 \oplus P_1 \cong Q \cong D_2 \oplus P_1$ as claimed. □

§2: The Grothendieck group of a ring:

Given a ring Λ, we denote by $\mathcal{P}(\Lambda)$ the isomorphism classes of finitely generated nonzero projective modules over Λ. It should cause no confusion if we identify the projective module P with its isomorphism class in $\mathcal{P}(\Lambda)$. Evidently $\mathcal{P}(\Lambda)$ is a commutative semigroup under direct sum

$$\mathcal{P}(\Lambda) \times \mathcal{P}(\Lambda) \to \mathcal{P}(\Lambda) \quad ; \quad (P,\, Q) \mapsto P \oplus Q.$$

We denote by $K_0(\Lambda)$ the universal abelian group obtained from $\mathcal{P}(\Lambda)$; that is, the elements of $K_0(\Lambda)$ are symbols $[P, Q]$ with $P, Q \in \mathcal{P}(\Lambda)$, subject to the following rule where $R \in \mathcal{P}(\Lambda)$

$$[P, Q] = [P', Q'] \iff P \oplus Q' \oplus R \;\cong\; P' \oplus Q \oplus R.$$

Addition is given by $[P_1, Q_1] + [P_2, Q_2] = [P_1 \oplus P_2, Q_1 \oplus Q_2]$ with $0 = [P, P]$ and $-[P, Q] = [Q, P]$. Now suppose $[P_1, Q_1] \in K_0(\Lambda)$. Then $[P_1, Q_1] = [P_1 \oplus P_2, Q_1 \oplus P_2]$ for any $P_2 \in \mathcal{P}(\Lambda)$. Choosing P_2 such that $P_1 \oplus P_2 \cong \Lambda^n$ and putting $Q = Q_1 \oplus P_2$ we see that:

(2.1) Every element of $K_0(\Lambda)$ can be represented in the form $[\Lambda^m, Q]$.

There is a semigroup homomorphism $\nu : \mathcal{P}(\Lambda) \to K_0(\Lambda)$ defined by the assignment $P \mapsto [P \oplus \Lambda, \Lambda]$. This has a more convenient description if instead we construct $K_0(\Lambda)$ by first adjoining the zero module to produce a monoid $\mathcal{P}_+(A)$ and then taking its universal abelian group. The correspondence $[P, Q] \mapsto [P, Q]$ is an isomorphism of the first definition onto the second. However, with the extra symbol '0' we now have $[P \oplus \Lambda, \Lambda] = [P, 0]$ so we can write ν in the form $\nu(P) = [P, 0]$.

A commutative additive semigroup $\mathcal{S}$ is said to have the *cancellation property* when $s + t = s' + t \Longrightarrow s = s'$ for all s, s', $t \in \mathcal{S}$. We note that

(2.2) $\nu : \mathcal{P}(\Lambda) \to K_0(\Lambda)$ is injective $\iff \mathcal{P}(\Lambda)$ is a cancellation semigroup.

The construction $\Lambda \mapsto K_0(\Lambda)$ is functorial on the category of rings. To see this, note that if $f : \Lambda \to \Omega$ is a ring homomorphism and P is a finitely generated projective Λ-module then $f_*(P) = P \otimes_{f(\Lambda)} \Omega \in \mathcal{P}(\Omega)$ so that the correspondence $P \mapsto f_*(P)$ induces a semigroup homomorphism $f_* : \mathcal{P}(\Lambda) \to \mathcal{P}(\Omega)$. Likewise, the assignment $[P, Q] \mapsto [f_*(P), f_*(Q)]$ defines a group homomorphism $f_* : K_0(\Lambda) \to K_0(\Omega)$. We proceed to review some standard examples:

(2.3) $K_0(\Lambda) \cong \mathbb{Z}$ generated by $[\Lambda, 0]$ if Λ is either a division ring or a commutative principal ideal domain.

We write $M_n(\Lambda)$ for the ring of $n \times n$ matrices over Λ. We define

$$\mathfrak{R}_n = \{(\lambda_1, \ldots, \lambda_n) \mid \lambda_i \in \Lambda\};$$

Morita's Theorem ([43] p. 165) then shows that:

(2.4) The mapping $P \mapsto P \otimes_\Lambda \mathfrak{R}_n$ defines an isomorphism $K_0(\Lambda) \xrightarrow{\cong} K_0(M_n(\Lambda))$.

We note that $K_0(M_n(\Lambda))$ is generated by $\mathfrak{R}_n$ and $M_n(\Lambda) \cong \underbrace{\mathfrak{R}_n \oplus \cdots \oplus \mathfrak{R}_n}_{n}$; thus, in contradistinction to (2.3), $[M_n(\Lambda), 0]$ does *not* generate $K_0(M_n(\Lambda))$ when $n \geq 2$. It is straightforward to show that K_0 commutes with products thus:

(2.5) $K_0(\Lambda_1 \times \Lambda_2) \cong K_0(\Lambda_1) \times K_0(\Lambda_2)$.

Consequently

(2.6) $K_0(M_{d_1}(D_1) \times \cdots \times M_{d_m}(D_m)) \cong \mathbb{Z}^{(m)}$ if each $K_0(D_i) \cong \mathbb{Z}$.

Let $\mathrm{rad}(\Lambda)$ denote the Jacobson radical of the ring Λ; the Nakayama-Bourbaki Lemma ([8] p. 85, [43] p. 183) shows that

(2.7) $K_0(\Lambda) \cong K_0(\Lambda/\mathrm{rad}(\Lambda))$ if $\mathrm{rad}(\Lambda)$ is a nilpotent ideal in Λ.

A right ideal $\mathfrak{J}$ is *simple* when it contains no proper nonzero submodule. We say that Λ is *finitely semisimple* when Λ decomposes as a direct sum of simple right ideals $\Lambda = \mathfrak{J}_1 \dot{+} \cdots \dot{+} \mathfrak{J}_N$. A theorem of Wedderburn ([50] p. 49) then shows:

(2.8) Λ is finitely semisimple $\iff \Lambda \cong M_{d_1}(D_1) \times \cdots \times M_{d_m}(D_m)$

for some sequence $(D_i, d_i)_{1 \leq i \leq m}$ of division algebras D_i and positive integers d_i.

Hence:

(2.9) If Λ is a finitely semisimple ring then $K_0(\Lambda) \cong \mathbb{Z}^{(m)}$ where m is the number of simple factors of Λ.

More generally, if Λ is an Artinian ring then $\Lambda/\mathrm{rad}(\Lambda)$ is finitely semisimple. Hence $\Lambda/\mathrm{rad}(\Lambda) \cong M_{d_1}(D_1) \times \cdots \times M_{d_m}(D_m)$ for some division rings $D_1, \ldots, D_m$. Furthermore, $\mathrm{rad}(\Lambda)$ is nilpotent ([52], p. 81) and so, by (2.7):

(2.10) If Λ is an Artinian ring then $K_0(\Lambda) \cong \mathbb{Z}^{(m)}$ where m is the number of simple factors of $\Lambda/\mathrm{rad}(\Lambda)$.

§3: The reduced Grothendieck group:

$K_0(\Lambda)$ has a subgroup $T(\Lambda) = \{[\Lambda^m, \Lambda^n] \mid m, n \in \mathbb{Z}_+\}$. We define the reduced projective class group $\widetilde{K}_0(\Lambda)$ as the quotient

$$\widetilde{K}_0(\Lambda) = K_0(\Lambda)/T(\Lambda).$$

Observe that if the class $[S, 0]$ is zero in $\widetilde{K}_0(\Lambda)$ then $S \oplus \Lambda^a \cong \Lambda^b$ for some positive integers a, b. Such modules S are said to be *stably free*; the question of whether stably free modules are actually free is a significant one in the subsequent development.

If $f : \Lambda \to \Omega$ is a ring homomorphism then $f_*(T(\Lambda)) \subset T(\Omega)$ so that f also induces a homomorphism $f_* : \widetilde{K}_0(\Lambda) \to \widetilde{K}_0(\Omega)$. Thus we have a functorial exact sequence

(3.1) $$0 \to T(\Lambda) \to K_0(\Lambda) \to \widetilde{K}_0(\Lambda) \to 0.$$

In many of the cases we encounter the sequence (3.1) splits. To see how this can happen, recall that the ring Λ has the invariant basis number property when

(IBN) $$\Lambda^m \cong \Lambda^n \iff m = n.$$

This is a very mild restriction on rings ([12]) and is satisfied in all the cases we shall encounter. In fact:

(3.2) If Λ satisfies IBN then $T(\Lambda) \cong \mathbb{Z}$ with generator $[\Lambda, 0]$.

Moreover if Λ and Ω both satisfy IBN then any ring homomorphism $i : \Omega \to \Lambda$ induces an isomorphism $i_* : T(\Omega) \xrightarrow{\simeq} T(\Lambda)$. We note that:

(3.3) If every finitely generated projective Ω-module is free then $K_0(\Omega) = T(\Omega)$.

This is the case, for example, if Ω is a commutative principal ideal domain. Recall that Λ *retracts onto* Ω when there exist ring homomorphisms $r : \Lambda \to \Omega$, $i : \Omega \to \Lambda$ such that $r \circ i = \mathrm{Id}_\Omega$. We then have:

Proposition 3.4: *If Λ retracts onto a commutative principal ideal domain then $T(\Lambda) \cong \mathbb{Z}$ and (3.1) splits to give an isomorphism $K_0(\Lambda) \cong \mathbb{Z} \oplus \widetilde{K}_0(\Lambda)$.*

Proof. The requirement that Λ retracts onto a commutative principal ideal domain Ω implies that Λ has IBN ([12]). Thus $T(\Lambda) \cong T(\Omega) \cong \mathbb{Z}$. As $K_0(\Omega) = T(\Omega)$ we can replace (3.1) by the exact sequence

$$0 \to K_0(\Omega) \xrightarrow{i_*} K_0(\Lambda) \to \widetilde{K}_0(\Lambda) \to 0.$$

If $r : \Lambda \to \Omega$ is a ring homomorphism such that $r \circ i = \mathrm{Id}_\Omega$ the induced homomorphism $r_* : K_0(\Lambda) \to K_0(\Omega)$ splits the above sequence on the left giving the required isomorphism $K_0(\Lambda) \cong K_0(\Omega) \oplus \widetilde{K}_0(\Lambda) \cong \mathbb{Z} \oplus \widetilde{K}_0(\Lambda)$. □

§4: Projective modules over $R[G]$:

If R is a commutative ring then, for any group G, the augmentation homomorphism $\epsilon : R[G] \to R$ composed with the canonical inclusion $i : R \to R[G]$ gives $\epsilon \circ i = \mathrm{Id}_R$; thus $R[G]$ retracts onto R. Consequently,

(4.1) If R is a commutative PID then $K_0(R[G]) \cong \mathbb{Z} \oplus \widetilde{K}_0(R[G])$ for any group G.

In what follows we assume that G is a nontrivial finite group. If $\mathbb{F}$ is a field the theorem of Maschke ([50], p. 51) shows that the group algebra $\mathbb{F}[G]$ is finitely semisimple provided the characteristic of $\mathbb{F}$ is coprime to the cardinal $|G|$ of G; interpreted in terms of Wedderburn's Theorem this becomes:

Theorem 4.2: *Let G be a finite group and $\mathbb{F}$ a field whose characteristic is coprime to the cardinal $|G|$ of G; then there is an isomorphism of $\mathbb{F}$-algebras*

$$\mathbb{F}[G] \cong M_{d_1}(D_1) \times \cdots \times M_{d_m}(D_m)$$

for some finite dimensional division algebras D_i and positive integers d_i.

Here the integer m is an invariant of both $\mathbb{F}$ and G. Likewise for a given field $\mathbb{F}$ the sequence $(D_i, d_i)_{1 \le i \le m}$ is uniquely determined by G up to isomorphism and order. It is conventional first to consider the case where $\mathbb{F} = \mathbb{C}$, the field of complex numbers. For us this case has limited interest as the only finite dimensional complex division algebra is $\mathbb{C}$ itself and we get:

(4.3) $$\mathbb{C}[G] \cong M_{d_1}(\mathbb{C}) \times \cdots \times M_{d_n}(\mathbb{C}).$$

Elementary representation theory shows that, in this case, n is the number of conjugacy classes of G and that $\sum_{i=1}^n d_i^2 = |G|$. Slightly more interesting is the case when $\mathbb{F} = \mathbb{R}$. Then we have an isomorphism

(4.4) $$\mathbb{R}[G] \cong M_{d_1}(D_1) \times \cdots \times M_{d_m}(D_m)$$

where each D_i is either $\mathbb{R}$, $\mathbb{C}$ and $\mathbb{H}$, the ring of Hamiltonian quaternions, these being the only the finite dimensional real division algebras. If we further progress to the case $\mathbb{F} = \mathbb{Q}$ and consider Wedderburn decompositions

$$\textbf{(4.5)} \qquad \mathbb{Q}[G] \cong M_{e_1}(D_1) \times \cdots \times M_{e_\mu}(D_\mu)$$

the situation is further complicated by the circumstance that there are now infinitely many isomorphically distinct finite dimensional rational division algebras D_i from which to choose. Not all of these actually feature in such a decomposition; for example, each D_i must then support a positive involution ([1], [29]). Nevertheless, by varying the group G, it is known that infinitely many do occur ([19]).

In this book we are primarily concerned with representation theory of finite groups over the ring of integers. In this connection, it is tempting to regard the rational representation theory of a finite group G as a reasonable first approximation to its integral representation theory. There are nevertheless significant differences. Whilst every module over $\mathbb{Q}[G]$ is projective this is far from the case over $\mathbb{Z}[G]$. If G is nontrivial then $K_0(\mathbb{Q}[G])$ is free abelian of rank ≥ 2. In contrast, (3.4) shows that

$$K_0(\mathbb{Z}[G]) \cong \mathbb{Z} \oplus \widetilde{K}_0(\mathbb{Z}[G]).$$

Moreover, we have the following theorem of Swan ([60], p. 573):

(4.6) $\widetilde{K}_0(\mathbb{Z}[G])$ is finite for any finite group G.

The difference is explained by the fact that $\mathbb{Q}[G]$ itself decomposes as a nontrivial direct sum of projective modules, all of which have $\mathbb{Q}$-rank less than $\mathbb{Q}[G]$; whereas

(4.7) $\mathbb{Z}[G]$ is indecomposable.

The indecomposability of $\mathbb{Z}[G]$ is an immediate consequence of the following special case of Swan's 'local freeness theorem.' ([14], vol. 1, p. 676):

(4.8) Let G be a finite group and let P be a finitely generated projective module over $\mathbb{Z}[G]$; then $P \otimes_{\mathbb{Z}} \mathbb{Q}$ is free over $\mathbb{Q}[G]$.

There is a further difference with regard to cancellation of modules. If M, X are modules over a ring Λ we shall say that M *cancels* X when

$$M \oplus X \cong N \oplus X \implies M \cong N$$

for any Λ-module N. Over a finitely semisimple algebra every module is projective and any module M cancels any other module X. However, this fails in general over integral group rings $\mathbb{Z}[G]$. Nevertheless, under suitable restrictions on the group G, there is a weaker form of cancellation due to Jacobinski ([28]) which is nevertheless extremely useful. Thus we shall say that M has *$\mathfrak{J}$-cancellation*(3) whenever M cancels any direct summand of any $M^{(m)}$; that is

$$(\mathfrak{J}) \qquad \text{If } X \oplus Y \cong M^{(m)} \quad \text{then} \quad M \oplus X \cong N \oplus X \implies M \cong N.$$

Let R be a commutative integral domain of characteristic zero with field of fractions k, let $\mathfrak{A}$ be k-algebra of finite dimension and let Λ be an R-subalgebra of $\mathfrak{A}$. We say that Λ is an *R-order* Λ in $\mathfrak{A}$ when its underlying R-module is finitely generated and free and such that $\Lambda \otimes_R k = \mathfrak{A}$. Thus for a finite group G the integral group ring $\mathbb{Z}[G]$ is a $\mathbb{Z}$-order in $\mathbb{Q}[G]$.

In order to state Jacobinski's Cancellation Theorem we first observe that if k is a finite algebraic extension of $\mathbb{Q}$ and A is a finite dimensional semisimple k-algebra then $A \otimes_{\mathbb{Q}} \mathbb{R}$ is a semisimple $\mathbb{R}$ algebra. We say that G satisfies the *Eichler condition* (cf [17], [62]) when in the decomposition (4.4), D_i is commutative whenever $d_i = 1$; that is, when no simple factor of $\mathbb{R}[G]$ is isomorphic to $\mathbb{H}$. It follows that:

(4.9) Any finite abelian group G satisfies the Eichler condition.

If k is an algebraic number field we say that an order Λ in a finite dimensional semisimple k-algebra A is an *Eichler order* when $A \otimes_{\mathbb{Q}} \mathbb{R}$ satisfies the Eichler condition. More generally, if M is any Λ-lattice then $\mathrm{End}_\Lambda(M)$ is also an order in a finite dimensional semisimple k-algebra and we say that M is an *Eichler lattice* when $\mathrm{End}_\Lambda(M)$ satisfies the Eichler condition. The following criterion for cancellation is due to Jacobinski [28].

(4.10) Let Λ be an order in a semisimple algebra over a finite algebraic extension k of $\mathbb{Q}$. If M is an Eichler lattice over Λ then M has $\mathfrak{J}$-cancellation.

As an immediate corollary we have:

(4.11) If Λ is an Eichler order then any Λ-lattice M has $\mathfrak{J}$-cancellation.

(3) The Gothic $\mathfrak{J}$ is chosen to connote 'Jacobinski'.

Now suppose that Λ is an Eichler order and that S is a stably free Λ-module so that for some positive integers $\Lambda^{(m)} \oplus \Lambda^{(n)} \cong S \oplus \Lambda^{(n)}$. Let N be an integer such that $n \leq mN$. Then $\Lambda^{(n)}$ is a direct summand of $(\Lambda^{(m)})^{(N)} = \Lambda^{(mN)}$ and so, by $\mathfrak{J}$-cancellation, $\Lambda^{(m)} \cong S$. Hence:

(4.12) If Λ is an Eichler order then any stably free Λ-module is free.

For the rest of this section we take $\Lambda = \mathbb{Z}[G]$ where G is a finite group which satisfies the Eichler condition. Suppose that P and P' are finitely generated projective Λ-modules such that $P \oplus \Lambda^{(n)} \cong P' \oplus \Lambda^{(n)}$. As $\widetilde{K}_0(\Lambda)$ is finite we may choose $d \geq 1$ such that $d[P] = 0 \in \widetilde{K}_0(\Lambda)$. Then $P^{(d)}$ is stably free and hence, by (4.12), $P^{(d)} \cong \Lambda^{(m)}$ for some $m \geq 1$. Again choose N so that $n \leq mN$. Then $\Lambda^{(n)}$ is a direct summand of $(\Lambda^{(m)})^{(N)} = P^{(dN)}$ and so $P \cong P'$ by $\mathfrak{J}$-cancellation. Hence:

(4.13) Suppose that $\Lambda = \mathbb{Z}[G]$ satisfies the Eichler condition and let $P \in \mathcal{P}(\Lambda)$; then P cancels $\Lambda^{(n)}$ for any $n \geq 1$.

Now suppose that P, P' and Q are finitely generated projective Λ-modules such that $P \oplus Q \cong P' \oplus Q$. For all $d \geq 1$ we thus have $P \oplus Q^{(d)} \cong P' \oplus Q^{(d)}$. Taking d to be the order of Q in $\widetilde{K}_0(\mathbb{Z}[G])$ then $Q^{(d)}$ is stably free and hence, by (4.12), $Q^{(d)} = \Lambda^{(n)}$ for some n. Thus $P \oplus \Lambda^{(n)} \cong P' \oplus \Lambda^{(n)}$ and so, by (4.13), $P \cong P'$; that is:

(4.14) If G satisfies the Eichler condition then $\mathcal{P}(\mathbb{Z}[G])$ is a cancellation semigroup.

The local freeness theorem (4.8) has the consequence that any finitely generated projective module P over $\mathbb{Z}[G]$ has a well defined rank namely

$$\mathrm{rk}(P) = \mathrm{rk}_{\mathbb{Q}[G]}(P \otimes_{\mathbb{Z}} \mathbb{Q}).$$

A further theorem of Swan [60] simplifies the description of projective modules over $\mathbb{Z}[G]$ as follows:

(4.15) Let $P \in \mathcal{P}(\mathbb{Z}[G])$; if $\mathrm{rk}(P) \geq 2$ then there exists $P_1 \in \mathcal{P}(\mathbb{Z}[G])$ with $\mathrm{rk}(P_1) = 1$ such that $P \cong P_1 \oplus \mathbb{Z}[G]^{(n-1)}$.

If G satisfies the Eichler condition then (4.14) shows that this description is unique:

(4.16) Suppose $P_1 \oplus \mathbb{Z}[G]^{(n-1)} \cong P_1' \oplus \mathbb{Z}[G]^{(n-1)}$ where $P_1, P_1' \in \mathcal{P}(\mathbb{Z}[G])$; if G satisfies the Eichler condition then $P_1 \cong P_1'$.

§5: Steinitz' Theorem:

In what follows A will denote a commutative integral domain and $\mathfrak{I}(A)$ will denote the set of nonzero ideals in A. We define a relation '$\leftrightarrow$' on $\mathfrak{I}(A)$ by

$$I \leftrightarrow J \iff aI = bJ$$

for some nonzero a, $b \in A$. It is straightforward to see that

(5.1) '$\leftrightarrow$' is an equivalence relation on $\mathfrak{I}(A)$.

We denote by $\mathbf{I}(A)$ the set of equivalence classes of $\mathfrak{I}(A)$ under '$\leftrightarrow$'. It is elementary to show:

(5.2) I is a principal ideal in $A \iff I \leftrightarrow A$.

If $I, J \in \mathfrak{I}(A)$ we recall that the product $I \cdot J$ is defined to be the set of finite sums

$$I \cdot J = \left\{ \sum_{r=1}^{N} a_r b_r \quad \text{where } a_r \in I, b_r \in J \right\}$$

Evidently $I \cdot J \in \mathfrak{I}(A)$ and we have:

(5.3) $$I \cdot A = I$$

(5.4) $$I \cdot J = J \cdot I$$

(5.5) $$I \cdot (J \cdot K) = (I \cdot J) \cdot K$$

The following is also elementary:

(5.6) $$(I_1 \leftrightarrow J_1) \wedge (I_2 \leftrightarrow J_2) \implies I_1 \cdot J_1 \leftrightarrow I_2 \cdot J_2.$$

Denoting by $[I]$ the class of I in $\mathbf{I}(A)$, the pairing $(I, J) \mapsto I \cdot J$ gives a multiplication:

$$\mathbf{I}(A) \times \mathbf{I}(A) \to \mathbf{I}(A) \quad ; \quad ([I], [J]) \mapsto [I \cdot J]$$

with respect to which $\mathbf{I}(A)$ becomes a commutative monoid in which the identity element is the class of A and thereby of any principal ideal. The construction is functorial in the following sense; suppose that A, B are commutative integral domains and $i : A \to B$ is an injective ring homomorphism. Then i induces a mapping

$$i_* : \mathfrak{I}(A) \to \mathfrak{I}(B) \quad ; \quad i_*(J) = i(J) \cdot B.$$

Moreover, it is clear that:

(5.7) $$J_1 \leftrightarrow J_2 \implies i_*(J_1) \leftrightarrow i_2(J_2).$$

Hence the correspondence $[J] \mapsto [i(J) \cdot B]$ defines a monoid homomorphism

(5.8) $$i_* : \mathbf{I}(A) \to \mathbf{I}(B).$$

We say that $I \in \mathfrak{I}(A)$ is *invertible* when $I \cdot J$ is principal for some $J \in \mathfrak{I}(A)$. We note that if $I \leftrightarrow J$ then I is invertible if and only if J is invertible. We denote by $Cl(A)$ the subset of $\mathbf{I}(A)$ consisting of invertible classes. Clearly

(5.9) $Cl(A)$ is a group contained within the monoid $\mathbf{I}(A)$.

Finally, suppose that I is an invertible ideal in A so that for some nonzero ideal J in A, $I \cdot J$ is a principal ideal $I \cdot J = (a)$. Then

$$\begin{aligned}(i(I) \cdot B) \cdot (i(J) \cdot B) &= ((i(I) \cdot i(J)) \cdot B) \\ &= i(I \cdot J) \cdot B \\ &= (i(a))\end{aligned}$$

so that $i(I) \cdot B$ is also invertible. Hence i_* induces a group homomorphism

(5.10) $$i_* : Cl(A) \to Cl(B).$$

For the rest of this section we restrict A to be a *Dedekind domain*; that is, we further assume that A is Noetherian and integrally closed in its field of fractions and that every prime ideal of A is maximal. In this context, we denote elements of $\mathfrak{I}(A)$, by lower case Gothic letters; $\mathfrak{a}, \mathfrak{b}, \mathfrak{c}, \ldots$. The theorem of Steinitz ([21], [59]) may be summarized in the statements (5.11)–(5.16) below:

(5.11) Every $\mathfrak{a} \in \mathfrak{I}(A)$ is invertible.

(5.12) If $\mathfrak{a}, \mathfrak{b} \in \mathfrak{I}(A)$ then $\mathfrak{a} \oplus \mathfrak{b} \cong A \oplus \mathfrak{ab}$.

Choosing $\mathfrak{b}$ so that $\mathfrak{ab} \cong A$ we see that $\mathfrak{a} \oplus \mathfrak{b} \cong A \oplus A$; hence:

(5.13) If $\mathfrak{a} \in \mathfrak{I}(A)$ then $\mathfrak{a}$ is a projective A-module.

Now take Q to be a nonzero finitely generated projective A-module; if $\mathbb{K}$ is the field of fractions of A then for some positive integer $n = \mathrm{rk}(Q)$, the *rank* of Q, we have:

$$Q \otimes_A \mathbb{K} \cong \mathbb{K}^n.$$

(5.14) If $\mathrm{rk}(Q) = 1$ then $Q \cong \mathfrak{a}$ for some $\mathfrak{a} \in \mathfrak{I}(A)$.

(5.15) If $\mathrm{rk}(Q) = n > 1$ then $Q \cong A^{(n-1)} \oplus \mathfrak{a}$ for some $\mathfrak{a} \in \mathfrak{I}(A)$.

(5.16) $A^m \oplus \mathfrak{a} \cong A^n \oplus \mathfrak{b} \iff (m = n) \wedge (\mathfrak{a} \cong \mathfrak{b})$.

As an immediate consequence we have:

(5.17) If A is a Dedekind domain then $\mathcal{P}(A)$ is a cancellation semigroup.

It follows that $K_0(A) \cong \mathbb{Z} \oplus Cl(A)$ when A is a Dedekind domain and hence:

(5.18) If A is a Dedekind domain then $\widetilde{K}_0(A) \cong Cl(A)$.

§6: Artinian rings are weakly Euclidean:

Given a ring Λ we denote by $M_n(\Lambda)$ the ring of $n \times n$-matrices over Λ. For each $n \geq 2$, $M_n(\Lambda)$ has the canonical Λ-basis $\epsilon(i,j)_{1 \leq i,j \leq n}$ given by $\epsilon(i,j)_{r,s} = \delta_{ir}\delta_{js}$. The elementary invertible matrices $E(i,j;\lambda)$ $(\lambda \in \Lambda)$ and $D(i,\delta)$ $(\delta \in \Lambda^*)$ which perform row and column operations are expressed in terms of the basic matrices as follows:

$$E(i,j;\lambda) = I_n + \lambda\epsilon(i,j) \quad (i \neq j);$$

$$D(i,\delta) = I_n + (\delta - 1)\epsilon(i,i).$$

For $n \geq 2$ we put $D_n(\Lambda) = \{D(1,\delta) : \delta \in \Lambda^*\}$; that is

$$D_n(\Lambda) = \left\{ \begin{bmatrix} \delta & & & 0 \\ & 1 & & \\ & & \ddots & \\ 0 & & & 1 \end{bmatrix} \right\}$$

Denote by $GL_n(\Lambda)$ the group of invertible $n \times n$-matrices over Λ and by $E_n(\Lambda)$ the subgroup of $GL_n(\Lambda)$ generated by the matrices $E(i,j;\lambda)$; then $D_n(\Lambda)$ normalises $E_n(\Lambda)$ and we define the *restricted linear group* $GE_n(\Lambda)$ to be the subgroup of $GL_n(\Lambda)$ given as the internal product

$$GE_n(\Lambda) = D_n(\Lambda) \cdot E_n(\Lambda).$$

The constructions GL_n, GE_n, E_n are all functorial under ring homomorphisms. If $\pi : A \to B$ is a surjective ring homomorphism the induced map $\pi_* : E_n(A) \to E_n(B)$ is also surjective. However $\pi_* : GE_n(A) \to GE_n(B)$ is not surjective unless the induced homomorphism on units $\pi_* : A^* \to B^*$ is surjective. In this case we say that ring homomorphism $\pi : A \to B$ has the *lifting property for units.* Then we have:

(6.1) Let $\pi : A \to B$ be a surjective ring homomorphism with the lifting property for units; then $\pi_* : GE_n(A) \to GE_n(B)$ is surjective.

In general $GE_n(\Lambda)$ is a proper subgroup of $GL_n(\Lambda)$ and Λ is said to be *weakly Euclidean*(4) when $GE_n(\Lambda) = GL_n(\Lambda)$ for all $n \geq 2$; that is, when for all $n \geq 2$, any $X \in GL_d(\Lambda)$ can be written as a product

$$X = E_1 \cdot \dots \cdot E_n \cdot \Delta_n(\lambda).$$

It is straightforward to see that:

(6.2) Let A, B be weakly Euclidean rings; then $A \times B$ is weakly Euclidean.

(6.3) If D is a division ring then $M_m(D)$ is weakly Euclidean for any integer $m \geq 1$.

(6.4) Let $D_1, \dots, D_m$ be division rings; then $M_{d_1}(D_1) \times \dots \times M_{d_m}(D_m)$ is weakly Euclidean for any positive integers $d_1, \dots, d_m$.

A ring homomorphism $\varphi : A \to B$ has the *lifting property for units* when the induced map $\phi_* : A^* \to B^*$ is surjective. We say φ has the *strong lifting property for units* when in addition the following holds for $\alpha \in A$;

$$\alpha \in A^* \iff \varphi(\alpha) \in B^*.$$

(4) The terminology arises from the classical theorem of H.J.S. Smith [58] which we may state as saying that an integral domain with a Euclidean algorithm is weakly Euclidean.

(6.5) Let $\varphi : A \to B$ be a surjective ring homomorphism; if $\mathrm{Ker}(\varphi)$ is nilpotent then φ has the strong lifting property for units.

Elsewhere ([35], Prop. 2.43, p. 21) we have shown:

(6.6) Let $\varphi : A \to B$ be a surjective ring homomorphism where B is weakly Euclidean; if φ has the strong lifting property for units then A is also weakly Euclidean.

Thus we have:

(6.7) Let $\varphi : A \to B$ be a surjective ring homomorphism with nilpotent kernel; if B is weakly Euclidean then A is also weakly Euclidean.

Theorem 6.8: *If A is an Artinian ring then A is weakly Euclidean.*

Proof. The radical $\mathrm{rad}(A)$ of the Artinian ring A is nilpotent (cf [52] p. 81). Moreover

$$A/\mathrm{rad}(A) \cong M_{d_1}(D_1) \times \cdots \times M_{d_m}(D_m)$$

for some division rings $D_1, \ldots, D_m$. The conclusion now follows from (6.4). □

Any finite ring is trivially Artinian; hence we have:

(6.9) If A is a finite ring then A is weakly Euclidean.

§7: Milnor's fibre square description of projective modules:

By a *fibre square* $\widehat{\mathfrak{L}}$ we mean a commutative diagram of ring homomorphisms

$$\widehat{\mathfrak{L}} = \left\{ \begin{array}{ccc} \Lambda & \stackrel{\pi_-}{\to} & \Lambda_- \\ \downarrow \pi_+ & & \downarrow \varphi_- \\ \Lambda_+ & \stackrel{\varphi_+}{\to} & \Lambda_0 \end{array} \right.$$

in which $\pi_+ \times \pi_-$ maps Λ isomorphically onto the fibre product

$$\Lambda_+ \times_\varphi \Lambda_- = \{(x_+, x_-) \in \Lambda_+ \times \Lambda_- : \varphi_+(x_+) = \varphi_-(x_-)\}.$$

By a *corner* $\mathfrak{L} = (\Lambda_+, \Lambda_-, \Lambda_0, \varphi_+, \varphi_-)$ we mean a diagram of ring homomorphisms:

$$\mathfrak{L} = \begin{cases} & \Lambda_- \\ & \downarrow \varphi_- \\ \Lambda_+ \stackrel{\varphi_+}{\rightarrow} & \Lambda_0. \end{cases}$$

Given a corner $\mathfrak{L} = (\Lambda_+, \Lambda_-, \Lambda_0, \varphi_+, \varphi_-)$ we construct a canonical fibre square by taking $\Lambda = \Lambda_+ \times_\varphi \Lambda_-$ and π_+, π_- to be the projections. By a module over a corner $\mathfrak{L}$ we mean a triple $\mathcal{M} = (M_+, M_-, \alpha)$ where M_+, M_- are modules over Λ_+ Λ_- respectively such that $M_+ \otimes \Lambda_0 \cong M_- \otimes \Lambda_0$ and where $\alpha : M_- \otimes \Lambda_0 \to M_+ \otimes \Lambda_0$ is a specific Λ_0-isomorphism. If $\mathcal{N} = (N_+, N_-, \beta)$ is also a module over $\mathfrak{L}$ then by an $\mathfrak{L}$-morphism $f : \mathcal{M} \to \mathcal{N}$ we mean a pair $f = (f_+, f_-)$ where $f_\sigma : M_\sigma \to N_\sigma$ is a homomorphism over Λ_σ such that the following commutes:

$$\begin{array}{ccc} M_- \otimes \Lambda_0 & \stackrel{f_- \otimes \mathrm{Id}}{\longrightarrow} & N_- \otimes \Lambda_0 \\ \downarrow \alpha & & \downarrow \beta \\ M_+ \otimes \Lambda_0 & \stackrel{f_+ \otimes \mathrm{Id}}{\longrightarrow} & N_+ \otimes \Lambda_0 \end{array}$$

There is a category $\mathcal{M}od_{\mathfrak{L}}$ whose objects are modules over the corner $\mathfrak{L}$ and whose morphisms are as described above. We say that an $\mathfrak{L}$-module $\mathcal{M} = (M_+, M_-; \alpha)$ is *finitely generated* when M_σ is finitely generated over Λ_σ for $\sigma = +, -$; we say that $\mathcal{M} = (M_+, M_-; \alpha)$ is *locally projective* when M_+, M_- are projective over Λ_+, Λ_- respectively. Observe that $\mathcal{M}od_{\mathfrak{L}}$ has coproducts given by

$$\mathcal{M} \oplus \mathcal{N} = (M_+ \oplus N_+, M_- \oplus N_-;\ \alpha \oplus \beta)\ \ ;\ \ f \oplus g = (f_+ \oplus g_+, f_- \oplus g_-).$$

For an $\mathfrak{L}$-module $\mathcal{M} = (M_+, M_-;\ \alpha)$ the pair (M_+, M_-) is called the *local type* of $\mathcal{M}$. Clearly isomorphic $\mathfrak{L}$-modules have the same local type. We denote by $\mathcal{L}(M_+, M_-)$ the set of isomorphism classes of $\mathfrak{L}$-modules of local type (M_+, M_-) and put

$$\mathrm{Iso}(M_- \otimes \Lambda_0, M_+ \otimes \Lambda_0) = \left\{ \begin{array}{c} \alpha : M_- \otimes \Lambda_0 \to M_+ \otimes \Lambda_0 \ \text{such} \\ \text{that } \alpha \text{ is an } \Lambda_0 \text{ isomorphism} \end{array} \right\}.$$

There is a two-sided action

$$\begin{aligned} \mathrm{Aut}_{\Lambda_+}(M_+) \times \mathrm{Iso}(M_- \otimes \Lambda_0, M_+ \otimes \Lambda_0) & \\ \times \mathrm{Aut}_{\Lambda_-}(M_-) & \to \mathrm{Iso}(M_- \otimes \Lambda_0, M_+ \otimes \Lambda_0) \\ (h_+, \alpha, h_-) & \mapsto [h_+] \circ \alpha \circ [h_-] \end{aligned}$$

where $[h_\sigma] = h_\sigma \otimes 1 : M_\sigma \otimes \Lambda_0 \to M_\sigma \otimes \Lambda_0$. There is evidently a surjective mapping

$$\begin{aligned} \natural : \mathrm{Iso}(M_- \otimes \Lambda_0, M_+ \otimes \Lambda_0) \to & \quad \mathcal{L}(M_+, M_-) \\ \natural(\alpha) & = [(M_+, M_-, \alpha)]. \end{aligned}$$

We obtain the following classification of $\mathfrak{L}$-modules within a local type:

Proposition 7.1: *$\natural$ induces a bijection*

$$\natural : \mathrm{Aut}_{\Lambda_+}(M_+)\backslash\mathrm{Iso}(M_- \otimes \Lambda_0, M_+ \otimes \Lambda_0)/\mathrm{Aut}_{\Lambda_-}(M_-) \longrightarrow \mathcal{L}(M_+, M_-).$$

If M is a module over the ring Λ then $M \otimes_{\varphi_-\pi_-} \Lambda_0 \equiv M \otimes_{\varphi_+\pi_+} \Lambda_0$ and there is a canonical isomorphism $(M \otimes_{\pi_-} \Lambda_-) \otimes_{\varphi_-} \Lambda_0 \overset{\natural}{\to} (M \otimes_{\pi_+} \Lambda_+) \otimes_{\varphi_+} \Lambda_0$ making the following commute:

$$\begin{array}{ccc} (M \otimes_{\pi_-} \Lambda_-) \otimes_{\varphi_-} \Lambda_0 & \overset{\natural}{\to} & (M \otimes_{\pi_+} \Lambda_+) \otimes_{\varphi_+} \Lambda_0 \\ \downarrow \nu_- & & \downarrow \nu_+ \\ M \otimes_{\varphi_-\pi_-} \Lambda_0 & \overset{\mathrm{Id}}{\to} & M \otimes_{\varphi_+\pi_+} \Lambda_0 \end{array}$$

Thus an Λ-module M gives rise to a module $(M \otimes_{\pi_+} \Lambda_+, M \otimes_{\pi_+} \Lambda_+; \natural)$ over $\mathfrak{L}$. If $f : M \to N$ is a Λ-homomorphism then putting $f_\sigma = f \otimes \mathrm{Id} : M \otimes_{\pi_\sigma} \Lambda_\sigma \to N \otimes_{\pi_\sigma} \Lambda_\sigma$, the correspondences $M \mapsto (M \otimes_{\pi_+} \Lambda_+, M \otimes_{\pi_+} \Lambda_+; \natural); f \mapsto (f_+, f_-)$ determine a functor $r : \mathcal{M}od_\Lambda \to \mathcal{M}od_{\mathfrak{L}}$. We say that the corner $\mathcal{L}$ satisfies the *Milnor condition* when either φ_+ or φ_- is surjective. Milnor's classification (cf [46]) of projective modules over a fibre product can be stated thus:

Theorem 7.2: *If the corner $\mathfrak{L}$ satisfies the Milnor condition then the functor r induces a 1-1 correspondence*

$$r : \left\{ \begin{array}{c} \text{Isomorphism classes} \\ \text{of finitely generated} \\ \textit{projective } \Lambda\textit{-modules} \end{array} \right\} \rightarrow \left\{ \begin{array}{c} \text{Isomorphism classes of} \\ \text{finitely generated } \textit{locally} \\ \textit{projective } \mathfrak{L}\textit{-modules} \end{array} \right\}.$$

We say that $\mathcal{M}$ is *locally free*[5] *with respect to* $\mathfrak{L}$ when $\mathcal{M} \cong (\Lambda_+^n, \Lambda_-^n, \alpha)$ for some $\alpha \in GL_n(\Lambda_0)$. The classification of modules within a local type given by (7.1) becomes simpler for locally free modules. Given a corner $\mathfrak{L}$ we denote by $\mathcal{LF}_n(\mathfrak{L})$ the isomorphism classes of locally free modules of rank n:

$$\mathcal{LF}_n(\mathfrak{L}) = \left\{ \begin{array}{c} \text{isomorphism classes of } \mathfrak{L}\text{-modules of the form} \\ (\Lambda_+^n, \Lambda_-^n; \alpha) \ \text{ where } \ \alpha \in \mathrm{Iso}(\Lambda_-^n \otimes \Lambda_0, \Lambda_+^n \otimes \Lambda_0) \end{array} \right\}.$$

A group valued functor $\mathcal{G}$: **Rings** $\rightarrow$ **Groups** applied to a corner $\mathfrak{L}$ gives a diagram of group homomorphisms

$$\mathcal{G}(\mathfrak{L}) = \left\{ \begin{array}{ccc} & & \mathcal{G}(\Lambda_-) \\ & & \downarrow \mathcal{G}(\varphi_-) \\ \mathcal{G}(\Lambda_+) & \stackrel{\mathcal{G}(\varphi_+)}{\rightarrow} & \mathcal{G}(\Lambda_0). \end{array} \right.$$

We write $\overline{\mathcal{G}}(\mathfrak{L}) = \mathrm{Im}\,\mathcal{G}(\varphi_+)\backslash\mathcal{G}(\Lambda_0)/\mathrm{Im}\,\mathcal{G}(\varphi_-)$. The set $\overline{\mathcal{G}}(\mathfrak{L})$ has a distinguished point, denoted by $*$, namely the class of the identity from $\mathcal{G}(\Lambda_0)$. Here we confine our attention to the functors $\mathcal{G} = GL_n$. For $k \geq 1$ the stabilization operator

$$X \mapsto \begin{pmatrix} X & 0 \\ 0 & \mathrm{I}_k \end{pmatrix}$$

(5) We caution the reader that all our subsequent uses of the term *locally free* will be in this sense. In particular, we shall not mean *'free after localization'* as in, for example, [63].

induces mappings $s_{n,k} : \overline{GL_n}(\mathfrak{L}) \rightarrow \overline{GL_{n+k}}(\mathfrak{L})$ satisfying $s_{n,m} = s_{n+k,m-k} \circ s_{n,k}$ for $1 \leq k < m$. We note that the following diagram commutes

$$\begin{array}{ccc} \overline{GL_n}(\mathfrak{L}) & \xrightarrow{s_{n,k}} & \overline{GL_{n+k}}(\mathfrak{L}) \\ \nu_n \downarrow & & \nu_{n+k} \downarrow \\ \mathcal{LF}_n(\mathfrak{L}) & \xrightarrow{\sigma_{n,k}} & \mathcal{LF}_{n+k}(\mathfrak{L}) \end{array} \qquad \textbf{(7.3)}$$

where $\sigma_{n,k} : \mathcal{LF}_n(\mathfrak{L}) \rightarrow \mathcal{LF}_{n+k}(\mathfrak{L})$ is the stabilization operator induced from the correspondence $\mathcal{P} \mapsto \mathcal{P} \oplus (\Lambda_+^k, \Lambda_-^k; I_k)$.

Identifying $\Lambda_+ \otimes \Lambda_0 = \Lambda_0 = \Lambda_- \otimes \Lambda_0$ we have $\mathrm{Iso}(\Lambda_-^n \otimes \Lambda_0, \Lambda_+^n \otimes \Lambda_0) = GL_n(\Lambda_0)$; then (7.3) gives a bijection:

(7.4) $\nu_n : \overline{GL_n}(\mathfrak{L}) = GL_n(\Lambda_+)\backslash GL_n(\Lambda_0)/GL_n(\Lambda_-) \xrightarrow{\simeq} \mathcal{LF}_n(\mathfrak{L})$.

Finally consider a Milnor square of ring homomorphisms:

$$\mathfrak{L} = \left\{ \begin{array}{ccc} \Lambda & \xrightarrow{\pi_-} & \Lambda_- \\ \downarrow \pi_+ & & \downarrow \varphi_- \\ \Lambda_+ & \xrightarrow{\varphi_+} & \Lambda_0. \end{array} \right.$$

Theorem 7.5: *If $\mathcal{P}(\Lambda)$ is a cancellation semigroup and Λ_0 is commutative and weakly Euclidean then the stabilization mapping $\sigma_{n,1} : \overline{GL_1}(\mathfrak{L}) \xrightarrow{\simeq} \overline{GL_n}(\mathfrak{L})$ is bijective for all $n \geq 2$.*

Proof. As Λ_0 is commutative and weakly Euclidean then the stabilization mapping

$$\sigma_{n,1} : \overline{GE_1}(\mathfrak{L}) \longrightarrow \overline{GE_n}(\mathfrak{L})$$

is bijective for all $n \geq 2$. Surjectivity of $\sigma_{n,1} : \overline{GL_1}(\mathfrak{L}) \xrightarrow{\simeq} \overline{GL_n}(\mathfrak{L})$ now follows as $\overline{GL_n}(\mathfrak{L})$ is a natural quotient of $\overline{GE_n}(\mathfrak{L})$ and $\overline{GL_1}(\mathfrak{L}) \equiv \overline{GE_1}(\mathfrak{L})$. To show that $\sigma_{n,1}$ is injective take $c_1, c_2 \in \overline{GL_1}(\mathfrak{L})$ and let P_i be the projective module which is locally free of rank 1 relative to $\mathfrak{L}$ and which is represented by c_i. If $\sigma_{n,1}(c) = \sigma_{n,1}(c') \in \overline{GL_n}(\mathfrak{L})$ then

$$P \oplus \Lambda^{n-1} \cong P' \oplus \Lambda^{n-1}.$$

As $\mathcal{P}(\Lambda)$ is a cancellation semigroup then $P_1 \cong P_2$ and so $c_1 = c_2$. Hence $\sigma_{n,1}$ is injective and this completes the proof. □

§8: The Milnor exact sequence:

For any ring A we write $GL(A) = \lim\limits_{\rightarrow} GL_n(A)$ and $E(A) = \lim\limits_{\rightarrow} E_n(A)$. We recall the observation of Whitehead ([45]) that

(8.1) $E(A)$ is the commutator subgroup of $GL(A)$.

Hence $GL(A)/E(A)$ is the abelianisation, $GL(A)^{ab}$, of $GL(A)$ and we put

$$K_1(A) = GL(A)/E(A) = GL(A)^{ab}.$$

Now take $\mathcal{L}$ to be a fibre square of ring homomorphisms

$$\mathcal{L} = \left\{ \begin{array}{ccc} \Lambda & \stackrel{\pi_-}{\rightarrow} & \Lambda_- \\ \downarrow \pi_+ & & \downarrow \varphi_- \\ \Lambda_+ & \stackrel{\varphi_+}{\rightarrow} & \Lambda_0 \end{array} \right.$$

where $\mathcal{L}$ satisfies Milnor's condition which, without loss, we may suppose takes the form that φ_+ is surjective. Then $\mathcal{L}$ induces an exact sequence

$$\textbf{(8.2)} \qquad K_0(\Lambda) \stackrel{\binom{\pi_-}{\pi_+}}{\longrightarrow} K_0(\Lambda_-) \oplus K_0(\Lambda_+) \stackrel{(\varphi_-, \varphi_+)}{\longrightarrow} K_0(\Lambda_0) \rightarrow 0.$$

$\mathcal{L}$ also induces a fibre square of group homomorphisms

$$GL(\mathcal{L}) = \left\{ \begin{array}{ccc} GL(\Lambda) & \stackrel{\pi_-}{\rightarrow} & GL(\Lambda_-) \\ \downarrow \pi_+ & & \downarrow \varphi_- \\ GL(\Lambda_+) & \stackrel{\varphi_+}{\rightarrow} & GL(\Lambda_0) \end{array} \right.$$

and thereby an exact sequence of abelian groups

$$GL^{ab}(\Lambda) \stackrel{\binom{\pi_-}{\pi_+}}{\longrightarrow} GL^{ab}(\Lambda_-) \times GL^{ab}(\Lambda_+) \stackrel{(\varphi_-, \varphi_+)}{\longrightarrow} GL^{ab}(\Lambda_0).$$

Under the identification $GL^{ab}(\Lambda) \cong K_1(\Lambda)$ we may re-write this as

$$\textbf{(8.3)} \qquad K_1(\Lambda) \stackrel{\binom{\pi_-}{\pi_+}}{\longrightarrow} K_1(\Lambda_-) \oplus K_1(\Lambda_+) \stackrel{(\varphi_-, \varphi_+)}{\longrightarrow} K_1(\Lambda_0).$$

As $\varphi_+ : \Lambda_+ \to \Lambda_0$ is surjective then $\varphi_+ : E(\Lambda_+) \to E(\Lambda_0)$ is also surjective enabling the identification

$$\begin{aligned}&\mathrm{Coker}\left(K_1(\Lambda_-) \oplus K_1(\Lambda_+) \overset{(\varphi_-,\varphi_+)}{\longrightarrow} K_1(\Lambda_0)\right)\\ &\qquad \overset{\simeq}{\longrightarrow} \varinjlim \left(GL_n(\Lambda_+)\backslash GL_n(\Lambda_0)/GL_n(\Lambda_-)\right).\end{aligned}$$

However Milnor's construction of §7 gives the alternative interpretation

$$\begin{aligned}&\varinjlim \left(GL_n(\Lambda_+)\backslash GL_n(\Lambda_0)/GL_n(\Lambda_-)\right)\\ &\qquad \overset{\simeq}{\longrightarrow} \mathrm{Ker}\left(K_0(\Lambda) \overset{\binom{\pi_-}{\pi_+}}{\longrightarrow} K_0(\Lambda_-) \oplus K_0(\Lambda_+)\right).\end{aligned}$$

Using this to connect (8.2) and (8.3) we obtain Milnor's exact sequence

(8.4) $K_1(\Lambda_+) \oplus K_1(\Lambda_-) \overset{\varphi_*}{\to} K_1(\Lambda_0) \overset{\partial}{\to} K_0(\Lambda) \overset{\pi_*}{\to} K_0(\Lambda_-) \oplus K_0(\Lambda_+) \overset{\varphi_*}{\twoheadrightarrow} K_0(\Lambda_0)$.

In particular, the structure of the group $K_1(\Lambda_0)$ plays a significant role in our calculations.

(8.5) If A is weakly Euclidean then $K_1(A) = A^*/(A^* \cap E(A))$.

It can be shown (cf [13], [35], p. 28–29) that

(8.6) If A is commutative then $A^* \cap E(A) = \{1\}$.

By (6.9), any finite ring is weakly Euclidean. Consequently, we have:

(8.7) If A is a finite commutative ring then $K_1(A) = A^*$.

In all the cases we shall meet, Λ_0 will be a finite commutative ring so that the Milnor exact sequence takes the form

$$\textbf{(8.8)}\quad 0 \to \Lambda_0^*/\mathrm{Im}(\varphi_*) \xrightarrow{\partial} K_0(\Lambda) \xrightarrow{\pi_*} K_0(\Lambda_-) \oplus K_0(\Lambda_+) \xrightarrow{\varphi_*} K_0(\Lambda_0) \to 0.$$

§9: Projective modules over $\mathbb{Z}[C_p]$:

To illustrate the foregoing and for future reference, we classify projective modules over $\mathbb{Z}[C_p]$ where p is an odd prime. This was first accomplished by Rim [55]. We follow Milnor's proof [46], enhanced by the observation that $\mathcal{P}(\mathbb{Z}[C_p])$ is a cancellation semigroup. Writing $\mathbb{Z}[C_p] = \mathbb{Z}[x]/(x^p - 1)$ then the factorization

$$x^p - 1 = (x-1)(x^{p-1} + \cdots + x + 1)$$

gives a fibre product decomposition

$$\textbf{(9.1)}\qquad \begin{array}{ccc} \mathbb{Z}[C_p] & \xrightarrow{\pi_+} & \mathbb{Z}[x]/(x^{p-1} + \ldots + x + 1) \\ \downarrow{\scriptstyle \pi_-} & & \downarrow{\scriptstyle \varphi_+} \\ \mathbb{Z}[x]/(x-1) & \xrightarrow{\varphi_-} & \mathbb{F}_p \end{array}$$

where π_-, π_+ are the obvious projections and φ_-, φ_+ represent reduction mod p. Now take $\zeta_p = \exp(2\pi i/p)$ and put $\mathbb{Z}(\zeta_p) = \mathrm{span}_{\mathbb{Z}}\{\zeta_p^r \mid r \in \mathbb{N}\}$. As $\sum_{r=0}^{p-1} \zeta_p^r = 0$ then

$$\mathbb{Z}(\zeta_p) = \left\{ \sum_{r=0}^{p-2} a_r \zeta_p^r \;\mid\; a_r \in \mathbb{Z} \right\}$$

and the correspondence $x \mapsto \zeta_p$ induces a ring isomorphism

$$\mathbb{Z}[x]/(x^{p-1} + \cdots + x + 1) \xrightarrow{\simeq} \mathbb{Z}(\zeta_p).$$

As $\mathbb{Z}[x]/(x-1) \cong \mathbb{Z}$ we may rewrite (9.1) above as

(9.2)
$$\begin{array}{ccc} \mathbb{Z}[C_p] & \xrightarrow{\pi_+} & \mathbb{Z}(\zeta_p) \\ \downarrow{\pi_-} & & \downarrow{\varphi_+} \\ \mathbb{Z} & \xrightarrow{\varphi_-} & \mathbb{F}_p \end{array}$$

In the Milnor exact sequence associated to (9.2)

$K_1(\mathbb{Z}) \oplus K_1(\mathbb{Z}(\zeta_p)) \stackrel{\varphi_*}{\to} K_1(\mathbb{F}_p) \stackrel{\partial}{\to} K_0(\mathbb{Z}[C_p]) \stackrel{\pi_*}{\to} K_0(\mathbb{Z}) \oplus K_0(\mathbb{Z}(\zeta_p)) \stackrel{\varphi_*}{\to}$
$K_0(\mathbb{F}_p) \to 0$

one see easily that the induced map on units $\varphi_* : \mathbb{Z}^* \times \mathbb{Z}(\zeta_p)^* \to \mathbb{F}_p^*$ is surjective. By (8.7) $K_1(\mathbb{F}_p) = \mathbb{F}_p^*$. Hence the induced map $K_1(\mathbb{Z}) \oplus K_1(\mathbb{Z}(\zeta_p)) \stackrel{\varphi_*}{\to} K_1(\mathbb{F}_p)$ is also surjective and the above exact sequence reduces to a short exact sequence

(9.3) $$0 \to K_0(\mathbb{Z}[C_p]) \stackrel{\pi_*}{\to} K_0(\mathbb{Z}) \oplus K_0(\mathbb{Z}(\zeta_p)) \stackrel{\varphi_*}{\to} K_0(\mathbb{F}_p) \to 0.$$

To proceed, we observe the following elementary proposition, the proof of which is left to the reader:

(9.4) Let $0 \to A \xrightarrow{\binom{i}{j}} B \oplus C \xrightarrow{(\xi,\eta)} D \to 0$ be an exact sequence of abelian groups; if $\xi : B \to D$ is an isomorphism then $j : A \to C$ is also an isomorphism.

In the sequence (9.3), $(\varphi_-)_* : K_0(\mathbb{Z}) \stackrel{\varphi_*}{\to} K_0(\mathbb{F}_p)$ is an isomorphism. It follows from (9.4) that

(9.5) $(\pi_+)_* : K_0(\mathbb{Z}[C_p]) \to K_0(\mathbb{Z}(\zeta_p))$ is an isomorphism.

We claim that:

Proposition 9.6: $\pi_+ : \mathcal{P}(\mathbb{Z}[C_p]) \to \mathcal{P}(\mathbb{Z}(\zeta_p))$ *is an isomorphism of semi-groups.*

Proof. Let $\xi \in \mathcal{P}(\mathbb{Z}(\zeta_p))$. By Steinitz' Theorem we may express ξ as the class of $\mathbb{Z}(\zeta_p)^{(n)} \oplus \mathbf{b}$ for some nonzero ideal $\mathbf{b}$ of $\mathbb{Z}(\zeta_p)$. By Milnor's construction there is a projective module Q over $\mathbb{Z}[C_p]$ obtained by glueing over $\mathrm{Id}_{\mathbb{F}_p}$ in the diagram

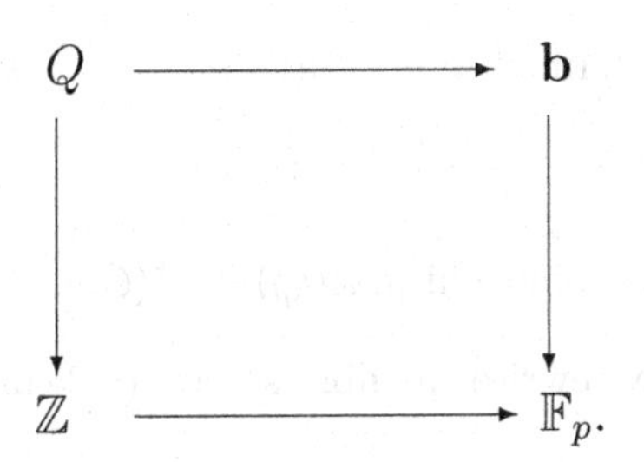

Then $\pi_+(Q) \cong \mathbf{b}$ so that $\pi_+(\mathbb{Z}[C_p]^{(n)} \oplus Q) \cong \mathbb{Z}(\zeta_p)^{(n)} \oplus \mathbf{b}$ and π_+ is surjective. To complete the proof we must show that π_+ is also injective. Thus consider the commutative diagram

$$\begin{array}{ccc} \mathcal{P}(\mathbb{Z}[C_p]) & \xrightarrow{\pi_+} & \mathcal{P}(\mathbb{Z}(\zeta_p)) \\ \nu_1 \downarrow & & \downarrow \nu_2 \\ K_0(\mathbb{Z}[C_p]) & \xrightarrow{(\pi_+)_*} & K_0(\mathbb{Z}(\zeta_p)) \end{array}$$

in which ν_1, ν_2 denote the respective canonical mappings. As C_p satisfies the Eichler condition then ν_1 is injective by (4.14) and $(\pi_+)_*$ is an isomorphism by (9.5). Hence $\pi_+ : \mathcal{P}(\mathbb{Z}[C_p]) \to \mathcal{P}(\mathbb{Z}(\zeta_p))$ is injective as required. □

For future reference we note that the ring $\mathbb{Z}(\zeta_p)$ has a somewhat different description as follows; let $\epsilon : \mathbb{Z}[C_p] \to \mathbb{Z}$ denote the augmentation homomorphism defined by

$$\epsilon\left(\sum_{r=0}^{p-1} a_r x^r\right) = \sum_{r=0}^{p-1} a_r$$

and take the exact sequence $0 \to I(C_p) \xrightarrow{i} \mathbb{Z}[C_p] \xrightarrow{\epsilon} \mathbb{Z} \to 0$ where $I(C_p) = \mathrm{Ker}(\epsilon)$. The dual then takes the form

(9.7) $$0 \to \mathbb{Z} \xrightarrow{\epsilon^*} \mathbb{Z}[C_p] \xrightarrow{i^*} I^*(C_p) \to 0.$$

As $\mathrm{Im}(\epsilon^*)$ is an ideal in $\mathbb{Z}[C_p]$ then $I^*(C_p)$ has a natural ring structure. However $\epsilon^*(1) = \sum_{r=0}^{p-1} x^r$ so that we have isomorphisms

$$I^*(C_p) \;\cong\; \mathbb{Z}[C_p]/\mathrm{Im}(\epsilon^*) \;\cong\; \mathbb{Z}[x]/(x^{p-1} + \cdots + x + 1) \;\cong\; \mathbb{Z}(\zeta_p).$$

Hence:

(9.8) There is a ring isomorphism $\mathbb{Z}(\zeta_p) \cong I^*(C_p)$.

Consequently we may rewrite the fibre square (9.2) in the form

(9.9)
$$\begin{array}{ccc} \mathbb{Z}[C_p] & \xrightarrow{\;i^*\;} & I^*(C_p) \\ \downarrow{\scriptstyle \epsilon} & & \downarrow{\scriptstyle \varphi_+} \\ \mathbb{Z} & \xrightarrow{\;\varphi_-\;} & \mathbb{F}_p \end{array}$$

Chapter Two

Homological algebra

Linear algebra over a field is rendered tractable by the fact that every module over a field is free ; that is, has a spanning set of linearly independent vectors. Over more general rings, when a module M is not free we make a first approximation to its being free by taking a surjective homomorphism $\varphi : F \to M$ where F is free. A free Λ-resolution of M of finite type is then an exact sequence

$$(\mathcal{F}) = (\cdots \stackrel{\partial_{n+1}}{\to} F_n \stackrel{\partial_n}{\to} F_{n-1} \cdots \cdots \stackrel{\partial_1}{\to} F_0 \stackrel{\partial_0}{\to} M \to 0)$$

in which each F_r is free of finite rank over Λ. More generally, we consider exact sequences

$$(\mathcal{P}) = (\cdots \stackrel{\partial_{n+1}}{\to} P_n \stackrel{\partial_n}{\to} P_{n-1} \cdots \cdots \stackrel{\partial_1}{\to} P_0 \stackrel{\partial_0}{\to} M \to 0)$$

in which each P_r is finitely generated projective over Λ. Cohomology is essentially the study of the invariants of such projective resolutions.

§10: Cochain complexes and cohomology:

We recall that, by a *chain complex* we mean a collection $C_* = (C_r, \partial_r^C)_{r \in \mathbf{Z}}$ where for each r, $\partial_r^C : C_r \to C_{r-1}$ is a homomorphism of abelian groups such that $\partial_r^C \circ \partial_{r+1}^C = 0$. We find it more convenient to work primarily with the dual notion, obtained by reversing the direction of the mappings ∂_r^C. Thus by a *cochain complex* we mean a collection $C^* = (C^r, \partial_C^r)_{r \in \mathbf{Z}}$ where for each r, $\partial_C^r : C^r \to C^{r+1}$ is a homomorphism of abelian groups such that $\partial_C^r \circ \partial_C^{r-1} = 0$.

If $C^* = (C^r, \partial_C^r)_{r \in \mathbf{Z}}$ $D^* = (D^r, \partial_D^r)_{r \in \mathbf{Z}}$ are cochain complexes then by a *cochain mapping* $f : C^* \to D^*$ we mean a collection $f = (f^r)_{r \in \mathbf{Z}}$ of

Λ-homomorphisms $f^r : C^r \to D^r$ such that, for each r, the following diagram commutes

$$\begin{array}{ccc} C^r & \xrightarrow{\partial_C^r} & C^{r+1} \\ f^r \downarrow & & f^{r+1} \downarrow \\ D^r & \xrightarrow{\partial_D^r} & D^{r+1} \end{array}$$

Evidently there is a category whose objects are cochain complexes and whose morphisms are cochain mappings. It is conventional to write $\mathfrak{Z}^r(C^*) = \mathrm{Ker}(\partial_C^r)$ and $\mathfrak{B}^r(C^*) = \mathrm{Im}(\partial_C^{r-1})$. The relation, $\partial^r \circ \partial^{r-1} = 0$ implies that $\mathfrak{B}^r(C^*) \subset \mathfrak{Z}^r(C^*)$. The *cohomology* $H^*(C)$ of C^* is then the graded abelian group $\{H^r(C^*)\}_{r\in\mathbb{Z}}$ where

(10.1) $$H^r(C^*) = \mathfrak{Z}^r(C^*)/\mathfrak{B}^r(C^*).$$

The correspondence $C^* \mapsto H^r(C^*)$ is functorial for if $f : C^* \to D^*$ is a cochain mapping then $f^r(\mathfrak{Z}_C^r) \subset \mathfrak{Z}_D^r$ and $f^r(\mathfrak{B}_C^r) \subset \mathfrak{B}_D^r$. Consequently for each $r \in \mathbf{Z}$, f^r induces a mapping $f^r : \mathfrak{Z}^r(C^*)/\mathfrak{B}^r(C^*) \to \mathfrak{Z}^r(D^*)/\mathfrak{B}^r(D^*)$; that is,

(10.2) $$f_C^r : H^r(C^*) \to H^*(D^*).$$

A sequence of cochain complexes $0 \to A^* \xrightarrow{i} B^* \xrightarrow{p} C^* \to 0$ is said to be exact when $0 \to A^n \xrightarrow{i} B^n \xrightarrow{p} C^n \to 0$ is exact for all n; that is, we have a commutative diagram as follows in which the rows are exact:

$$\begin{array}{ccccccccc} 0 & \longrightarrow & A^{n-1} & \xrightarrow{i^{n-1}} & B^{n-1} & \xrightarrow{p^{n-1}} & C^{n-1} & \longrightarrow & 0 \\ & & \downarrow \partial_A^{n-1} & & \downarrow \partial_B^{n-1} & & \downarrow \partial_C^{n-1} & & \\ 0 & \longrightarrow & A^n & \xrightarrow{i^n} & B^n & \xrightarrow{p^n} & C^n & \longrightarrow & 0 \\ & & \downarrow \partial_A^n & & \downarrow \partial_B^n & & \downarrow \partial_C^n & & \\ 0 & \longrightarrow & A^{n+1} & \xrightarrow{i^{n+1}} & B^{n+1} & \xrightarrow{p^{n+1}} & C^{n+1} & \longrightarrow & 0 \end{array}$$

We begin by constructing a homomorphism $\widetilde{\delta} : \mathcal{Z}^n(C) \to H^{n+1}(A)$. Thus choose $\mathbf{c} \in \mathcal{Z}^n(C)$ and define a subset $\widehat{\delta}(\mathbf{c})$ of $\mathcal{Z}^{n+1}(A)$ by

$$\widehat{\delta}(\mathbf{c}) = \{\mathbf{a} \in \mathcal{Z}^{n+1}(A) | i^{n+1}(\mathbf{a}) = \partial_B^n(\beta) \text{ for some } \beta \in B^n \text{ such that } p^n(\beta) = \mathbf{c}\}.$$

It is straightforward to see that

(10.3) $$\widehat{\delta}(\mathbf{c}) \neq \emptyset \text{ for each } \mathbf{c} \in \mathcal{Z}^n(C).$$

Furthermore, denoting by $[\mathbf{a}]$ the class $[\mathbf{a}] = \mathbf{a} + \mathrm{Im}(\partial_A^n) \in H^{n+1}(A)$ then

(10.4) $$\text{If } \mathbf{a}, \mathbf{a}' \in \widehat{\delta}(\mathbf{c}) \text{ then } [\mathbf{a}] = [\mathbf{a}'].$$

Thus we obtain a homomorphism $\widetilde{\delta} : \mathcal{Z}^n(C) \to H^{n+1}(A)$ by

(10.5) $$\widetilde{\delta}(\mathbf{c}) = [\mathbf{a}] \text{ when } \mathbf{a} \in \widehat{\delta}(\mathbf{c}).$$

Moreover $0 \in \widehat{\delta}(\mathbf{c})$ if $\mathbf{c} \in \mathrm{Im}(\partial_C^{n-1})$. Hence $\widetilde{\delta}$ is identically zero when restricted to $\mathfrak{B}_C^n = \mathrm{Im}(\partial_C^{n-1})$ and so induces a Λ-homomorphism $\delta : H^n(C) \to H^{n+1}(A)$. It is a matter of standard technique to show that if $0 \to A^* \xrightarrow{i} B^* \xrightarrow{p} C^* \to 0$ is an exact sequence of cochain complexes then

(10.6) $H^n(A) \xrightarrow{i^*} H^n(B) \xrightarrow{p^*} H^n(C) \xrightarrow{\delta} H^{n+1}(A) \xrightarrow{i^*} H^{n+1}(B)$ is exact.

The connecting homomorphism δ is natural in the following sense; suppose given a commutative diagram of cochain mappings

(10.7) $$\begin{pmatrix} 0 \to & A^* & \xrightarrow{i} & B^* & \xrightarrow{p} & C^* & \to 0 \\ & \downarrow f_A & & \downarrow f_B & & \downarrow f_C & \\ 0 \to & \widetilde{A}^* & \xrightarrow{\widetilde{i}} & \widetilde{B}^* & \xrightarrow{\widetilde{p}} & \widetilde{C}^* & \to 0 \end{pmatrix}$$

in which both rows are exact. To each row there corresponds a connecting homomorphism thus: $\delta : H^*(C) \to H^{*+1}(A); \widetilde{\delta} : H^*(\widetilde{C}) \to H^{*+1}(\widetilde{A})$. These are natural in the sense that for each n the following square commutes:

(10.8) $$\begin{array}{ccc} H^n(C) & \xrightarrow{\delta} & H^{n+1}(A) \\ \downarrow f_C^n & & \downarrow f_A^{n+1} \\ H^n(\widetilde{C}) & \xrightarrow{\widetilde{\delta}} & H^{n+1}(\widetilde{A}). \end{array}$$

As a consequence, a commutative diagram of cochain complexes with exact rows, as in (10.7), gives rise to a commutative ladder for each n as follows:

(10.9) $$\left\{ \begin{array}{ccccccccc} H^n(A) & \xrightarrow{i_*} & H^n(B) & \xrightarrow{p_*} & H^n(C) & \xrightarrow{\delta} & H^{n+1}(A) & \xrightarrow{i^*} & H^{n+1}(B) \\ \downarrow f_A^n & & \downarrow f_B^n & & \downarrow f_C^n & & \downarrow f_A^{n+1} & & \downarrow f_B^{n+1} \\ H^n(\widetilde{A}) & \xrightarrow{\widetilde{i}^*} & H^n(\widetilde{B}) & \xrightarrow{\widetilde{p}^*} & H^n(\widetilde{C}) & \xrightarrow{\widetilde{\delta}} & H^{n+1}(\widetilde{A}) & \xrightarrow{\widetilde{i}^*} & H^{n+1}(\widetilde{B}). \end{array} \right.$$

§11: The cohomology theory of modules:

We begin by recalling briefly the basics of the Eilenberg-Maclane cohomology theory ([10], [42]). Let M be a Λ-module; a *resolution* of M is an exact sequence of Λ-homomorphisms

$$\mathbf{A} = (\cdots \xrightarrow{\partial^A_{n+1}} A_n \xrightarrow{\partial^A_n} A_{n-1} \xrightarrow{\partial^A_{n-1}} \cdots\cdots \xrightarrow{\partial^A_1} A_0 \xrightarrow{\epsilon} M \to 0),$$

abbreviated to $\mathbf{A} = (A_* \to M)$. We say that a resolution $\mathbf{A}$ is *free* (resp. *projective*) when each A_r is a free (resp. *projective*) Λ-module.

(11.1) Every Λ-module has a free resolution.

As free modules are projective we see also that:

(11.2) Every Λ-module has a projective resolution.

Free resolutions will be denoted thus $\mathbf{F} = (F_* \to M)$ and projective resolutions thus $\mathbf{P} = (P_* \to M)$.

If $\mathbf{A} = (A_* \to M_1)$, $\mathbf{B} = (B_* \to M_2)$ are resolutions and $f : M_1 \to M_2$ is a module homomorphism then by a morphism of resolutions, $\varphi : \mathbf{A} \to \mathbf{B}$ *over* f we mean a collection (φ_r) of Λ-homomorphisms completing a commutative diagram

$$\begin{array}{c}\mathbf{A}\\ \downarrow\varphi\\ \mathbf{B}\end{array} = \begin{pmatrix} \cdots\to & A_n \to & A_{n-1}\to & \cdots & \cdots & \to & A_0\to & M_1\to & 0\\ & \downarrow\varphi_n & \downarrow\varphi_{n-1} & & & & \downarrow\varphi_0 & \downarrow f & \\ \cdots\to & B_n\to & B_{n-1}\to & \cdots & \cdots & \to & B_0\to & M_2\to & 0 \end{pmatrix}.$$

Such a morphism is called a *lifting* of f. Given a Λ-homomorphism $f : M_1 \to M_2$, morphisms $\varphi, \psi : \mathbf{A} \to \mathbf{B}$ over f are *homotopic over* f (written $\varphi \simeq_f \psi$) when there exists a collection $\eta = (\eta_r)_{r\geq 0}$ of Λ-homomorphisms $\eta_r : A_r \to B_{r+1}$ such that

(i) $\varphi_0 - \psi_0 = \partial_1\eta_0$ and (ii) $\varphi_r - \psi_r = \partial_{r+1}\eta_r + \eta_{r-1}\partial_r$ for all $r \geq 1$.

Let $f : M_1 \to M_2$ be a Λ -homomorphism; if $\mathbf{P} = (P_* \to M_1)$ is a projective resolution and $\mathbf{B} = (B_* \to M_2)$ is a resolution then there exists a lifting $\tilde{f} : \mathbf{P} \to \mathbf{B}$ of f. Moreover any two liftings of f are then homotopic over f.

For the rest of this chapter we shall abbreviate $\mathrm{Hom}_\Lambda(-,-)$ simply to $\mathrm{Hom}(-,-)$. If M, N are Λ-modules then choosing a projective resolution $\mathbf{P} = (P_* \to M)$ of M we construct a cochain complex (P_r^N, ∂_r^*) thus :

$$P_r^N = \begin{cases} \mathrm{Hom}(P_r, N) & \text{for } r \geq 0 \\ 0 & \text{for } r < 0 \end{cases}$$

where for $r \geq 0$, $\partial_r^* : P_r^N \to P_{r+1}^N$ is the induced map $\partial_r^*(\alpha) = \alpha \circ \partial_r$. We denote by $H^*_{\mathbf{P}}(M, N)$ the cohomology of this cochain complex ; that is

$$H^k_{\mathbf{P}}(M, N) \cong \frac{\mathrm{Ker}(\mathrm{Hom}(P_k, N) \stackrel{\partial^*_{k+1}}{\to} \mathrm{Hom}(P_{k+1}, N))}{\mathrm{Im}(\mathrm{Hom}(P_{k-1}, N) \stackrel{\partial^*_k}{\to} \mathrm{Hom}(P_k, N))} \quad \text{for } k \geq 1.$$

Here $H^0_{\mathbf{P}}(M, N) = \mathrm{Ker}(\mathrm{Hom}(P_0, N) \stackrel{\partial^*_1}{\to} \mathrm{Hom}(P_1, N))$ and $H^k_{\mathbf{P}}(M, N) = 0$ for $k < 0$. If $\mathbf{Q} = (Q_* \to L)$ is a projective resolution of L, $f : M \to L$ is a homomorphism and $\tilde{f} : \mathbf{P} \to \mathbf{Q}$ is a lifting over f then the induced maps f_r^* : $\mathrm{Hom}(Q_r, N) \to \mathrm{Hom}(P_r, N)$ give rise to homomorphisms in cohomology $f_{\mathbf{PQ}} : H^n_{\mathbf{Q}}(L, N) \to H^n_{\mathbf{P}}(M, N)$ which are independent of the particular lifting of f. Given a projective resolution $\mathbf{R} \to K$ and a homomorphism g : $L \to K$ then $(g \circ f)_{\mathbf{PR}} : H^n_{\mathbf{R}}(K, N) \to H^n_{\mathbf{P}}(M, N)$ satisfies the transitivity property:

$$(g \circ f)^*_{\mathbf{PR}} = f^*_{\mathbf{PQ}} g^*_{\mathbf{QR}}.$$

If $\mathbf{P}$, $\mathbf{Q}$ are projective resolutions of M and $H^*_{\mathbf{P}}(M, N)$ (resp. $H^*_{\mathbf{Q}}(M, N)$) is the cohomology computed using $\mathbf{P}$ (resp. $\mathbf{Q}$) then there is a *transition homomorphism* $t_{\mathbf{PQ}} = \mathrm{Id}^*_{\mathbf{PQ}} : H^*_{\mathbf{Q}}(M, N) \to H^*_{\mathbf{P}}(M, N)$. If $\mathbf{R}$ is also a projective resolution of M then

$$t_{\mathbf{PR}} = t_{\mathbf{PQ}} \circ t_{\mathbf{QR}}$$

whilst in the special case where $\mathbf{Q} = \mathbf{P}$ then $t_{\mathbf{PP}} = \mathrm{Id} : H^*_{\mathbf{P}}(M, N) \to H^*_{\mathbf{P}}(M, N)$. It follows that each transition homomorphism $t_{\mathbf{PQ}}$ is an isomorphism and $t^{-1}_{\mathbf{PQ}} = t_{\mathbf{QP}}$. In particular the isomorphism type of $H^n_{\mathbf{P}}(M, N)$ is independent of the particular projective resolution $\mathbf{P}$ of M; moreover:

(11.3) There is a natural equivalence of functors

$$\natural_{\mathbf{P}} : \mathrm{Hom}(-, N) \xrightarrow{\simeq} H^0_{\mathbf{P}}(-, N).$$

We eliminate the subscript $\mathbf{P}$ from $H^n_{\mathbf{P}}$ by making, for each module M, a specific choice of projective resolution $\mathbf{P}(M)$ for M. We then write

$$H^*(M,N) = H^*_{\mathbf{P}(M)}(M,N).$$

If $f : L \to M$ is a homomorphism we write $f^* = f^*_{\mathbf{P}(M)\mathbf{P}(L)} : H^n(M,N) \to H^n(L,N)$. As the projective module P has the resolution $(\cdots \to 0 \to 0 \to P \stackrel{\mathrm{Id}}{\to} P \to 0)$ then:

(11.4) If P is projective then, for any $n \geq 1$, $H^n(P,C) = 0$ for any module C.

§12: The exact sequences in cohomology:

Let $\mathcal{E} = (0 \to A \stackrel{i}{\to} B \stackrel{p}{\to} C \to 0)$ be an exact sequence of Λ-modules. If P is a projective module it is straightforward to see that the induced sequence

$$0 \to \mathrm{Hom}(P,A) \stackrel{i_*}{\to} \mathrm{Hom}(P,B) \stackrel{p_*}{\to} \mathrm{Hom}(P,C) \to 0$$

is exact. More generally, if $P_* \to M$ is a projective resolution of M then we obtain an exact sequence of cochain complexes

(12.1) $$0 \to \mathrm{Hom}(P_*,A) \stackrel{i_*}{\to} \mathrm{Hom}(P_*,B) \stackrel{p_*}{\to} \mathrm{Hom}(P_*,C) \to 0.$$

Applying the cohomology functor to (12.1) gives a long exact sequence

(12.2) $\to H^{n-1}(M,C) \stackrel{\partial}{\to} H^n(M,A) \stackrel{i_*}{\to} H^n(M,B) \stackrel{p_*}{\to} H^n(M,C) \stackrel{\partial}{\to} H^{n+1}(M,A) \to$

in which the terms $H^r(M,-)$ are zero for $r < 0$. In view of the natural equivalence $H^0(M,-) \cong \mathrm{Hom}(M,-)$ the initial portion of the sequence can be written :

$$0 \to \mathrm{Hom}(M,A) \stackrel{i_*}{\to} \mathrm{Hom}(M,B) \stackrel{p_*}{\to} \mathrm{Hom}(M,C) \stackrel{\partial}{\to} H^1(M,A)$$
$$\stackrel{i_*}{\to} H^1(M,B) \stackrel{p_*}{\to} H^1(M,C) \to \cdots$$

The above *direct exact sequence* is covariant with respect to the exact sequence $\mathcal{E}$. In addition, as we now proceed to establish, there is a *reverse exact sequence* which is contravariant with respect to $\mathcal{E}$. Thus suppose given a commutative diagram of Λ-modules and homomorphisms

$$\begin{array}{ccccccccc} 0 \to & X & \stackrel{j}{\to} & Y & \stackrel{\pi}{\to} & Z \to & 0 \\ & \downarrow \epsilon_A & & \downarrow \epsilon_B & & \downarrow \epsilon_C & \\ 0 \to & A & \stackrel{i}{\to} & B & \stackrel{p}{\to} & C \to & 0 \end{array}$$

in which both rows are exact and ϵ_A, ϵ_B, ϵ_C are all surjective. It then follows that:

Lemma 12.3:

(i) $j(\mathrm{Ker}(\epsilon_A) \subset \mathrm{Ker}(\epsilon_B)$;
(ii) $\pi(\mathrm{Ker}(\epsilon_B) \subset \mathrm{Ker}(\epsilon_C)$;
(iii) *the sequence* $0 \to \mathrm{Ker}(\epsilon_A) \stackrel{j}{\to} \mathrm{Ker}(\epsilon_B) \stackrel{\pi}{\to} \mathrm{Ker}(\epsilon_C) \to 0$ *is exact.*

Proof. (i) and (ii) are clear. For (iii) $j : \mathrm{Ker}(\epsilon_A) \to \mathrm{Ker}(\epsilon_B)$ is injective as it is the restriction of the injective mapping $j : X \to Y$, and since $\pi \circ j = 0$ then evidently

$$\mathrm{Im}(j : \mathrm{Ker}(\epsilon_A) \to \mathrm{Ker}(\epsilon_B)) \subset \mathrm{Ker}(\pi : \mathrm{Ker}(\epsilon_B) \to \mathrm{Ker}(\epsilon_C)).$$

Now suppose $y \in \mathrm{Ker}(\epsilon_B)$ satisfies $\pi(y) = 0$. Choose $x \in X$ such that $j(x) = y$. Then $i\epsilon_A(x) = \epsilon_B j(x) = \epsilon_B(y) = 0$. However i is injective so that $\epsilon_A(x) = 0$. As required there exists $x \in \mathrm{Ker}(\epsilon_A)$ such that $j(x) = y$. To complete the proof, suppose $z \in \mathrm{Ker}(\epsilon_C)$ and choose $y_1 \in Y$ such that $\pi(y_1) = z$. Then

$$p\epsilon_B(y_1) = \epsilon_C \pi(y_1) = \epsilon_C(z) = 0,$$

so that $\epsilon_B(y_1) \in \mathrm{Ker}(p) = \mathrm{Im}(i)$. Now choose $a \in A$ such that $i(a) = \epsilon_B(y_1)$, and choose $x \in X$ such that $\epsilon_A(x) = a$. Then $\epsilon_B j(x) = i\epsilon_A(x) = \epsilon_B(y_1)$. Put $y = y_1 - j(x)$. Then $y \in \mathrm{Ker}(\epsilon_B)$ and $\pi(y) = \pi(y_1) - \pi j(x) = \pi(y_1) = z$. □

Lemma 12.4: *Let* $0 \to A \stackrel{i}{\to} B \stackrel{p}{\to} C \to 0$ *be an exact sequence of* Λ*-modules with surjective homomorphisms* $\epsilon_A : P \to A$ *and* $\epsilon_C : Q \to C$ *where* P, Q *are projective. Then there exists a surjective homomorphism* $\epsilon_B : P \oplus Q \to B$ *making the following diagram commute*

$$\begin{array}{ccccccccc} 0 \to & P & \stackrel{j}{\to} & P \oplus Q & \stackrel{\pi}{\to} & Q \to & 0 \\ & \downarrow \epsilon_A & & \downarrow \epsilon_B & & \downarrow \epsilon_C & \\ 0 \to & A & \stackrel{i}{\to} & B & \stackrel{p}{\to} & C \to & 0 \end{array}$$

where $j(x) = (x, 0)$ *and* $\pi(x, y) = y$.

Proof. Since $p : B \to C$ is surjective and Q is projective then there exists a homomorphism $\widetilde{\epsilon}_C : Q \to B$ making the following diagram commute.

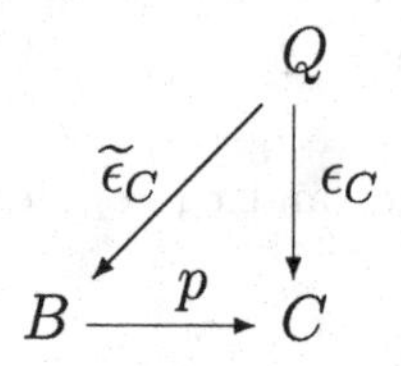

Put $\epsilon_B = (i\epsilon_A, \widetilde{\epsilon}_C) : P \oplus Q \to B$. It is easy to check that the diagram

$$\begin{array}{ccccc} 0 \to & P \xrightarrow{j} & P \oplus Q \xrightarrow{\pi} & Q \to & 0 \\ & \downarrow \epsilon_A & \downarrow \epsilon_B & \downarrow \epsilon_C & \\ 0 \to & A \xrightarrow{i} & B \xrightarrow{p} & C \to & 0 \end{array}$$

commutes. To show ϵ_B is surjective let $b \in B$ and choose $z \in Q$ such that $\epsilon_C(z) = p(b)$. Then $b - \widetilde{\epsilon}_C(z) \in \mathrm{Ker}(p) = \mathrm{Im}(i)$. Choose $a \in A$ such that $i(a) = b - \widetilde{\epsilon}_C(z)$ and choose $x \in P$ such that $\epsilon_A(x) = a$. Then $\epsilon_B(x, z) = b$ and ϵ_B is surjective. □

Theorem 12.5: *Let $\mathcal{E} = (0 \to A \xrightarrow{i} B \xrightarrow{p} C \to 0)$ be an exact sequence of Λ-modules with projective resolutions $P_* \to A$ and $R_* \to C$. Then there is a projective resolution $Q_* \to B$ and an exact sequence of complexes $0 \to P_* \to Q_* \to R_* \to 0$ covering $\mathcal{E}$.*

Proof. Let $(P_* \to A)$ and $(R_* \to C)$ be projective resolutions. For each n write $D_n^A = \mathrm{Im}(\partial_n^A)(= \mathrm{Ker}(\partial_{n-1}^A) \subset P_{n-1})$ and consider the canonical factorization of ∂_n^A through its image

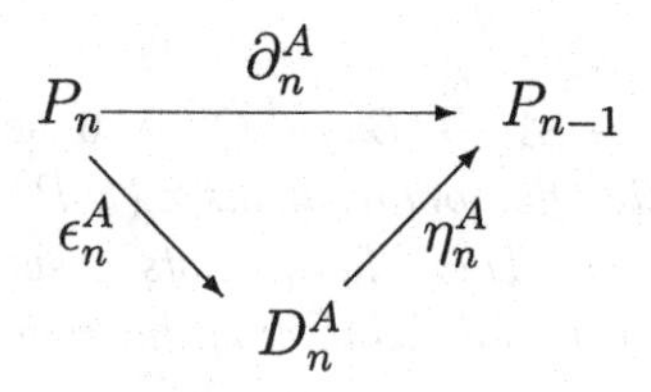

where η_n^A is inclusion and $\epsilon_n^A(x) = \partial_n^A(x)$. Likewise there is a canonical factorization of ∂_n^C. When convenient we will suppress the symbols η_n^A, η_n^C denoting inclusions. For inductive purposes it is convenient to write

$D_0^A = A$, $\epsilon_{-1}^A = \epsilon^A$ and $D_0^C = C$, $\epsilon_{-1}^C = \epsilon^C$. We shall consider commutative diagrams $\mathcal{D}_r$ of the form

$$\mathcal{D}_r = \left(\begin{array}{ccccccccc} & & 0 & & 0 & & 0 & & \\ & & \downarrow & & \downarrow & & \downarrow & & \\ 0 & \longrightarrow & D_r^A & \xrightarrow{i_r} & D_r^B & \xrightarrow{p_r} & D_r^C & \longrightarrow & 0 \\ & & \downarrow \eta_r^A & & \downarrow \eta_r^B & & \downarrow \eta_r^C & & \\ 0 & \longrightarrow & P_r & \xrightarrow{j_r} & P_r \oplus R_r & \xrightarrow{\pi_r} & R_r & \longrightarrow & 0 \\ & & \downarrow \epsilon_{r-1}^A & & \downarrow \epsilon_{r-1}^B & & \downarrow \epsilon_{r-1}^C & & \\ 0 & \longrightarrow & D_{r-1}^A & \xrightarrow{i_{r-1}} & D_{r-1}^B & \xrightarrow{p_{r-1}} & D_{r-1}^C & \longrightarrow & 0 \\ & & \downarrow & & \downarrow & & \downarrow & & \\ & & 0 & & 0 & & 0 & & \end{array}\right)$$

with exact rows and columns, where $j_r(x) = (x, 0)$ and $\pi_r(x, y) = y$, and where η_r^A, η_r^B, η_r^C, are set-theoretic inclusions, so that, in particular, i_r is the restriction of j_r and p_r is the restriction of π_r. We argue by induction. The induction base will be to construct $\mathcal{D}_1$; the induction step will be to construct $\mathcal{D}_n$ from $\mathcal{D}_{n-1}$. It is more convenient to treat the induction step first. Suppose $\mathcal{D}_{n-1}$ is constructed. From it we may extract the top row $\mathcal{E}_{n-1} = (0 \longrightarrow D_{n-1}^A \xrightarrow{i_{n-1}} D_{n-1}^B \xrightarrow{p_{n-1}} D_{n-1}^C \longrightarrow 0)$; in addition we have surjective homomorphisms $\epsilon_n^A : P_n \to D_{n-1}^A, \epsilon_n^C : R_n \to D_{n-1}^C$ where P_n, R_n are projective. We may therefore apply (12.4) to obtain a commutative diagram

$$\mathcal{D}_n = \left(\begin{array}{ccccccccc} & & 0 & & 0 & & 0 & & \\ & & \downarrow & & \downarrow & & \downarrow & & \\ 0 & \longrightarrow & D_n^A & \xrightarrow{i_n} & D_n^B & \xrightarrow{p_n} & D_n^C & \longrightarrow & 0 \\ & & \downarrow \eta_n^A & & \downarrow \eta_n^B & & \downarrow \eta_n^C & & \\ 0 & \longrightarrow & P_n & \xrightarrow{j_n} & P_n \oplus R_n & \xrightarrow{\pi_n} & R_n & \longrightarrow & 0 \\ & & \downarrow \epsilon_{n-1}^A & & \downarrow \epsilon_{n-1}^B & & \downarrow \epsilon_{n-1}^C & & \\ 0 & \longrightarrow & D_{n-1}^A & \xrightarrow{i_{n-1}} & D_{n-1}^B & \xrightarrow{p_{n-1}} & D_{n-1}^C & \longrightarrow & 0 \\ & & \downarrow & & \downarrow & & \downarrow & & \\ & & 0 & & 0 & & 0 & & \end{array}\right)$$

where $\epsilon_n^B : P_n \oplus R_n \to D_{n-1}^B$ is surjective, $D_n^B = \mathrm{Ker}(\epsilon_n^B)$ and where η_n^B is the inclusion of D_n^B in $P_n \oplus R_n$. Thus all columns are exact. Also by (12.3) the sequence $0 \to D_n^A \xrightarrow{i_n} D_n^B \xrightarrow{p_n} D_n^C \to 0$ is exact so that $\mathcal{D}_n$ is a commutative diagram of the required type. With only a slight modification the same argument now establishes the induction base, starting with the exact sequence $\mathcal{E}_0 = (0 \to A \stackrel{i}{\to} B \stackrel{p}{\to} C \to 0)$ (corresponding to $n = 1$) described for purposes of formal agreement as

$$\mathcal{E}_0 = (0 \to D_0^A \xrightarrow{i_0} D_0^B \xrightarrow{p_0} D_0^C \to 0).$$

We now put $Q_n = P_n \oplus R_n$ and define $\partial_n^B : Q_n \to Q_{n-1}$ by $\partial_n^B = \eta_n^B \circ \epsilon_n^B$. It is straightforward to check that

$$Q_* \to B = (\cdots \to Q_n \stackrel{\partial_n^B}{\to} Q_{n-1} \to \cdots \to Q_1 \stackrel{\partial_1^B}{\to} Q_0 \stackrel{\epsilon^B}{\to} B \to 0)$$

is a projective resolution of B. Now consider the following commutative diagram in which at each level $Q_n = P_n \oplus R_n$.

$$\text{(12.6)} \qquad \begin{array}{ccccccccc}
 & & \vdots & & \vdots & & \vdots & & \\
0 & \longrightarrow & P_n & \xrightarrow{j_n} & Q_n & \xrightarrow{\pi_n} & R_n & \longrightarrow & 0 \\
 & & \downarrow \partial_n^A & & \downarrow \partial_n^B & & \downarrow \partial_n^C & & \\
0 & \longrightarrow & P_{n-1} & \xrightarrow{j_{n-1}} & Q_{n-1} & \xrightarrow{\pi_{n-1}} & R_{n-1} & \longrightarrow & 0 \\
 & & \downarrow \partial_{n-1}^A & & \downarrow \partial_{n-1}^B & & \downarrow \partial_{n-1}^C & & \\
 & & \vdots & & \vdots & & \vdots & & \\
 & & \vdots & & \vdots & & \vdots & & \\
 & & \downarrow \partial_1^A & & \downarrow \partial_1^B & & \downarrow \partial_1^C & & \\
0 & \longrightarrow & P_0 & \xrightarrow{j_0} & Q_0 & \xrightarrow{\pi_0} & R_0 & \longrightarrow & 0 \\
 & & \downarrow \epsilon^A & & \downarrow \epsilon^B & & \downarrow \epsilon^C & & \\
0 & \longrightarrow & A & \xrightarrow{i} & B & \xrightarrow{p} & C & \longrightarrow & 0 \\
 & & \downarrow & & \downarrow & & \downarrow & & \\
 & & 0 & & 0 & & 0 & &
\end{array}$$

Clearly we have an exact sequence of projective resolutions $0 \to P_* \to Q_* \to R_* \to 0$ covering the initial exact sequence $0 \to A \to B \to C \to 0$.

□

We note that in the above, whilst $Q_n = P_n \oplus R_n$ in general $\partial_n^B \neq \partial_n^A \oplus \partial_N^C$. Given the exact sequence $0 \to P_* \to Q_* \to R_* \to 0$ just constructed then for any Λ-module N we obtain an exact sequence of cochain complexes

(12.7) $$0 \to \mathrm{Hom}_\Lambda(R_*, N) \to \mathrm{Hom}_\Lambda(Q_*, N) \to \mathrm{Hom}_\Lambda(P_*, N) \to 0$$

Applying the cohomology functor we get the *reverse exact sequence*

(12.8) $$\overset{i^*}{\to} H^{n-1}(A,N) \overset{\delta}{\to} H^n(C,N) \overset{p^*}{\to} H^n(B,N) \overset{i^*}{\to} H^n(A,N) \overset{\delta}{\to} H^{n+1}(C,N) \overset{p^*}{\to}$$

Here $H^r(-, N) = 0$ for $r < 0$; moreover, as $H^0(-, N) \cong \mathrm{Hom}(-, N)$ then the initial portion of the sequence can be written:

$$0 \to \mathrm{Hom}(C,N) \overset{p^*}{\to} \mathrm{Hom}(B,N) \overset{i^*}{\to} \mathrm{Hom}(A,N) \overset{\delta}{\to} H^1(C,N)$$
$$\overset{p^*}{\to} H^1(B,N) \overset{i^*}{\to} H^1(A,N) \to \cdots$$

§13: Module extensions:

If $A, C \in \mathcal{M}\mathrm{od}_\Lambda$, $\mathcal{E}\mathrm{xt}^1_\Lambda(A, C)$ will denote the class of exact sequences of Λ-homomorphisms of the form

$$\mathcal{E} = \big(0 \to \ C \to \ B \to \ A \to \ 0\big).$$

If $\mathcal{E}, \mathcal{F} \in \mathcal{E}\mathrm{xt}^1(A, C)$, recall that $\mathcal{E}$ and $\mathcal{F}$ are *congruent*, written '$\mathcal{E} \equiv \mathcal{F}$', when there is commutative diagram of Λ-homomorphisms of the form

$$\begin{matrix} \mathcal{E} \\ \downarrow \varphi \\ \mathcal{F} \end{matrix} = \begin{pmatrix} 0 \to & C \to & E_0 \to & A \to & 0 \\ & \downarrow \mathrm{Id} & \downarrow \varphi_0 & \downarrow \mathrm{Id} & \\ 0 \to & C \to & F_0 \to & A \to & 0 \end{pmatrix}.$$

By the Five Lemma, congruence is an equivalence relation on $\mathcal{E}\mathrm{xt}^1(A, C)$. We denote by $\mathrm{Ext}^1(A, C)$ the set of equivalence classes in $\mathcal{E}\mathrm{xt}^1(A, C)$ under '$\equiv$'. Given Λ-modules A, C we denote by $\mathcal{T}$ the *trivial extension*

$$\mathcal{T} = (0 \to C \overset{i_C}{\to} C \oplus A \overset{\pi_A}{\to} A \to 0)$$

where $i_C(c) = \binom{c}{0}$ and $\pi_A\binom{c}{a} = a$. An extension $\mathcal{E} = (0 \to C \overset{j}{\to} X \overset{p}{\to} A \to 0)$ is said to *split* when it is congruent to the trivial extension; that is, when there exists an isomorphism $\varphi : X \to C \oplus A$ making the following diagram commute:

$$\begin{array}{ccccccc} 0 \to & C & \xrightarrow{j} & X & \xrightarrow{p} & A & \to 0 \\ & \downarrow \mathrm{Id}_C & & \downarrow \varphi & & \downarrow \mathrm{Id}_A & \\ 0 \to & C & \xrightarrow{i} & C \oplus A & \xrightarrow{\pi} & A & \to 0. \end{array}$$

To repeat (1.2) we see that

(13.1) $\mathcal{E}$ splits $\Longleftrightarrow \mathcal{E}$ splits on the right $\Longleftrightarrow \mathcal{E}$ splits on the left.

To proceed we recall some natural constructions on $\mathcal{E}\mathrm{xt}^1(A, C)$.

Pushout: Let A, C_1, C_2 be Λ-modules; if $f : C_1 \to C_2$ is a Λ-homomorphism and $\mathcal{E} = (0 \to \ C_1 \xrightarrow{i} \ X \xrightarrow{\eta} \ A \to \ 0) \in \mathcal{E}\mathrm{xt}^1(A, C_1)$ we put

$$f_*(\mathcal{E}) = \left(0 \to \ C_2 \xrightarrow{j} \ \varinjlim(f, i) \xrightarrow{\epsilon} \ A \to \ 0\right),$$

where $\varinjlim(f, i) = (C_2 \oplus E_0)/\mathrm{Im}(f \times -i)$ denotes the colimit and j is the injection $j : C_2 \to \varinjlim(f, i); j(x) = [x, 0]$. The correspondence $\mathcal{E} \mapsto f_*(\mathcal{E})$ determines the 'pushout' functor $f_* : \mathcal{E}\mathrm{xt}^1(A, C_1) \to \mathcal{E}\mathrm{xt}^1(A, C_2)$. If in addition $g : C_2 \to C_3$, it is straightforward to see that $(g \circ f)_*(\mathcal{E}) = g_* f_*(\mathcal{E})$. Furthermore, there is a natural transformation $\nu_f : \mathrm{Id} \to f_*$ obtained as follows:

$$\begin{array}{c} \mathcal{E} \\ \downarrow \nu_f \\ f_*(\mathcal{E}) \end{array} = \left(\begin{array}{ccccc} 0 \to & C_1 \xrightarrow{i} & X \to & A \to & 0 \\ & \downarrow f & \downarrow \nu & \downarrow \mathrm{Id} & \\ 0 \to & C_2 \to & \varinjlim(f, i) \to & A \to & 0 \end{array} \right)$$

where $\nu : X \to \varinjlim(f, i)$ is the mapping $\nu(x) = [0, x]$. The correspondence $\mathcal{E} \mapsto f_*(\mathcal{E})$ gives a mapping $f_* : \mathcal{E}\mathrm{xt}^1(A, C) \to \mathcal{E}\mathrm{xt}^1(A, C')$. Moreover

$$(g \circ f)_*(\mathcal{E}) \equiv g_* f_*(\mathcal{E})$$

for any homomorphism $g : C' \to C''$. Generalizing the statement of (1.2) for right splittings we have a criterion for a pushout extension to be trivial.

Proposition 13.2: *Let $\mathcal{E} = (0 \to A \xrightarrow{i} B \xrightarrow{p} C \to 0)$ be an exact sequence of Λ-modules and let $\alpha : A \to N$ be a Λ-homomorphism; then the following two statements are equivalent:*

(i) $\alpha_*(\mathcal{E})$ *splits;*

(ii) *there exists a homomorphism* $\hat{\alpha} : B \to N$ *making the following diagram commute:*

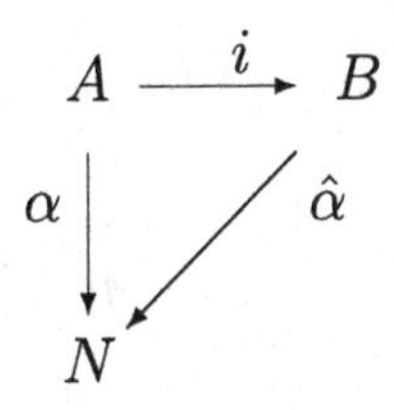

Pullback: Consider an extension $\mathcal{E} = (0 \to C \to X \xrightarrow{\eta} A_2 \to 0) \in \mathcal{E}\mathrm{xt}^1(A_2, C)$ for Λ-modules A_1, A_2, C. If $f : A_1 \to A_2$ is a Λ-homomorphism we put

$$f^*(\mathcal{E}) = (0 \to C \to \varprojlim(\eta, f) \xrightarrow{\epsilon} A_1 \to 0),$$

where $\varprojlim(\eta, f)$ is the fibre product $\varprojlim(\eta, f) = X \underset{\eta, f}{\times} A_1 = \{(x, y) : \eta(x) = f(y)\}$ and $\epsilon : \varprojlim(\eta, f) \to A_1$ is the projection $\epsilon(x, y) = y$. If $g : A_2 \to A_3$ is a Λ -homomorphism, it is straightforward to see that $(g \circ f)^*(\mathcal{E}) \cong f^* \circ g^*(\mathcal{E})$. We obtain the 'pullback functor' $f^* : \mathcal{E}\mathrm{xt}^1(A_2, C) \to \mathcal{E}\mathrm{xt}^1(A_1, C)$ via the correspondence $\mathcal{E} \mapsto f^*(\mathcal{E})$. Moreover, there is a natural transformation $\mu_f : f^* \to \mathrm{Id}$ defined by

$$\begin{matrix} f^*(\mathcal{E}) \\ \downarrow \mu_f \\ \mathcal{E} \end{matrix} = \begin{pmatrix} 0 \to & C \to & \varprojlim(\eta, f) \to & A_1 \to & 0 \\ & \downarrow \mathrm{Id} & \downarrow \mu_0 & \downarrow f & \\ 0 \to & C \to & X \to & A_2 \to & 0 \end{pmatrix}$$

where $\mu_0 : \varprojlim(\eta, f) \to X$ is the projection $\mu_0(x, y) = x$.
Likewise if $f : A' \to A$ is a Λ-homomorphism the correspondence $\mathcal{E} \mapsto f^*(\mathcal{E})$ thus defines a mapping $f^* : \mathcal{E}\mathrm{xt}^1(A, C) \to \mathcal{E}\mathrm{xt}^1(A', C)$. Also

$$(g \circ f)^*(\mathcal{E}) \equiv f^* \circ g^*(\mathcal{E}).$$

Corresponding to (13.2) is the dual statement for a pullback extension to be trivial:

Proposition 13.3: *Let* $\mathcal{E} = (0 \to A \xrightarrow{i} B \xrightarrow{p} C \to 0)$ *be an exact sequence of* Λ*-modules and let* $\gamma : M \to C$ *be a* Λ*-homomorphism; then the following two statements are equivalent:*

(i) $\gamma^*(\mathcal{E})$ *splits;*

(ii) *there exists a homomorphism* $\tilde{\gamma} : M \to B$ *making the following diagram commute:*

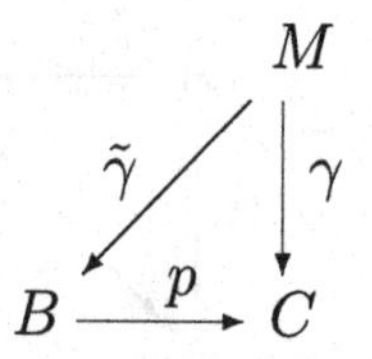

§14: The group structure on $\mathbf{Ext}^1$:

There is a natural group structure on $\mathrm{Ext}^1(A, B)$ which we proceed to describe; thus A_1, A_2, C_1, C_2 be Λ-modules and for $r = 1, 2$ let

$$\mathcal{E}(r) = \big(0 \to \quad C_r \to \quad X_r \to \quad A_r \to \quad 0\big) \in \mathcal{E}\mathrm{xt}^1(A_r, C_r).$$

Then $\mathcal{E}(1) \times \mathcal{E}(2) = (0 \to C_1 \times C_2 \to X_1 \times X_2 \to A_1 \times A_2 \to 0)$ is exact and gives a functorial pairing

$$\times : \mathcal{E}\mathrm{xt}^1(A_1, B_1) \times \mathcal{E}\mathrm{xt}^1(A_2, B_2) \to \mathcal{E}\mathrm{xt}^1(A_1 \oplus A_2, B_1 \oplus B_2).$$

For Λ-modules A, B_1, B_2 there is a functorial pairing, *external sum,*

$$\oplus : \mathcal{E}\mathrm{xt}^1(A, B_1) \times \mathcal{E}\mathrm{xt}^1(A, B_2) \to \mathcal{E}\mathrm{xt}^1(A, B_1 \oplus B_2)$$

given by $\mathcal{E}_1 \oplus \mathcal{E}_2 = \Delta^*(\mathcal{E}_1 \times \mathcal{E}_2)$. where $\Delta : A \to A \times A$ is the diagonal. The addition map $+ : B \times B \to B$ can also be regarded as a Λ-homomorphism

$$\alpha : B \oplus B \to B; \quad \alpha(b_1, b_2) = b_1 + b_2.$$

Let $\mathcal{E}_r \in \mathcal{E}\mathrm{xt}^1(A, B)$ for $r = 1, 2$, and define the '*Baer sum*' ([4], [42]) $\mathcal{E}_1 + \mathcal{E}_2$ by

$$\mathcal{E}_1 + \mathcal{E}_2 = \alpha_*(\mathcal{E}_1 \oplus \mathcal{E}_2)(= \alpha_*\Delta^*(\mathcal{E}_1 \times \mathcal{E}_2)).$$

This gives a functorial pairing

$$+ : \mathcal{E}\mathrm{xt}^1(A, B) \times \mathcal{E}\mathrm{xt}^1(A, B) \to \mathcal{E}\mathrm{xt}^1(A, B).$$

It is straightforward to see that congruence in $\mathcal{E}\mathrm{xt}^1$ is compatible with Baer sum.

From the identity on diagonal maps $(\Delta \times \mathrm{Id}) \circ \Delta = (\mathrm{Id} \times \Delta) \circ \Delta$ it follows that for $\mathcal{E}_1, \mathcal{E}_2, \mathcal{E}_3 \in \mathcal{E}\mathrm{xt}^1(A, B)$ then $\mathcal{E}_1 \oplus (\mathcal{E}_2 \oplus \mathcal{E}_3) \equiv (\mathcal{E}_1 \oplus \mathcal{E}_2) \oplus \mathcal{E}_3$. By associativity of addition (in B) there is a commutative diagram of morphisms in $\mathcal{E}\mathrm{xt}^1$

$$
\begin{array}{ccccc}
 & (\mathcal{E}_1 \oplus \mathcal{E}_2) \oplus \mathcal{E}_3 & \xrightarrow{\ \gamma\ } & \mathcal{E}_1 \oplus (\mathcal{E}_2 \oplus \mathcal{E}_3) & \\
\swarrow{\scriptstyle \nu \oplus 1} & \downarrow & & \downarrow & \searrow{\scriptstyle 1 \oplus \nu} \\
\alpha_*(\mathcal{E}_1 \oplus \mathcal{E}_2) \oplus \mathcal{E}_3 & \nu \circ (\nu \oplus 1) & & \nu \circ (1 \oplus \nu) & \mathcal{E}_1 \oplus \alpha_*(\mathcal{E}_2 \oplus \mathcal{E}_3) \\
\searrow{\scriptstyle \nu} & \downarrow & & \downarrow & \swarrow{\scriptstyle \nu} \\
 & \alpha_*(\alpha_*(\mathcal{E}_1 \oplus \mathcal{E}_2) \oplus \mathcal{E}_3) & \xrightarrow{\ c\ } & \alpha_*(\mathcal{E}_1 \oplus \alpha_*(\mathcal{E}_2 \oplus \mathcal{E}_3)) &
\end{array}
$$

in which each ν is an instance of the natural transformation $\nu_\alpha : \mathrm{Id} \to \alpha_*$ and in which γ and c are congruences. It follows immediately that:

Proposition 14.1: *If $\mathcal{E}_1, \mathcal{E}_2, \mathcal{E}_3 \in \mathcal{E}\mathrm{xt}^1(A, B)$ then*

$$\mathcal{E}_1 + (\mathcal{E}_2 + \mathcal{E}_3) \equiv (\mathcal{E}_1 + \mathcal{E}_2) + \mathcal{E}_3.$$

Commutativity of addition and the obvious congruence $\mathcal{E} \oplus \mathcal{F} \equiv \mathcal{F} \oplus \mathcal{E}$ show that:

Proposition 14.2: *If $\mathcal{E}, \mathcal{F} \in \mathcal{E}\mathrm{xt}^1(A, B)$ then $\mathcal{E} + \mathcal{F} \equiv \mathcal{F} + \mathcal{E}$.*

We denote by $\mathcal{T}$ the trivial extension $\mathcal{T} = (0 \to B \to B \oplus A \to A \to 0)$.

Proposition 14.3: *If $\mathcal{E} \in \mathcal{E}\mathrm{xt}^1(A, B)$ then $\mathcal{E} + \mathcal{T} \equiv \mathcal{E}$.*

We denote by $-\mathcal{E}$ the extension $-\mathcal{E} = (0 \to B \xrightarrow{i} X \xrightarrow{-p} A \to 0)$.

Proposition 14.4: *If $\mathcal{E} \in \mathcal{E}\mathrm{xt}^1(A, B)$ then $\mathcal{E} + (-\mathcal{E}) \equiv \mathcal{T}$.*

Observe that we have a congruence

$$
\begin{array}{ccccccc}
0 \to & B & \xrightarrow{-j} & X & \xrightarrow{p} & A \to & 0 \\
 & \downarrow \mathrm{Id} & & \downarrow -\mathrm{Id} & & \downarrow \mathrm{Id} & \\
0 \to & B & \xrightarrow{j} & X & \xrightarrow{-p} & A \to & 0
\end{array}
$$

so that the additive inverse of $\mathcal{E}$ is equally well represented by the extension

$$(0 \to B \xrightarrow{-j} X \xrightarrow{p} A \to 0).$$

Corollary 14.5: $\mathrm{Ext}^1(A, B)$ *is an abelian group with respect to Baer sum.*

Observe that if Q is projective then any exact sequence $(0 \to N \to X \to Q \to 0)$ splits. However $0 \in \mathrm{Ext}^1(Q, N)$ is defined by the split sequence so that:

(14.6) Let Q be a projective module over Λ; then $\mathrm{Ext}^1(Q, N) = 0$ for any Λ-module N.

If $f : A_1 \to A_2$ is a Λ-homomorphism the correspondence $\mathcal{E} \mapsto f^*(\mathcal{E})$ gives a functor $f^* : \mathcal{E}\mathrm{xt}^1(A_2, N) \to \mathcal{E}\mathrm{xt}^1(A_1, N)$ and f induces a homomorphism of abelian groups $f^* : \mathrm{Ext}^1(A_2, B) \to \mathrm{Ext}^1(A_1, B)$ via the congruence $f^*(\mathcal{E}_1 + \mathcal{E}_2) \equiv f^*(\mathcal{E}_1) + f^*(\mathcal{E}_2)$. Similarly, if $g : B_1 \to B_2$ is a Λ-homomorphism the correspondence $\mathcal{E} \mapsto g_*(\mathcal{E})$ gives a functor $g_* : \mathcal{E}\mathrm{xt}^1(A, B_1) \to \mathcal{E}\mathrm{xt}^1(A, B_2)$ and g induces a homomorphism of abelian groups $g_* : \mathrm{Ext}^1(A, B_1) \to \mathrm{Ext}^1(A, B_2)$ as $g_*(\mathcal{E}_1 + \mathcal{E}_2) \equiv g_*(\mathcal{E}_1) + g_*(\mathcal{E}_2)$.

§15: The cohomological interpretation of Ext^1:

We first give a model for the cohomology group $H^1(M, N)$ which confers certain advantages at the cost of a slight degree of unconventionality. A short exact sequence $\mathcal{P} = (0 \to K \xrightarrow{i} P \xrightarrow{p} M \to 0)$ in which P is projective is called a *projective* 0-*complex*. Given such a projective 0-complex we define

$$H^1_{\mathcal{P}}(M, N) = \frac{\mathrm{Hom}(K, N)}{\mathrm{Im}\big(\mathrm{Hom}(P, N) \xrightarrow{i^*} \mathrm{Hom}(K, N)\big)}.$$

We first show that this construction is functorial on morphisms of projective 0-complexes. Specifically, given a morphism of projective 0-complexes

$$\begin{matrix} \mathcal{P} \\ \downarrow \varphi = \\ \mathcal{Q} \end{matrix} \begin{pmatrix} 0 \to & K \xrightarrow{i} & P \xrightarrow{\epsilon} & M \to & 0 \\ & \downarrow \varphi_+ & \downarrow \varphi_0 & \downarrow \varphi_- & \\ 0 \to & K' \xrightarrow{j} & Q \xrightarrow{\eta} & M' \to & 0 \end{pmatrix}$$

we construct an induced morphism $\varphi^* : H^1_{\mathcal{Q}}(M', N) \to H^1_{\mathcal{P}}(M, N)$. To do this observe that the induced map $(\varphi_+)^* : \mathrm{Hom}(K', N) \to \mathrm{Hom}(K, N)$ has the property that

$$\mathrm{Im}\big(\mathrm{Hom}(Q,N) \xrightarrow{j^*} \mathrm{Hom}(K',N)\big) \subset \mathrm{Im}\big(\mathrm{Hom}(P,N) \xrightarrow{i^*} \mathrm{Hom}(K,N)\big)$$

since if $\alpha = j^*(\beta)$ for $\beta \in \mathrm{Hom}(Q,N)$ then $\varphi_+^*(\alpha) = \varphi_+^* j^*(\beta) = i^*\varphi_0^*(\beta)$. We define

$$\varphi^* : H^1_{\mathcal{Q}}(M',N) \to H^1_{\mathcal{P}}(M,N)$$

to be the homomorphism induced from $\varphi_+^* : \mathrm{Hom}(K',N) \to \mathrm{Hom}(K,N)$. It is straightforward to check that this construction is functorial on morphisms of projective 0-complexes; that is, if $\varphi : \mathcal{P} \to \mathcal{Q}$, $\psi : \mathcal{Q} \to \mathcal{R}$ are morphisms as follows :

$$\begin{matrix} \mathcal{P} \\ \downarrow \varphi \\ \mathcal{Q} \\ \downarrow \psi \\ \mathcal{R} \end{matrix} = \begin{pmatrix} 0 \to & K \to & P \to & M \to & 0 \\ & \downarrow \varphi_+ & \downarrow \varphi_0 & \downarrow \varphi_- & \\ 0 \to & K' \to & Q \to & M' \to & 0 \\ & \downarrow \psi_+ & \downarrow \psi_0 & \downarrow \psi_- & \\ 0 \to & K'' \to & R \to & M'' \to & 0 \end{pmatrix}$$

then $(\psi\varphi)^* = \varphi^*\psi^*$. Next we show that φ^* depends only upon φ_-.

Proposition 15.1: *If $\varphi : \mathcal{P} \to \mathcal{Q}$ has the property that $\varphi_0 = 0$ then $\varphi^* = 0$.*

Proof. We have a commutative diagram

$$\begin{matrix} 0 \to & K \xrightarrow{i} & P \xrightarrow{\epsilon} & M \to & 0 \\ & \downarrow \varphi_+ & \downarrow \varphi_0 & \downarrow 0 & \\ 0 \to & K' \xrightarrow{j} & Q \xrightarrow{\eta} & M' \to & 0 \end{matrix}$$

from which it is clear that $\mathrm{Im}(\varphi_0) \subset \mathrm{Ker}(\eta) = \mathrm{Im}(j)$. Put $\tilde{\varphi} = j^{-1}\varphi_0 : P \to K'$ so that the following diagram commutes:

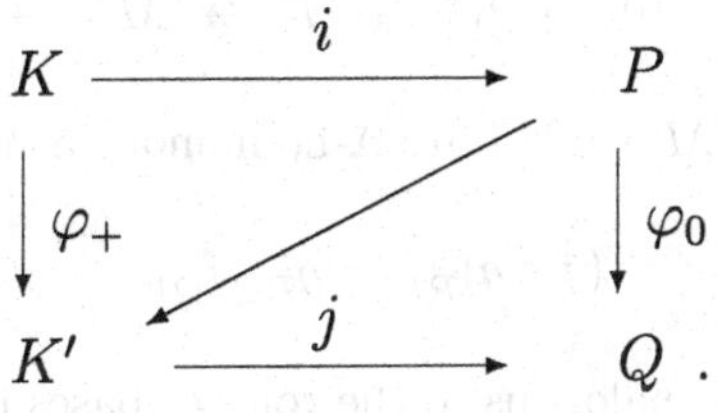

$$\begin{array}{ccc} K & \xrightarrow{\quad i \quad} & P \\ \downarrow \varphi_+ & \swarrow & \downarrow \varphi_0 \\ K' & \xrightarrow{\quad j \quad} & Q\ . \end{array}$$

As $\varphi_+^*(\beta) = i^*(\beta\tilde{\varphi}) \in \mathrm{Im}(\mathrm{Hom}(P,N) \to \mathrm{Hom}(K,N))$ for $\beta \in \mathrm{Hom}(K',N)$ it follows that that $\varphi^* = 0$. □

Corollary 15.2: *If $\varphi_1, \varphi_2 : \mathcal{P} \to \mathcal{Q}$ satisfy $(\varphi_1)_- = (\varphi_2)_-$ then $(\varphi_1)^* = (\varphi_2)^*$.*

Proof. For then $(\varphi_1 - \varphi_2)_- = 0$ so that $(\varphi_1 - \varphi_2)^* = 0$ and so $\varphi_1^* = \varphi_2^*$. □

We can free this construction from global dependence upon the category of projective 0-complexes, at least in part ; suppose that $\mathcal{P}$, $\mathcal{Q}$ are projective 0-complexes

$$\mathcal{P} = (0 \to K \overset{i}{\to} P \overset{\epsilon}{\to} M \to 0)$$

$$\mathcal{Q} = (0 \to K' \overset{j}{\to} Q \overset{\eta}{\to} M' \to 0)$$

and that $f : M \to M'$ is a Λ-homomorphism. By the universal property of projective modules we may construct a lifting $\hat{f}$ of f thus:

$$\begin{matrix} \mathcal{P} \\ \downarrow \hat{f} \\ \mathcal{Q} \end{matrix} = \begin{pmatrix} 0 \to & K \overset{i}{\to} & P \overset{\epsilon}{\to} & M \to & 0 \\ & \downarrow f_+ & \downarrow f_0 & \downarrow f & \\ 0 \to & K' \overset{j}{\to} & Q \overset{\eta}{\to} & M' \to & 0 \end{pmatrix}.$$

We may then define $f^*_{\mathcal{PQ}} : H^1_{\mathcal{Q}}(M', N) \to H^1_{\mathcal{P}}(M, N)$ by $f^*_{\mathcal{PQ}} = (\hat{f})^*$. This is meaningful since if $\tilde{f}$ is another lifting of f then $(\hat{f} - \tilde{f})_+ = 0$ and so $(\tilde{f})^* = (\hat{f})^*$. The construction $f \mapsto f_{\mathcal{PQ}}$ is functorial in the following sense; suppose that $\mathcal{P}$, $\mathcal{Q}$, $\mathcal{R}$ are projective 0-complexes

$$\mathcal{P} = (0 \to K \to P \to M \to 0)$$

$$\mathcal{Q} = (0 \to K' \to Q \to M' \to 0)$$

$$\mathcal{R} = (0 \to K'' \to R \to M'' \to 0)$$

and $g : M \to M'$, $f : M' \to M''$ are Λ-homomorphisms ; then

$$(f \circ g)^*_{\mathcal{PR}} = g^*_{\mathcal{PQ}} f^*_{\mathcal{QR}}.$$

The suffices $\mathcal{P}$, $\mathcal{Q}$ are analogous to the roles of bases in the change of basis formula in linear algebra. In the case of projective 0-complexes over the same module M thus

$$\mathcal{P} = (0 \to K \to P \to M \to 0)$$
$$\mathcal{P}' = (0 \to K' \to P' \to M \to 0)$$

there is the important special case of lifting the identity Id_M; in this case we will write $\tau_{\mathcal{P}'\mathcal{P}} = (\mathrm{Id}_M)^*_{\mathcal{P}'\mathcal{P}}$. Although this notation suppresses the module M it should cause no confusion. We obtain the general transformation rule.

$$f^*_{\mathcal{P}'\mathcal{Q}'} = \tau_{\mathcal{P}'\mathcal{P}} f^*_{\mathcal{P}\mathcal{Q}} \tau_{\mathcal{Q}\mathcal{Q}'}.$$

We proceed by showing that the model we have just given for H^1 is isomorphic to the standard model. Specifically, we will show:

Proposition 15.3: *If* $\mathcal{P} = (0 \to K \to P \to M \to 0)$ *is a projective* 0*-complex and* $\mathbf{Q} = (\cdots \to Q_n \to Q_{n-1} \to \cdots \to Q_0 \to M \to 0)$ *is a projective resolution then there exists an isomorphism* $\natural_{\mathbf{Q}\mathcal{P}} : H^1_{\mathcal{P}}(M,N) \to H^1_{\mathbf{Q}}(M,N)$ *which is natural in the sense that the following diagrams commute:*

$$\begin{array}{ccc} H^1_{\mathcal{Q}}(M,N) & \xrightarrow{\natural_{\mathbf{Q}\mathcal{Q}}} & H^1_{\mathbf{Q}}(M,N) \\ \downarrow \tau_{\mathcal{P}\mathcal{Q}} & & \downarrow t_{\mathbf{PQ}} \\ H^1_{\mathcal{P}}(M,N) & \xrightarrow{\natural_{\mathbf{P}\mathcal{P}}} & H^1_{\mathbf{P}}(M,N) \end{array}$$

Proof. Extend the projective 0-complex $\mathcal{P}$ to a projective resolution

$$\mathbf{P} = (\cdots \to P_n \xrightarrow{\partial_n} P_{n-1} \xrightarrow{\partial_{n-1}} \cdots \xrightarrow{\partial_3} P_2 \xrightarrow{\partial_2} P_1 \xrightarrow{\partial_1} P_0 \xrightarrow{\epsilon} M \to 0)$$

in which $P_0 = P$ and such that, in the factorization

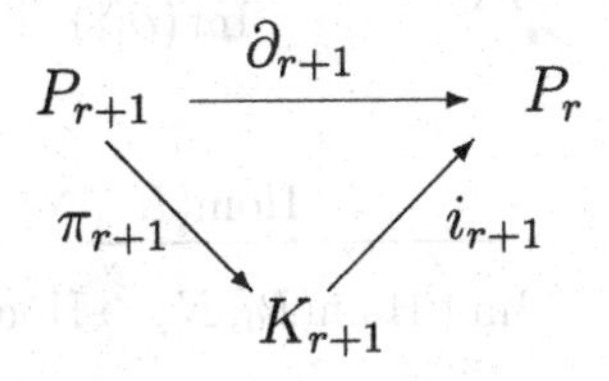

where π_{r+1} is surjective and i_{r+1} is injective, then $K_1 = K$ and $i_1 = i$. In particular, we have a pair of short exact sequences

$$0 \to K_1 \xrightarrow{i_1} P_0 \xrightarrow{\epsilon} M \to 0; \quad 0 \to K_2 \xrightarrow{i_2} P_1 \xrightarrow{\pi_1} K_1 \to 0.$$

From the exact sequence $0 \to \mathrm{Hom}(K_n, N) \xrightarrow{\pi_n^N} \mathrm{Hom}(P_n, N) \xrightarrow{i_n^N} \cdots$ we see that π_r^N is injective for $r = 1, 2$. Now we have a diagram in which the row is exact and the triangle commutes:

$$\begin{array}{ccccccc} 0 & \longrightarrow & \mathrm{Hom}(K_1, N) & \xrightarrow{\pi_1^N} & \mathrm{Hom}(P_1, N) & \xrightarrow{i_2^N} & \mathrm{Hom}(K_2, N) \\ & & & & & \searrow^{\partial_2^N} & \downarrow \pi_2^N \\ & & & & & & \mathrm{Hom}(P_2, N) \end{array}$$

As π_2^N is injective and $\partial_2^N = \pi_2^N \circ i_2^N$ then $\mathrm{Ker}(\partial_2^N) = \mathrm{Ker}(i_2^N)$. By exactness we have an isomorphism $\pi_1^N : \mathrm{Hom}(K_1, N) \to \mathrm{Ker}(\partial_2^N)$. As $\pi_1^N : \mathrm{Hom}(K_1, N) \to \mathrm{Hom}(P_1, N)$ is injective it induces an isomorphism

$$\pi_1^N : \frac{\mathrm{Hom}(K_1, N)}{\mathrm{Im}\left(\mathrm{Hom}(P_0, N) \xrightarrow{i_1^N} \mathrm{Hom}(K, N)\right)}$$

$$\xrightarrow{\simeq} \frac{\mathrm{Im}(\mathrm{Hom}(K_1, N) \xrightarrow{\pi_1^N} \mathrm{Hom}(P_1, N))}{\mathrm{Im}(\mathrm{Hom}(P_0, N) \xrightarrow{\pi_1^N i_1^N} \mathrm{Hom}(P_1, N))}.$$

As above $\mathrm{Ker}(\partial_2^N) = \mathrm{Im}(\mathrm{Hom}(K_1, N) \xrightarrow{\pi_1^N} \mathrm{Hom}(P_1, N))$ and $\partial_1^N = \pi_1^N i_1^N$ so that π_1^N induces an isomorphism, denoted by $\natural_{\mathbf{P}\mathcal{P}}$, thus

$$\natural_{\mathbf{P}\mathcal{P}} = \pi_1^N : \frac{\mathrm{Hom}(K_1, N)}{\mathrm{Im}\left(\mathrm{Hom}(P_0, N) \xrightarrow{i_1^N} \mathrm{Hom}(K, N)\right)} \longrightarrow \frac{\mathrm{Ker}(\partial_2^N)}{\mathrm{Im}(\partial_1^N)}.$$

However

$$H^1_{\mathbf{P}}(M, N) = \frac{\mathrm{Ker}(\partial_2^N)}{\mathrm{Im}\,(\partial_1^N)}$$

and

$$H^1_{\mathcal{P}}(M, N) = \frac{\mathrm{Hom}(K_1, N)}{\mathrm{Im}\left(\mathrm{Hom}(P_0, N) \xrightarrow{i_1^N} \mathrm{Hom}(K, N)\right)}$$

so that $\natural_{\mathbf{P}\mathcal{P}}$ gives an isomorphism $\natural_{\mathbf{P}\mathcal{P}} : H^1_{\mathcal{P}}(M,N) \longrightarrow H^1_{\mathbf{P}}(M,N)$. In general if $\mathbf{Q}$ is an arbitrary projective resolution of M we define

$$\natural_{\mathbf{Q}\mathcal{P}} = t_{\mathbf{QP}}\natural_{\mathbf{P}\mathcal{P}} : H^1_{\mathcal{P}}(M,N) \to H^1_{\mathbf{Q}}(M,N)$$

where $\mathbf{P}$ is a projective resolution of M extending $\mathcal{P}$. That this is independent of $\mathbf{P}$ can be seen easily from the fact that if $\mathbf{P}'$ is also a projective resolution of M extending $\mathcal{P}$ then the following diagram commutes:

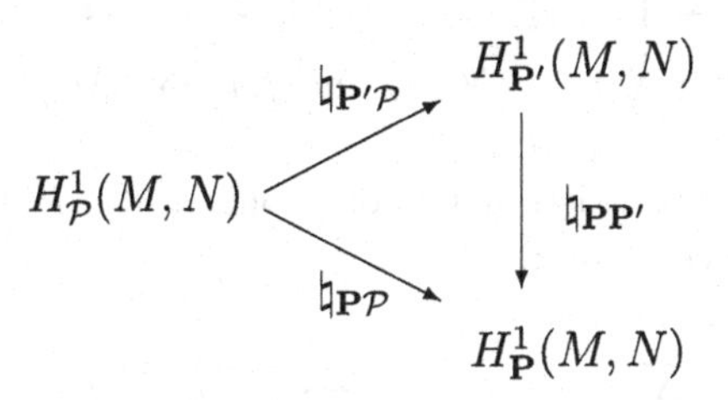

This completes the proof. □

Let $\mathcal{P} = (0 \to K \to P \to M \to 0)$ be a projective 0-complex. Given a Λ-module N and a homomorphism $f : K \to N$ we may form the pushout sequence thus:

$$\begin{matrix} \mathcal{P} \\ \downarrow \nu_f \\ f_*(\mathcal{P}) \end{matrix} = \begin{pmatrix} & \to & K \xrightarrow{i} & P \xrightarrow{\epsilon} & M \to & 0 \\ & & \downarrow f & \downarrow \natural & \downarrow \mathrm{Id} & \\ & 0 \to & N \to & \varinjlim(f,i) \to & M \to & 0 \end{pmatrix}.$$

It follows that we have a group homomorphism $\nu_{\mathcal{P}} : \mathrm{Hom}(K,N) \to \mathrm{Ext}^1(M,N)$ given by $\nu_{\mathcal{P}}(f) = f_*(\mathcal{P})$. If $\alpha \in \mathrm{Ext}^1(M,N)$ is represented by the extension

$$\alpha = (0 \to N \to A \to M \to 0)$$

then from the universal property of projectives there is a morphism of extensions lifting the identity of M thus:

$$\begin{matrix} \mathcal{P} \\ \downarrow g \\ \alpha \end{matrix} = \begin{pmatrix} 0 \to & K \xrightarrow{i} & P \xrightarrow{\epsilon} & M \to & 0 \\ & \downarrow g_+ & \downarrow g_0 & \downarrow \mathrm{Id} & \\ 0 \to & N \to & A \to & M \to & 0 \end{pmatrix}.$$

It follows by general nonsense that $(g_+)_*(\mathcal{P}) \equiv \alpha$ so that:

Proposition 15.4: $\nu_{\mathcal{P}} : \mathrm{Hom}(K,N) \to \mathrm{Ext}^1(M,N)$ *is surjective.*

Proposition 15.5: $\mathrm{Ker}(\nu_{\mathcal{P}}) = \mathrm{Im}(\mathrm{Hom}(P,N) \stackrel{i^*}{\to} \mathrm{Hom}(K,N))$.

Proof. Observe that $f \in \mathrm{Ker}(\nu_{\mathcal{P}})$ if and only if the pushout extension $f_*(\mathcal{P})$ splits. Thus suppose that $f_*(\mathcal{P})$ splits and consider the pushout morphism

$$\begin{matrix} \mathcal{P} \\ \downarrow \\ f_*(\mathcal{P}) \end{matrix} = \left(\begin{matrix} 0 \to & K \stackrel{i}{\to} & P \stackrel{\epsilon}{\to} & M \to & 0 \\ & \downarrow f & \downarrow \natural & \downarrow \mathrm{Id} & \\ 0 \to & N \stackrel{j}{\to} & \varinjlim(f,i) \stackrel{\eta}{\to} & M \to & 0 \end{matrix} \right).$$

Then by (13.2) there exists a homomorphism $\hat{f} : P \to N$ making the following diagram commute;

$$\begin{array}{ccc} K & \xrightarrow{i} & P \\ \downarrow f & \swarrow \hat{f} & \downarrow \natural \\ N & \xrightarrow{j} & \varinjlim(f,i) \end{array}$$

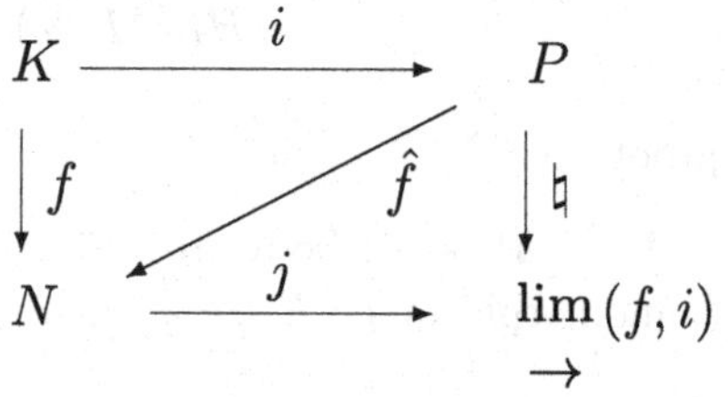

Then $f = i^*(\hat{f})$; equivalently $f \in \mathrm{Im}(\mathrm{Hom}(P,N) \stackrel{i^*}{\to} \mathrm{Hom}(K,N))$. □

As
$$H^1_{\mathcal{P}}(M,N) = \frac{\mathrm{Hom}(K,N)}{\mathrm{Im}(\mathrm{Hom}(P,N) \stackrel{i^*}{\to} \mathrm{Hom}(K,N))}$$

then denoting the map induced by $\nu_{\mathcal{P}}$ by the same symbol we see that:

Proposition 15.6: *$\nu_{\mathcal{P}}$ induces an isomorphism*

$$\nu_{\mathcal{P}} : H^1_{\mathcal{P}}(M,N) \xrightarrow{\simeq} \mathrm{Ext}^1(M,N).$$

This construction is compatible with the transition isomorphisms constructed above in the sense that the following diagram commutes:

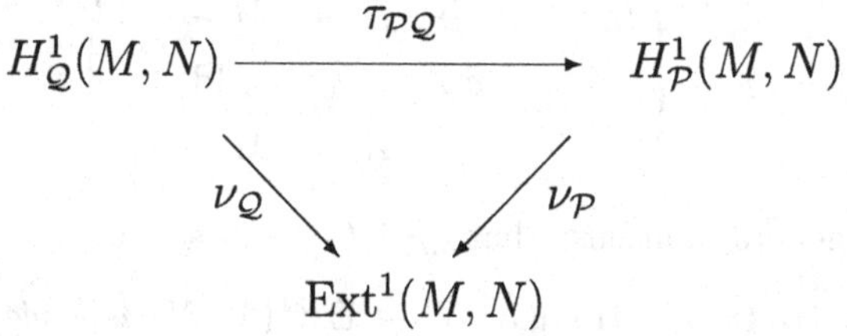

$$\begin{array}{ccc} H^1_{\mathcal{Q}}(M,N) & \xrightarrow{\tau_{\mathcal{P}\mathcal{Q}}} & H^1_{\mathcal{P}}(M,N) \\ & \searrow \nu_{\mathcal{Q}} \quad \nu_{\mathcal{P}} \swarrow & \\ & \mathrm{Ext}^1(M,N) & \end{array}$$

Up to this point we have preserved the notational distinction between cohomology $H^1(-,-)$ and the group of extensions $\mathrm{Ext}^1(-,-)$. In the light of (15.6) it will frequently seem pedantic to maintain the distinction.

Proposition 15.7: *For any module P over Λ the following are equivalent:*

(i) *P is projective;*
(ii) *$Ext^1(P,N) = 0$ for all Λ-modules N;*
(iii) *$H^1(P,N) = 0$ for all Λ-modules N.*

Proof. The cohomological interpretation of Ext^1 given in (15.6) shows that that (ii) $\iff$ (iii). If P is projective then any exact sequence $0 \to N \to X \to P \to 0$ splits and hence (i) $\implies$ (ii) holds. Finally suppose that P is a Λ-module with the property that $\mathrm{Ext}^1(P,N) = 0$ for all modules N and let $p : F \to P$ be a surjective homomorphism from a free module F. Putting $N = \mathrm{Ker}(p)$ it follows from the hypothesis $\mathrm{Ext}^1(P,N) = 0$ that the exact sequence

$$0 \to N \xrightarrow{j} F \xrightarrow{p} P \to 0$$

splits. Thus $F \cong P \oplus N$ so that, P being a direct summand of a free module is projective. Thus (ii) $\implies$ (i), so completing the proof. □

§16: The exact sequences of Ext^1:

Given an exact sequence of Λ-modules $\mathcal{E} = (0 \to A \xrightarrow{i} B \xrightarrow{p} C \to 0)$, there is a mapping $\delta : \mathrm{Hom}_\Lambda(A,N) \to \mathrm{Ext}^1(C,N)$, the *connecting homomorphism*, given by

$$\delta(\alpha) = [\alpha_*(\mathcal{E})].$$

It is straightforward to check that:

(16.1) $\delta : \mathrm{Hom}(A,N) \to \mathrm{Ext}^1(C,N)$ is a homomorphism of abelian groups.

It follows that we have a sequence of abelian groups

(*) $0 \to \mathrm{Hom}(C,N) \xrightarrow{p^*} \mathrm{Hom}(B,N) \xrightarrow{i^*} \mathrm{Hom}(A,N) \xrightarrow{\delta} \mathrm{Ext}^1(C,N) \xrightarrow{p^*} \mathrm{Ext}^1(B,N) \xrightarrow{i^*} \mathrm{Ext}^1(A,N).$

From (16.1) and (12.8) we obtain a commutative ladder of abelian groups

$$\begin{array}{ccccccccccc} 0 \to \mathrm{Hom}(C,N) & \overset{p^*}{\to} & \mathrm{Hom}(B,N) & \overset{i^*}{\to} & \mathrm{Hom}(A,N) & \overset{\delta}{\to} & H^1(C,N) & \overset{p^*}{\to} & H^1(B,N) & \overset{i^*}{\to} & H^1(A,N) \\ \downarrow \mathrm{Id} & & \downarrow \mathrm{Id} & & \downarrow \mathrm{Id} & & \downarrow \nu & & \downarrow \nu & & \downarrow \nu \\ 0 \to \mathrm{Hom}(C,N) & \overset{p^*}{\to} & \mathrm{Hom}(B,N) & \overset{i^*}{\to} & \mathrm{Hom}(A,N) & \overset{\delta}{\to} & \mathrm{Ext}^1(C,N) & \overset{p^*}{\to} & \mathrm{Ext}^1(B,N) & \overset{i^*}{\to} & \mathrm{Ext}^1(A,N) \end{array}$$

in which, by (15.6), the downarrows are isomorphisms. From (12.8) we see that

Theorem 16.2: *The sequence* $(*)$ *is exact for any* Λ*-module* N.

Given an exact sequence $\mathcal{E} = (0 \to A \overset{i}{\to} B \overset{p}{\to} C \to 0)$ there is a mapping $\partial : \mathrm{Hom}(M, C) \to \mathrm{Ext}^1(M, A)$ given by $\partial(\gamma) = [\gamma^*(\mathcal{E})]$. One sees easily that the connecting map $\partial : \mathrm{Hom}(M, C) \to \mathrm{Ext}^1(M, A)$ is additive. From this we obtain a sequence of homomorphisms of abelian groups:

(**) $0 \to \mathrm{Hom}(M, A) \overset{i_*}{\to} \mathrm{Hom}(M, B) \overset{p_*}{\to} \mathrm{Hom}(M, C) \overset{\partial}{\to} \mathrm{Ext}^1(M, A) \overset{i_*}{\to} \mathrm{Ext}^1(M, B) \overset{p_*}{\to} \mathrm{Ext}^1(M, C)$.

Theorem 16.3: *The sequence* **(**)** *is exact for any* Λ*-module* M.

The proof of (16.3) is dual to that of (16.2). We leave the details to the reader.

§17: Example, the cyclic group of order m:

We describe the cyclic group C_m of order m by means of the presentation

$$C_m = \langle y | y^m = 1 \rangle.$$

The trivial $\mathbb{Z}[C_m]$ module $\mathbb{Z}$ admits a free resolution of period two

$$\cdots \overset{\Sigma}{\to} \mathbb{Z}[C_m] \overset{y-1}{\to} \mathbb{Z}[C_m] \overset{\Sigma}{\to} \cdots \overset{y-1}{\to} \mathbb{Z}[C_m] \overset{\Sigma}{\to} \mathbb{Z}[C_m] \overset{y-1}{\to} \mathbb{Z}[C_m] \overset{\epsilon}{\to} \mathbb{Z} \to 0$$

where ϵ is the augmentation homomorphism and $\Sigma = \sum_{r=0}^{m-1} y^r \in \mathbb{Z}[C_m]$. If A is a module over $\mathbb{Z}[C_m]$ then y induces a $\mathbb{Z}[C_m]$ homomorphism $y_* : A \to A$, $y_*(a) = a \cdot y$ and hence we have induced homomorphisms

$$y_* - 1 : A \to A \quad \text{and} \quad \Sigma_* = \sum_{r=0}^{m-1} y_*^r : A \to A$$

Consequently

$$\textbf{(17.1)} \qquad \begin{cases} H^0(\mathbb{Z}, A) &= \operatorname{Ker}(y_* - 1) \\ H^{2k-1}(\mathbb{Z}, A) &= \operatorname{Ker}(\Sigma_*)/\operatorname{Im}(y_* - 1) \quad (1 \le k) \\ H^{2k}(\mathbb{Z}, A) &= \operatorname{Ker}(y_* - 1)/\operatorname{Im}(\Sigma_*) \quad (1 \le k) \end{cases} .$$

By change of emphasis, we have a periodic resolution of the augmentation ideal $I(C_m) = \operatorname{Ker}(\epsilon)$

$$\dots \xrightarrow{\Sigma} \mathbb{Z}[C_m] \xrightarrow{y-1} \mathbb{Z}[C_m] \xrightarrow{\Sigma} \dots \xrightarrow{y-1} \mathbb{Z}[C_m] \xrightarrow{\Sigma} \mathbb{Z}[C_m] \xrightarrow{y-1} I(C_m) \to 0$$

so that

$$\textbf{(17.2)} \qquad \begin{cases} H^0(I(C_m), A) &= \operatorname{Ker}(\Sigma_*) \\ H^{2k-1}(I(C_m), A) &= \operatorname{Ker}(y_* - 1)/\operatorname{Im}(\Sigma_*) \quad (1 \le k) \\ H^{2k}(I(C_m), A) &= \operatorname{Ker}(\Sigma_*)/\operatorname{Im}(y_* - 1) \quad (1 \le k) \end{cases} .$$

In particular, if A is a torsion free module on which C_m acts trivially then:

$$\textbf{(17.3)} \qquad \begin{cases} H^0(\mathbb{Z}, A) &= A \\ H^{2k-1}(\mathbb{Z}, A) &= 0 \qquad (1 \le k) \\ H^{2k}(\mathbb{Z}, A) &= A/mA \quad (1 \le k) \end{cases}$$

and

$$\textbf{(17.4)} \qquad \begin{cases} H^{2k}(I(C_m), A) &= 0 \qquad (0 \le k) \\ H^{2k+1}(I(C_m), A) &= A/mA \quad (0 \le k). \end{cases}$$

Chapter Three

The derived module category

In this chapter, and indeed for the remainder of this book, we will assume for convenience that *all rings under consideration are Noetherian.* Moreover, without further mention, *all modules will be assumed to be finitely generated.* Whilst in more general settings these conditions are prohibitively restrictive, nevertheless in the context in which we shall operate they are sufficient. Given a projective resolution

$$\cdots \stackrel{\partial_{n+1}}{\to} P_n \stackrel{\partial_n}{\to} P_{n-1} \cdots\cdots \stackrel{\partial_1}{\to} P_0 \stackrel{\partial_0}{\to} M \to 0$$

we consider its canonical decomposition into short exact sequences thus:

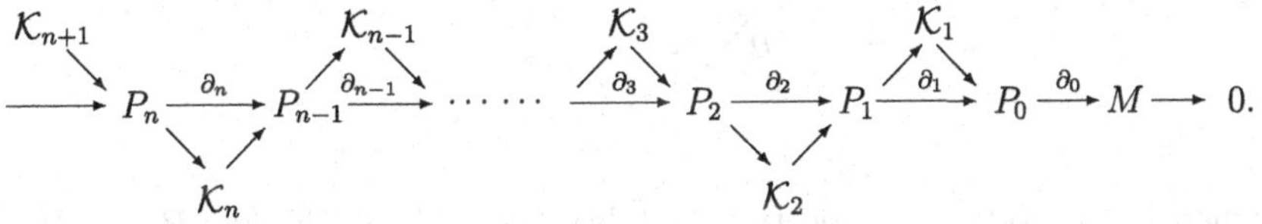

$$\begin{array}{ccccccccccccccc}
\mathcal{K}_{n+1} & & & & \mathcal{K}_{n-1} & & & & \mathcal{K}_3 & & & & \mathcal{K}_1 & & \\
& \searrow & & \nearrow & & \searrow & & \nearrow & & \searrow & & \nearrow & & \searrow & \\
\longrightarrow & & P_n & \xrightarrow{\partial_n} & P_{n-1} & \xrightarrow{\partial_{n-1}} & \cdots\cdots & \xrightarrow{\partial_3} & P_2 & \xrightarrow{\partial_2} & P_1 & \xrightarrow{\partial_1} & P_0 & \xrightarrow{\partial_0} M \longrightarrow & 0. \\
& & & \searrow & \nearrow & & & & & \searrow & \nearrow & & & & \\
& & & & \mathcal{K}_n & & & & & & \mathcal{K}_2 & & & &
\end{array}$$

Ideally we would prefer to consider only free resolutions. However, as will become apparent, the problems we encounter force us from the outset to consider the more general case of projective resolutions. Given another such projective resolution

$$\cdots \stackrel{\partial'_{n+1}}{\to} P'_n \stackrel{\partial'_n}{\to} P'_{n-1} \cdots\cdots \stackrel{\partial'_1}{\to} P'_0 \stackrel{\partial'_0}{\to} M \to 0$$

the corresponding module $\mathcal{K}'_n = \mathrm{Ker}(\partial'_{n-1})$ is *projectively equivalent* to $\mathcal{K}_n$ in the sense that $\mathcal{K}_n \oplus P \cong \mathcal{K}'_n \oplus P'$ for projective modules P, P'. Consequently the projective equivalence class $[\mathcal{K}_n]$ is independent of the particular choice of projective resolution of M. This class is called n^{th} *derivative of* M and is written

$$D_n(M) = [\mathcal{K}_n].$$

The invariants of these derivative modules give an alternative definition of module cohomology. These considerations may be made precise by working in the 'derived module category', the quotient of the category of Λ-modules obtained by quotienting out by morphisms which factorize through a projective.

At the risk of triviality, we point out that, when $\Lambda = \mathbb{Z}[G]$ is an integral group ring, it has been clear since the work of Fox (cf. [20]) that the derivative modules of free resolutions play a fundamental role in the homotopy theory of spaces with fundamental group G.

§18: The derived module category:

We denote by $\mathcal{M}od_\Lambda$ the category of finitely generated right modules over the Noetherian ring Λ. If $f : M \to N$ is a morphism in $\mathcal{M}od_\Lambda$ we say that f *factors through a projective module*, written '$f \approx 0$', when f can be written as a composite $f = \xi \circ \eta$ thus

$$\begin{array}{ccc} M & \xrightarrow{\quad f \quad} & N \\ & {\scriptstyle\eta}\searrow \quad \nearrow{\scriptstyle\xi} & \\ & P & \end{array}$$

where $P \in \mathcal{M}od_\Lambda$ is a projective module and $\eta : M \to P$ and $\xi : P \to N$ are homomorphisms over Λ. As projective modules are direct summands of free modules the condition '$f \approx 0$' is evidently equivalent to the requirement that $f : M \to N$ factors through a free module. We define

$$\mathrm{Hom}^0_\Lambda(M, N) = \{f \in \mathrm{Hom}_\Lambda(M, N) : f \approx 0\}.$$

On taking either ξ or η to be zero we see that $0 \in \mathrm{Hom}^0_\Lambda(M, N)$. Moreover, if $f, g : M \to N$ are Λ-homomorphisms and $f = \alpha \circ \beta$, $g = \gamma \circ \delta$ are factorizations through the projectives P, Q respectively ; then

$$f - g = (\alpha, \gamma)\begin{pmatrix} \beta \\ -\delta \end{pmatrix}$$

is a factorization of $f - g$ through the projective $P \oplus Q$. It follows that:

(18.1) $\mathrm{Hom}^0_\Lambda(M, N)$ is an additive subgroup of $\mathrm{Hom}_\Lambda(M, N)$.

We extend $\approx$ to a binary relation on $\mathrm{Hom}_\Lambda(M, N)$ by means of

$$f \approx g \Longleftrightarrow f - g \approx 0.$$

So extended, $\approx$ is an equivalence relation compatible with composition; that is given Λ-homomorphisms $f, f' : M_0 \to M_1$, $g, g' : M_1 \to M_2$ then:

(18.2) $f \approx f'$ and $g \approx g' \Longrightarrow g \circ f \approx g' \circ f'$.

We obtain the *derived module category*(6) $\mathcal{D}\mathrm{er}(\Lambda)$ whose objects are finitely generated right Λ-modules and where the set of morphisms $\mathrm{Hom}_{\mathcal{D}\mathrm{er}}(M, N)$ is defined by

$$\mathrm{Hom}_{\mathcal{D}\mathrm{er}}(M, N) = \mathrm{Hom}_\Lambda(M, N)/\mathrm{Hom}^0_\Lambda(M, N).$$

Since $\mathrm{Hom}^0_\Lambda(M, N)$ is a subgroup of $\mathrm{Hom}_\Lambda(M, N)$, it follows that:

(18.3) $\mathrm{Hom}_{\mathcal{D}\mathrm{er}}(M, N)$ has the natural structure of an abelian group(7).

Now take Λ-modules M and P and let $i_M : M \to M \oplus P$ and $i_p : M \to M \oplus P$ denote the canonical inclusions and $\pi_M : M \oplus P \to M$, $\pi_P : M \oplus P \to P$ the canonical projections. Then $\mathrm{Id}_M = \pi_M \circ i_M$ and $\mathrm{Id}_{M \oplus P} = i_M \circ \pi_M + i_P \circ \pi_P$. If P is projective then $\mathrm{Id}_{M \oplus P} \approx i_M \circ \pi_M$ and so i_M and π_M are mutually inverse isomorphisms in $\mathcal{D}\mathrm{er}(\Lambda)$; that is

(18.4) $M \oplus P \cong_{\mathcal{D}\mathrm{er}} M$ for any projective module P.

Let $\mathcal{A}\mathrm{b}$ denote the category of abelian groups and homomorphisms. Given an additive functor $E : \mathcal{M}\mathrm{od}(\Lambda) \to \mathcal{A}\mathrm{b}$ we say that E *descends to* $\mathcal{D}\mathrm{er}(\Lambda)$ when there is an additive functor $E : \mathcal{D}\mathrm{er}(\Lambda) \to \mathcal{A}\mathrm{b}$ such that $E = E \circ \natural$ where '$\natural$' denotes the quotient functor $\natural : \mathcal{M}\mathrm{od}(\Lambda) \to \mathcal{D}\mathrm{er}(\Lambda)$; it is straightforward to see that:

(18.5) E descends to $\mathcal{D}\mathrm{er}(\Lambda)$ if and only if $E(P) = 0$ for any projective module P.

Proof. The statement is true whether E is either covariant or contravariant. We give the proof when E is contravariant, the proof when E is covarariant being entirely analogous. Thus suppose that E descends to

(6) Referred to as the *stable module category* by some authors. However, this terminology is misleading for, as we shall see, the isomorphism criterion in this category is *not* stable equivalence but rather *projective equivalence* in the sense of MacLane.
(7) We point out that in the rest of this book, for Λ modules M, N, the terms $\mathrm{Hom}_\Lambda(M, N)$ and $\mathrm{Hom}_{\mathcal{D}\mathrm{er}}(M, N)$ will both be in frequent use and it will be important to distinguish between them.

$\mathcal{D}\mathrm{er}(\Lambda)$ and that P is projective. As $P \cong_{\mathcal{D}\mathrm{er}} 0$ by (18.4) then $E(P) \cong E(0) = 0$ so proving $(\Longrightarrow)$.

Conversely, suppose that $E(P) = 0$ when P is projective and let $f, g : M_1 \to M_2$ be Λ-homomorphisms such that $f \approx g$. Hence we have a factorization

$$\begin{array}{ccc} M_1 & \xrightarrow{f-g} & M_2 \\ & {}_{\alpha}\searrow \quad \nearrow_{\beta} & \\ & P & \end{array}$$

of $f - g$ through a projective P. Application of $E(-)$ yields a factorization

$$\begin{array}{ccc} E(M_2) & \xrightarrow{f^* - g^*} & E(M_1) \\ & {}_{\beta^*}\searrow \quad \nearrow_{\alpha^*} & \\ & E(P) & \end{array}$$

of $(f-g)^* = f^* - g^*$ through $E(P)$. As P is projective then by hypothesis $E(P) = 0$; thus if $f \approx g$ then $f^* = g^* : E(M_2) \to E(M_1)$, so completing the proof. □

It now follows directly from (14.6) that:

(18.6) For any Λ-module N, the correspondence $M \mapsto \mathrm{Ext}^1(M, N)$ defines a contravariant functor $\mathrm{Ext}^1(-, N) : \mathcal{D}\mathrm{er} \to \mathrm{Ab}$.

For Λ-modules M, N we denote by $H^k(M, N)$ the module cohomology of M with coefficients in N. If Q is projective then $H^k(Q, N) = 0$ whenever $k > 0$; consequently:

(18.7) For $k > 0$, the functor $H^k(-, N)$ descends to $\mathcal{D}\mathrm{er}(\Lambda)$.

As a criterion for Λ-modules M_1, M_2 to become isomorphic in $\mathcal{D}\mathrm{er}(\Lambda)$ we have:

Theorem 18.8: *Let M_1, M_2 be modules over Λ; then*

$$M_1 \cong_{\mathcal{D}\mathrm{er}} M_2 \Longleftrightarrow M_1 \oplus P_1 \cong_\Lambda M_2 \oplus P_2$$

for some projective modules P_1, P_2.

Proof. The implication ($\Longleftarrow$) is an easy deduction from (18.4). To show ($\Longrightarrow$), suppose $f : M_1 \to M_2$ is a Λ homomorphism which defines an isomorphism in $\mathcal{D}$er. Then by (18.7), $f^* : H^k(M_2, N) \xrightarrow{\simeq} H^k(M_1, N)$ is an isomorphism for all $k \geq 1$. Let $\eta : F \to M_2$ be a surjective Λ-homomorphism where F is a free module over Λ and let $p : M_1 \oplus F \to M_2$ be the Λ-homomorphism $p(x, y) = f(x) + \eta(y)$. Since both i and f are isomorphisms in $\mathcal{D}$er then it follows from the factorization

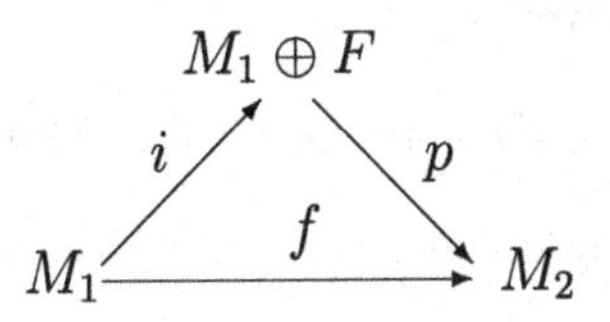

that p is also an isomorphism in $\mathcal{D}$er. Put $K = \mathrm{Ker}(p)$. Applying $H^k(-, K)$ to the exact sequence $0 \to K \xrightarrow{i} M_1 \oplus F \xrightarrow{p} M_2 \to 0$ we get an exact sequence

$$\begin{aligned} 0 \to \mathrm{Hom}_\Lambda(M_2, K) &\xrightarrow{p^*} \mathrm{Hom}_\Lambda(M_1 \oplus F, K) \xrightarrow{i^*} \mathrm{Hom}_\Lambda(K, K) \\ &\xrightarrow{\delta_*} H^1(M_2, K) \xrightarrow{p^*} H^1(M_1 \oplus F, K). \end{aligned}$$

Since p is an isomorphism in $\mathcal{D}$er, $p^* : H^1(M_2, K) \to H^1(M_1 \oplus F, K)$ is an isomorphism and the above exact sequence reduces to

$$0 \to \mathrm{Hom}_\Lambda(M_2, K) \xrightarrow{p^*} \mathrm{Hom}_\Lambda(M_1 \oplus F, K) \xrightarrow{i^*} \mathrm{Hom}_\Lambda(K, K) \to 0.$$

Choosing $r : M_1 \oplus F \to K$ such that $i^*(r) = \mathrm{Id}_K$ we see that $r \circ i = \mathrm{Id}_K$. In particular, r splits the sequence $0 \to K \xrightarrow{i} M_1 \oplus F \xrightarrow{p} M_2 \to 0$ on the left, so that $M_1 \oplus F \cong_\Lambda M_2 \oplus K$. Now F is projective since it is free. To establish the conclusion as stated it remains to show that K is projective.

Since p is an isomorphism in $\mathcal{D}$er then $p^* : H^k(M_2, N) \to H^k(M_1 \oplus F, N)$ is an isomorphism for all $k \geq 1$ and any coefficient module N. Now from the exact sequence

$$\begin{aligned} \ldots \xrightarrow{\delta_*} H^1(M_2, N) &\xrightarrow{p^*} H^1(M_1 \oplus F, N) \xrightarrow{i_*} H^1(K, N) \\ &\xrightarrow{\delta^*} H^2(M_2, N) \xrightarrow{p^*} H^2(M_1 \oplus F, N) \end{aligned}$$

we see easily that $H^1(K, N) = 0$ for all Λ-modules N. Thus K is projective by (15.7), and this completes the proof. $\square$

For Λ-modules M_1, M_2 we formalize the above isomorphism criterion by writing

(18.9) $$M_1 \approx M_2 \Longleftrightarrow M_1 \oplus P_1 \cong_\Lambda M_2 \oplus P_2$$

for some projective Λ-modules P_1, P_2. Clearly $\approx$ is an equivalence relation. Following MacLane (cf [42] p. 101), we then say that M_1, M_2 are *projectively equivalent.* Denoting by $\langle M \rangle$ the isomorphism class in $\mathcal{D}\mathrm{er}(\Lambda)$ of the Λ-module M we then have:

(18.10) $$\langle M \rangle = \{N \in \mathcal{M}\mathrm{od}_{\mathrm{fin}}(\Lambda) | M \approx N\}.$$

We obtain a useful characterization of projective modules which generalizes (15.7).

Proposition 18.11: *For any module P over Λ the following are equivalent:*

(i) *P is projective;*
(ii) $P \cong_{\mathcal{D}\mathrm{er}} 0$;
(iii) $\mathrm{Hom}_{\mathcal{D}\mathrm{er}}(P, N) = 0$ *for all Λ-modules N;*
(iv) $\mathrm{Hom}_{\mathcal{D}\mathrm{er}}(M, P) = 0$ *for all Λ-modules M;*
(v) $\mathrm{End}_{\mathcal{D}\mathrm{er}}(P) = 0$;
(vi) $\mathrm{Ext}^1(P, N) = 0$ *for all Λ-modules N;*
(vii) $H^1(P, N) = 0$ *for all Λ-modules N.*

Proof. It follows immediately from (18.8) that

(a) (i) $\Longleftrightarrow$ (ii).

The implications (iii) $\Longrightarrow$ (v) and (iv) $\Longrightarrow$ (v) are obvious. If (v) holds then $\mathrm{Id}_P \approx 0$. Moreover, if $\alpha : P \to N$ is a Λ-homomorphism then $\alpha = \alpha \circ \mathrm{Id}$ and hence $\alpha \approx 0$; thus we see that (v) $\Longrightarrow$ (iii). A similar proof shows that (v) $\Longrightarrow$ (iv) and hence

(b) (iii) $\Longleftrightarrow$ (iv) $\Longleftrightarrow$ (v).

The implication (ii) $\Longrightarrow$(iii) is likewise obvious. To show that (iii) $\Longrightarrow$ (i), suppose that $\mathrm{Hom}_{\mathcal{D}\mathrm{er}}(P, N) = 0$ for all Λ-modules N, let $\alpha : P \to N$ be a Λ-homomorphism and let $p : M \to N$ be a surjective Λ-homomorphism. To show P is projective we will show there exists a Λ-homomorphism $\widehat{\alpha} : P \to M$ making the following commute:

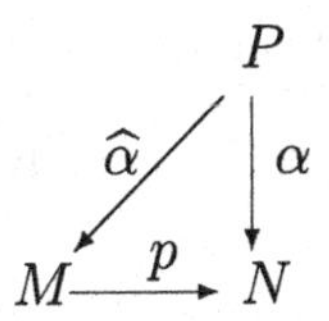

By hypothesis $\mathrm{Hom}_{\mathcal{D}\mathrm{er}}(P, N) = 0$ so that there exists a factorization $\alpha = \xi \circ \eta$ through a projective Q as in the diagram below. As p is surjective, there exists a homomorphism $\widehat{\xi} : Q \to M$ which makes the following diagram commute as indicated:

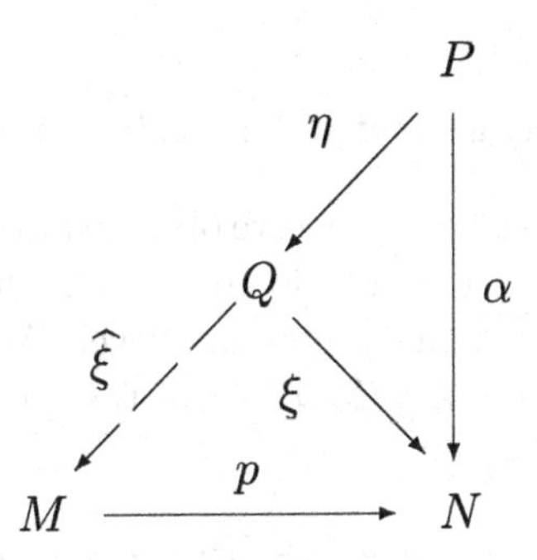

Hence taking $\widehat{\alpha} = \widehat{\xi} \circ \eta$ we see that $p \circ \widehat{\alpha} = \alpha$ which is the condition required to guarantee that P is projective. Thus (iii) $\Longrightarrow$ (i), so establishing the equivalences

(c) $$\text{(i)} \Longleftrightarrow \text{(ii)} \Longleftrightarrow \text{(iii)} \Longleftrightarrow \text{(iv)} \Longleftrightarrow \text{(v)}.$$

However the equivalences (i) $\Longleftrightarrow$ (vi) $\Longleftrightarrow$ (vii) simply re-state (15.7). □

§19: Stable equivalence and projective equivalence:

As we have seen, the isomorphism criterion in $\mathcal{D}\mathrm{er}(\Lambda)$ is given by projective equivalence '$\approx$'. There is an equivalence relation analogous to '$\approx$' which for us is more significant and with which it is frequently confused. We say the Λ-modules M_1, M_2 are *stably equivalent*, written $M_1 \sim M_2$ when for some integers $n_1, n_2 \geq 0$:

(19.1) $$M_1 \sim M_2 \Longleftrightarrow M_1 \oplus \Lambda^{n_1} \cong M_2 \oplus \Lambda^{n_2}.$$

For a Λ-module M, we denote by $[M]$ the corresponding *stable module*; that is:

(19.2) $$[M] = \{N \in \mathcal{M}\mathrm{od}(\Lambda) | M \sim N\}.$$

The following observation is obvious.

(19.3) M is finitely generated if and only if each $N \in [M]$ is finitely generated.

When M is a finitely generated Λ-module, the stable module $[M]$ has the structure of a graph in which the vertices are the isomorphism classes of modules $N \in [M]$ and where we draw an edge $N_1 \to N_2$ when $N_2 \cong N_1 \oplus \Lambda$. As Λ is Noetherian it is straightforward to show (cf [35] p.7) that if N is finitely generated then $N \oplus \Lambda^m \not\cong N$ when $m \geq 1$. Consequently $[M]$ contains no loops; hence:

(19.4) If M is a finitely generated Λ-module then $[M]$ is an infinite tree.

Additionally, the directed tree structure obtained from such a stable module $[M]$ has the property that its roots do not extend infinitely downwards ([35] p.7); thus $[M]$ has a well defined minimum level. As an example, take Λ to be the integral group ring $\Lambda = \mathbb{Z}[Q(28)]$ of the quaternionic group Q_{28} of order 28

$$Q_{28} = \langle x, y | x^7 = y^2; \quad xyx = y \rangle.$$

Following Swan [63], we may depict the stable class [0] by the tree

(19.5)

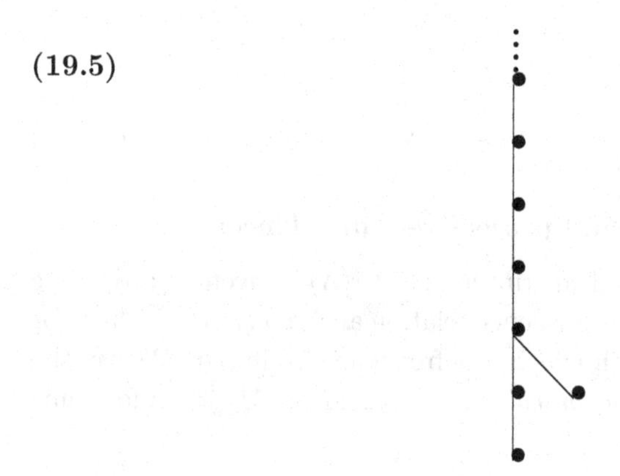

In particular, there is a non-free stably free module of rank 1. Clearly we have:

(19.6) $$M_1 \sim M_2 \Longrightarrow M_1 \approx M_2.$$

The converse is however false; in general, '$\approx$' is a coarser relation than '$\sim$'. In fact (cf [35] pp. 124–125)

(19.7) $\langle M\rangle$ has the structure of a directed graph, in general a disjoint union of trees, in which $[M]$ imbeds as a connected component.

It is useful to think of $[M]$ as a 'polarized state' of $\langle M\rangle$. As an example again take $\Lambda = \mathbb{Z}[Q_{28}]$. In this case $\widetilde{K}_0(\mathbb{Z}[Q_{28}]) \cong \mathbb{Z}/2$. Hence the isomorphism class $\langle 0\rangle$ of the zero module in $\mathcal{D}\mathrm{er}(\Lambda)$ is a graph with two connected components, described by the following diagram in which the imbedding of $[0]$ in $\langle 0\rangle$ is clear.

(19.8)

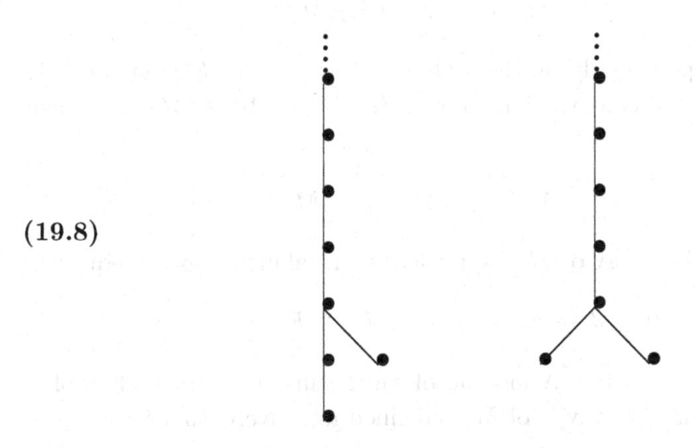

In particular, there are precisely four isomorphically distinct projective modules of rank 1 which, after stabilization via $- \oplus \Lambda$, become isomorphic in pairs.

§20: Syzygies and generalized syzygies:

Given a finitely generated Λ-module M there exists an exact sequence of Λ-homomorphisms

(20.1) $$0 \to J \to \Lambda^a \to M \to 0$$

for some positive integer a. To consider the extent to which J is determined by M suppose given another such exact sequence $0 \to J' \to \Lambda^b \to M \to 0$. It follows from Schanuel's Lemma, (1.4) that

(20.2) $$J' \oplus \Lambda^a \cong J \oplus \Lambda^b.$$

Thus if M is a finitely generated Λ-module there is a well defined stable module $\Omega_1(M)$, the *first syzygy* of M, determined by the rule that

(20.3) $$\Omega_1(M) = [J]$$

whenever J occurs in a short exact sequence of the form (20.1). Given a sequence of the form (20.1) then J is finitely generated as Λ is Noetherian. By (19.3) above it follows that every module in $\Omega_1(M)$ is finitely generated. Moreover, by modifying (20.1) to a sequence of the form $0 \to J \to \Lambda^{a+b} \to M \oplus \Lambda^b \to 0$ we note that

(20.4) $$\Omega_1(M \oplus \Lambda^a) = \Omega_1(M).$$

Thus $\Omega_1(M)$ depends only on the stability class $[M]$ of M. Consequently one may repeat the construction to form $\Omega_2(M) = \Omega_1(\Omega_1(M))$ and then iterate to form

(20.5) $$\Omega_n(M) = \Omega_1(\Omega_{n-1}(M))$$

$\Omega_n(M)$ is the nth-syzygy of M. Equivalently on taking an exact sequence

(20.6) $$0 \to J \to F_{n-1} \to \cdots \to F_1 \to F_0 \to M \to 0$$

in which each F_r is a free Λ module of finite rank, the stable class of J coincides with the nth-syzygy of M as defined iteratively; that is:

(20.7) $$\Omega_n(M) = [J]$$

Given exact sequences in which P, P' are projective

$$0 \to D \to P \to M \to 0; \quad 0 \to D' \to P' \to M \to 0$$

then Schanuel's Lemma implies that $D \oplus P' \cong D' \oplus P$. Otherwise expressed we see from (18.8) that $D \cong_{\mathcal{D}\mathrm{er}} D'$; that is, the projective equivalence class of D depends only upon M. We regard such a module D as a *first derivative* of M and put

(20.8) $$D_1(M) = \langle D \rangle.$$

The projective equivalence class $D_1(M)$ is called the *generalized first syzygy* of M. It is straightforward to see that if Q is projective then:

(20.9) $$D_1(M \oplus Q) = D_1(M).$$

We may again iterate the construction and define $D_n(M) = D_1(D_{n-1}(M))$. From the exact sequence $(0 \to D \to P \to M \to 0)$ we see that if M is projective then $P \cong D \oplus M$ so that, as P is projective, then D is also projective; hence by the isomorphism criterion (18.8) $D \cong_{\mathcal{D}\mathrm{er}(\Lambda)} 0$; that is, $D_1(M) = \langle 0 \rangle$ if M is projective. Iterating we see that

(20.10) If M is projective then $D_n(M) = \langle 0 \rangle$ for all $n \geq 1$.

Take an exact sequence $(0 \to D \xrightarrow{i} P \xrightarrow{\eta} M \to 0)$ in which P is projective and $(\cdots \xrightarrow{\partial_{n+1}} Q_n \xrightarrow{\partial_n} Q_{n-1} \xrightarrow{\partial_{n-1}} \cdots\cdots \xrightarrow{\partial_1} Q_0 \xrightarrow{\eta} D \to 0)$ is a projective resolution for D. On splicing the sequences together we obtain a projective resolution

$$\cdots \xrightarrow{\delta_{n+1}} P_n \xrightarrow{\delta_n} P_{n-1} \xrightarrow{\delta_{n-1}} \cdots \xrightarrow{\delta_2} P_1 \xrightarrow{\delta_1} P_0 \xrightarrow{\epsilon} M \to 0$$

for M where $P_0 = P$, $\delta_1 = i \circ \eta$ $P_r = Q_{r-1}$ for $r \geq 1$ and $\delta_r = \partial_{r-1}$ for $r \geq 2$. As a consequence we see that $H^{n+1}(M, N) = H^n(D, N)$. This is the phenomenon of *dimension shifting* which we may formalize as

(20.11) $$H^{n+1}(M, N) = H^n(D_1(M), N).$$

Inductively we see that

(20.12) $$H^{n+1}(M, N) = H^1(D_n(M), N).$$

Given the cohomological interpretation of $\mathrm{Ext}^1(-, N)$ it now follows that

(20.13) $$H^{n+1}(M, N) = \mathrm{Ext}^1(D_n(M), N).$$

Henceforth without any explicit appeal to Yoneda's cohomological interpretation of higher extensions ([42], [69]), we shall, whenever $n \geq 2$ and where it seems appropriate, adopt the standard convention of writing

$$\mathrm{Ext}^n(M, N) = H^n(M, N).$$

We shall maintain a clear distinction between $\Omega_n(M)$ and $D_n(M)$; $\Omega_n(M)$ is a stability class of finitely generated modules whereas $D_n(M)$ constitutes a single isomorphism class in $\mathcal{D}\mathrm{er}(\Lambda)$. As remarked in (19.7), $D_n(M)$ consists in general of a disjoint union of trees, in which $\Omega_n(M)$ is included as a single connected component.

§21: The corepresentation theorem for Ext^1:

Let Q be projective over Λ. We saw in (11.4) that $H^k(Q, N) = 0$ for all $k \geq 1$ and that, consequently, $H^k(-, N)$ descends to $\mathcal{D}\mathrm{er}(\Lambda)$. By contrast, in general $H^k(M, Q) \neq 0$. We say that M is *n-coprojective* when, for Q projective, $H^k(M, Q) = 0$ for $1 \leq k \leq n$. By the additivity of $H^k(M, -)$ it follows easily that:

(21.1) M is n-coprojective $\iff H^k(M, \Lambda) = 0$ for $1 \leq k \leq n$.

From (18.5) it follows that

(21.2) If M is n-coprojective then $H^k(M, -)$ descends to $\mathcal{D}\mathrm{er}(\Lambda)$ for $1 \leq k \leq n$.

Whilst from (20.13) we see that

(21.3) M is n-coprojective $\iff \mathrm{Ext}^1(D_r(M), \Lambda) = 0$ for $0 \leq r \leq n-1$.

In §12 with any exact sequence of Λ-modules $\mathcal{E} = (0 \to A \to B \xrightarrow{p} C \to 0)$ we associated the reverse exact sequence:

$$0 \to \mathrm{Hom}_\Lambda(C, N) \xrightarrow{p^*} \mathrm{Hom}_\Lambda(B, N) \xrightarrow{i^*} \mathrm{Hom}_\Lambda(A, N) \xrightarrow{\delta} \mathrm{Ext}^1(C, N)$$
$$\xrightarrow{p^*} \mathrm{Ext}^1(B, N) \xrightarrow{i^*} \mathrm{Ext}^1(A, N).$$

This sequence is functorial on the category $\mathcal{M}od_\Lambda$. With some extra hypotheses, it can be modified to give a similar sequence for $\mathcal{D}\mathrm{er}(\Lambda)$; we first consider the sequence

(21.4) $$\mathrm{Hom}_{\mathcal{D}\mathrm{er}}(C, N) \xrightarrow{p^*} \mathrm{Hom}_{\mathcal{D}\mathrm{er}}(B, N) \xrightarrow{i^*} \mathrm{Hom}_{\mathcal{D}\mathrm{er}}(A, N)$$

Proposition 21.5: *If C is 1-coprojective then the sequence (21.4) is exact.*

Proof. The exactness of the above reverse exact sequence implies that $i^* \circ p^* = 0$. Hence it suffices to show that $\mathrm{Ker}(i^*) \subset \mathrm{Im}(p^*)$. Thus suppose that $\beta \in \mathrm{Hom}_\Lambda(B, N)$ is such that $\beta \circ i$ factors through a projective Q.

$$\begin{array}{ccc} A & \xrightarrow{\beta \circ i} & N \\ & \eta \searrow \quad \nearrow \xi & \\ & Q & \end{array}$$

Next note that we have a commutative diagram of exact sequences

$$\begin{array}{ccccccc} 0 \longrightarrow & A \stackrel{i}{\longrightarrow} & B \stackrel{p}{\longrightarrow} & C \longrightarrow & 0 \\ & \downarrow \eta & \downarrow \nu & \downarrow \mathrm{Id}_C & \\ 0 \longrightarrow & Q \stackrel{j}{\longrightarrow} & \varprojlim(\eta, i) \stackrel{\pi}{\longrightarrow} & C \longrightarrow & 0 \end{array}$$

As C is 1-coprojective then $\mathrm{Ext}^1(C, Q) = 0$. Hence the extension $\eta_*(\mathcal{E})$ splits and there exists a homomorphism $\hat{\eta} : B \to Q$ such that $\hat{\eta} \circ i = \eta$. Then $(\beta - \xi\hat{\eta}) \circ i \equiv 0$ so that $\beta - \xi\hat{\eta}$ induces a homomorphism $B/\mathrm{Im}(i) \to N$. On identifying $B/\mathrm{Im}(i)$ with C, there is a unique homomorphism $\gamma : C \to N$ defined by the condition that

$$\gamma(y) = (\beta - \xi \circ \hat{\eta})(x) \quad \text{when} \quad p(x) = y \ \ (x \in B, y \in C).$$

Then $\beta - p^*(\gamma) = \xi \circ \hat{\eta}$; however $\xi \circ \hat{\eta}$ factors through Q so that $p^*(\gamma) \approx \beta$ and $[p^*(\gamma)] = [\beta] \in \mathrm{Hom}_{\mathcal{D}\mathrm{er}}(B, N)$. Hence as required we have $\mathrm{Ker}(i^*) \subset \mathrm{Im}(p^*)$. □

We also have an exact sequence

(21.6)
$$\mathrm{Ext}^1(C, N) \stackrel{p^*}{\to} \mathrm{Ext}^1(B, N) \stackrel{i^*}{\to} \mathrm{Ext}^1(A, N).$$

To connect (21.4) with (21.6) we first consider the possibility of factorizing δ through $\mathrm{Hom}_{\mathcal{D}\mathrm{er}}(A, N)$ as in the diagram below:

(21.7)
$$\begin{array}{ccc} \mathrm{Hom}_\Lambda(A, N) & \stackrel{\delta}{\longrightarrow} & \mathrm{Ext}^1(C, N) \\ & \natural \searrow \quad \nearrow \delta_* & \\ & \mathrm{Hom}_{\mathcal{D}\mathrm{er}}(A, N) & \end{array}$$

Proposition 21.8: *If C is 1-coprojective then the connecting homomorphism $\delta : \mathrm{Hom}_\Lambda(A, N) \to \mathrm{Ext}^1(C, N)$ given by $\delta(\alpha) = [\alpha_*(\mathcal{E})]$ factors through $\mathrm{Hom}_{\mathcal{D}\mathrm{er}}(A, N)$ as in (21.7); in particular, $\mathrm{Im}(\delta_*) = \mathrm{Im}(\delta)$.*

Proof. Suppose that $\alpha \in \mathrm{Hom}_\Lambda(A, N)$ factors through a projective Q:

$$\begin{array}{ccc} A & \stackrel{\alpha}{\longrightarrow} & N \\ & \eta \searrow \quad \nearrow \xi & \\ & Q & \end{array}$$

Then $\alpha_*(\mathcal{E}) = \xi_*\eta_*(\mathcal{E})$. As C is 1-coprojective then $\mathrm{Ext}^1(C,Q) = 0$ and hence $\eta_*(\mathcal{E})] = 0$; thus $\delta(\alpha) = [\alpha_*(\mathcal{E}) = [\xi_*\eta_*(\mathcal{E})] = \xi_*[\eta_*(\mathcal{E})] = \xi_*(0) = 0$. In particular, δ vanishes on $\mathrm{Hom}^0_\Lambda(A,N)$ so that δ factors through $\mathrm{Hom}_{\mathcal{D}\mathrm{er}}(A,N)$ thus:

$$\begin{array}{ccc} \mathrm{Hom}_\Lambda(A,N) & \xrightarrow{\quad\delta\quad} & \mathrm{Ext}^1(C,N) \\ & \natural\searrow \quad \nearrow\delta_* & \\ & \mathrm{Hom}_{\mathcal{D}\mathrm{er}}(A,N) & \end{array}$$

Hence $\mathrm{Im}(\delta_*) = \mathrm{Im}(\delta)$, as claimed. □

Thus if C is 1-coprojective we have a sequence

(21.9) $$\mathrm{Hom}_{\mathcal{D}\mathrm{er}}(B,N) \xrightarrow{i^*} \mathrm{Hom}_{\mathcal{D}\mathrm{er}}(A,N) \xrightarrow{\delta_*} \mathrm{Ext}^1(C,N)$$

Suppose that $[\alpha] \in \mathrm{Hom}_{\mathcal{D}\mathrm{er}}(A,N)$ is the class of $\alpha \in \mathrm{Hom}_\Lambda(A,N)$ and satisfies $\delta_*([\alpha]) = 0$. Then $\delta(\alpha) = 0$ and so there exists $\beta \in \mathrm{Hom}_\Lambda(B,N)$ such that $i^*(\beta) = \alpha$. If $[\beta]$ denotes the class of β in $\mathrm{Hom}_{\mathcal{D}\mathrm{er}}(B,N)$ then $i^*([\beta]) = [\alpha]$. Thus we see that $\mathrm{Ker}(\delta_*) \subset \mathrm{Im}(i^*)$ and hence:

(21.10) The sequence (21.9) is exact.

Collecting our results we obtain the *reverse exact sequence* in $\mathcal{D}\mathrm{er}(\Lambda)$, namely:

Proposition 21.11: *If $\mathcal{E} = (0 \to A \xrightarrow{i} B \xrightarrow{p} C \to 0)$ be an exact sequence over Λ in which C is 1-coprojective then for any Λ-module N we have an exact sequence*:

$$\mathrm{Hom}_{\mathcal{D}\mathrm{er}}(C,N) \xrightarrow{p^*} \mathrm{Hom}_{\mathcal{D}\mathrm{er}}(B,N) \xrightarrow{i^*} \mathrm{Hom}_{\mathcal{D}\mathrm{er}}(A,N) \xrightarrow{\delta_*} \mathrm{Ext}^1(C,N)$$
$$\xrightarrow{p^*} \mathrm{Ext}^1(B,N) \xrightarrow{i^*} \mathrm{Ext}^1(A,N).$$

Thus given an exact sequence $0 \to J \xrightarrow{i} P \xrightarrow{p} A \to 0$ in which A is 1-coprojective then for any Λ-module N we have an exact sequence

$$\mathrm{Hom}_{\mathcal{D}\mathrm{er}}(P,N) \xrightarrow{i_*} \mathrm{Hom}_{\mathcal{D}\mathrm{er}}(J,N) \xrightarrow{\delta} \mathrm{Ext}^1(A,N) \xrightarrow{p_*} \mathrm{Ext}^1(P,N).$$

If in addition P is projective then $\mathrm{Hom}_{\mathcal{D}\mathrm{er}}(P,N) \cong \mathrm{Ext}^1(P,N) = 0$ so that $\delta : \mathrm{Hom}_{\mathcal{D}\mathrm{er}}(J,N) \xrightarrow{\simeq} \mathrm{Ext}^1(A,N)$ is an isomorphism. As J represents $D_1(A)$ we obtain the following *corepresentation theorem* for $\mathrm{Ext}^1(A,-)$.

(21.12) If A is 1-coprojective there is a natural equivalence of functors

$$\delta : \mathrm{Hom}_{\mathcal{D}\mathrm{er}}(D_1(A), -) \xrightarrow{\simeq} \mathrm{Ext}^1(A, -).$$

By induction we obtain:

(21.13) If A is n-coprojective there is a natural equivalence of functors

$$\delta : \mathrm{Hom}_{\mathcal{D}\mathrm{er}}(D_n(A), -) \xrightarrow{\simeq} \mathrm{Ext}^n(A, -).$$

The sequence (21.11) above is the analogue for $\mathcal{D}\mathrm{er}(\Lambda)$ of the reverse exact sequence (12.8) of $\mathcal{E} = (0 \to A \xrightarrow{i} B \xrightarrow{p} C \to 0)$. As we now indicate, there is a corresponding an analogue of the direct sequence (12.2). To see this, first consider the sequence

$$\mathrm{Hom}_{\mathcal{D}\mathrm{er}}(M, A) \xrightarrow{i_*} \mathrm{Hom}_{\mathcal{D}\mathrm{er}}(M, B) \xrightarrow{p_*} \mathrm{Hom}_{\mathcal{D}\mathrm{er}}(M, C)$$

Clearly we have $p_* \circ i_* = 0$ so that $\mathrm{Im}(i_*) \subset \mathrm{Ker}(p_*)$. To show that $\mathrm{Ker}(p_*) \subset \mathrm{Im}(i_*)$ suppose $\beta \in \mathrm{Hom}_\Lambda(M, B)$ is such that $p \circ \beta$ factorizes through a projective Q thus:

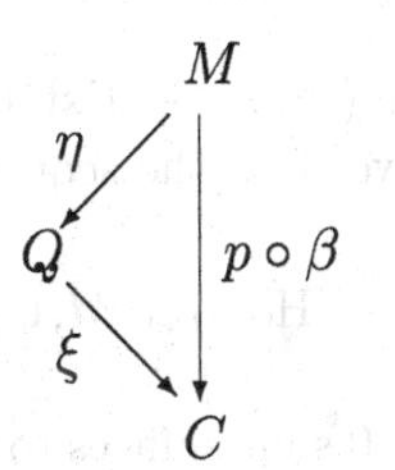

We must find $\alpha \in \mathrm{Hom}_\Lambda(M, A)$ such that $i_*(\alpha) \approx \beta$. Since Q is projective and p is surjective, there exists a homomorphism $\widehat{\xi} : Q \to B$ such that $\xi = p \circ \widehat{\xi}$; in particular the diagram below commutes:

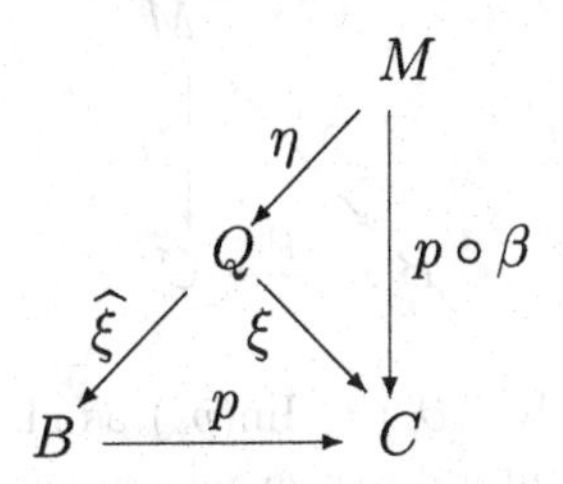

Now $\beta - \widehat{\xi} \circ \eta \in \mathrm{Hom}_\Lambda(M, \mathrm{Ker}(p))$ since $p(\beta - \widehat{\xi} \circ \eta) \equiv 0$. Since $\mathrm{Ker}(p) = \mathrm{Im}(i)$ we may define $\alpha = i^{-1}(\beta - \widehat{\xi} \circ \eta) \in \mathrm{Hom}_\Lambda(M, A)$. Then $\beta - i_*(\alpha) \approx 0$

by virtue of the identity $\beta - i_*(\alpha) = \widehat{\xi} \circ \eta$ and hence $i_*(\alpha) \approx \beta$. Thus the sequence

$$\mathrm{Hom}_{\mathcal{D}\mathrm{er}}(M, A) \xrightarrow{i_*} \mathrm{Hom}_{\mathcal{D}\mathrm{er}}(M, B) \xrightarrow{p_*} \mathrm{Hom}_{\mathcal{D}\mathrm{er}}(M, C)$$

is exact. Furthermore, it follows from (12.2) and (15.6) that the sequence

$$\mathrm{Ext}^1(M, A) \xrightarrow{i_*} \mathrm{Ext}^1(M, B) \xrightarrow{p_*} \mathrm{Ext}^1(M, C)$$

is also exact. To link the two sequences together first recall that the connecting homomorphism $\partial : \mathrm{Hom}_\Lambda(M, C) \to \mathrm{Ext}^1(M, A)$ is defined by $\partial(\gamma) = [\gamma^*(\mathcal{E})]$. If $\gamma = \xi \circ \eta$ is a factorization of γ through a projective Q then $\xi^*(\mathcal{E}) = 0 \in \mathrm{Ext}^1(Q, A)$ as Q is projective. Thus $\gamma^*(\mathcal{E}) = \eta^*\xi^*(\mathcal{E})$ is also trivial, so that $\partial(\gamma) = 0$ for $\gamma \in \mathrm{Hom}^0_\Lambda(M, C)$. Hence the connecting homomorphism ∂ of (12.2) induces a homomorphism $\partial_* : \mathrm{Hom}_{\mathcal{D}\mathrm{er}}(M, C) \to \mathrm{Ext}^1(M, A)$. It is clear from the above that

$$\mathrm{Im}(\partial_*) = \mathrm{Im}(\partial) = \mathrm{Ker}(i^* : \mathrm{Ext}^1(M, A) \to \mathrm{Ext}^1(M, B)).$$

Thus $\mathrm{Hom}_{\mathcal{D}\mathrm{er}}(M, C) \xrightarrow{\partial_*} \mathrm{Ext}^1(M, A) \xrightarrow{i_*} \mathrm{Ext}^1(M, B) \xrightarrow{p_*} \mathrm{Ext}^1(M, C)$ is well defined and exact. Finally we show the sequence

$$\mathrm{Hom}_{\mathcal{D}\mathrm{er}}(M, B) \xrightarrow{p_*} \mathrm{Hom}_{\mathcal{D}\mathrm{er}}(M, C) \xrightarrow{\partial_*} \mathrm{Ext}^1(M, A)$$

is exact. Evidently $\partial_* \circ p_* = 0$ so it suffices to show that $\mathrm{Ker}(\partial_*) \subset \mathrm{Im}(p_*)$. Suppose that $\gamma \in \mathrm{Hom}_\Lambda(M, C)$ satisfies $\partial(\gamma) = 0$; that is, $\gamma^*(\mathcal{E})$ splits. Thus there exists a homomorphism $\hat{\gamma} : M \to B$ making the following diagram commute:

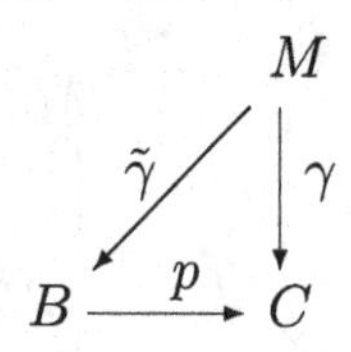

Then $p_*(\tilde{\gamma}) = \gamma$ so that $\mathrm{Ker}(\partial_*) \subset \mathrm{Im}(p_*)$ as claimed; to summarize, we have shown the existence of a direct exact sequence on the derived module category:

$$\mathrm{Hom}_{\mathcal{D}\mathrm{er}}(M,A) \xrightarrow{i_*} \mathrm{Hom}_{\mathcal{D}\mathrm{er}}(M,B) \xrightarrow{p_*} \mathrm{Hom}_{\mathcal{D}\mathrm{er}}(M,C) \xrightarrow{\partial_*} \mathrm{Ext}^1(M,A)$$
$$\xrightarrow{i_*} \mathrm{Ext}^1(M,B) \xrightarrow{p_*} \mathrm{Ext}^1(M,C).$$

We note that, unlike (21.11) the exactness of the above makes no demand of coprojectivity on any of the modules involved. A more subtle point however is that, in order for the terms $\mathrm{Ext}^1(M,A)$, $\mathrm{Ext}^1(M,B)$, $\mathrm{Ext}^1(M,C)$ to be functorial on $\mathcal{D}\mathrm{er}(\Lambda)$, we require M to be 1-coprojective. Thus we have:

Theorem 21.14: *Let $\mathcal{E} = (0 \to A \xrightarrow{i} B \xrightarrow{p} C \to 0)$ be an exact sequence of Λ-modules; then for any 1-coprojective Λ-module M there is an exact sequence*

$$\mathrm{Hom}_{\mathcal{D}\mathrm{er}}(M,A) \xrightarrow{i_*} \mathrm{Hom}_{\mathcal{D}\mathrm{er}}(M,B) \xrightarrow{p_*} \mathrm{Hom}_{\mathcal{D}\mathrm{er}}(M,C) \xrightarrow{\partial_*} \mathrm{Ext}^1(M,A)$$
$$\xrightarrow{i_*} \mathrm{Ext}^1(M,B) \xrightarrow{p_*} \mathrm{Ext}^1(M,C)$$

which is functorial on $\mathcal{D}\mathrm{er}(\Lambda)$.

§22: De-stabilization:

The following de-stabilization lemma is essential at a number of points:

Proposition 22.1: *Let $0 \to J \oplus Q_0 \xrightarrow{j} Q_1 \to M \to 0$ be an exact sequence of Λ-modules in which Q_0, Q_1 are projective; if M is 1-coprojective then $Q_1/j(Q_0)$ is projective.*

Proof. Let $i : J \to J \oplus Q_0$ and $\pi : J \oplus Q_0 \to J$ be the mappings $i(x) = (x, 0)$ and $\pi(x,y) = x$. Put $L = \varinjlim(i \circ \pi, j)$; then we have a commutative diagram

$$\begin{matrix} \mathcal{E} \\ \downarrow \nu(\alpha) \\ (i\circ\pi)_*(\mathcal{E}) \end{matrix} \quad = \quad \begin{pmatrix} 0 \to & J \oplus Q_0 \xrightarrow{j} & Q_1 \to & M \to & 0 \\ & \downarrow i \circ \pi & \downarrow \nu & \downarrow \mathrm{Id} & \\ 0 \to & J \oplus Q_0 \to & L \to & M \to & 0 \end{pmatrix}$$

where $\nu : Q_1 \to L = \varinjlim(i \circ \pi, j)$ is the natural map. As $H^1(M, Q_0) = 0$ we have isomorphisms $\pi_* : H^1(M, J \oplus Q_0) \to H^1(M, J)$ and $i_* : H^1(M, J) \to H^1(M, J \oplus Q_0)$. As $\pi_* \circ i_* = \mathrm{Id}$ then $i_* \circ \pi_* = \mathrm{Id} : H^1(M, J \oplus Q_0) \to H^1(M, J \oplus Q_0)$. Let $c = c_{\mathcal{E}} \in H^1(M, J \oplus Q_0)$ be the element classifying the extension $\mathcal{E}$. Then $(i \circ \pi)_*(\mathcal{E})$ is classified by $(i \circ \pi)_*(c) = c$. Thus $(i \circ \pi)_*(\mathcal{E})$ is congruent to $\mathcal{E}$, so that $L \cong Q_1$, and in particular, L is projective. Now put $S = \varinjlim(\pi, j)$. It is straightforward to check that $S = Q_1/j(Q_0)$, thus it suffices to show that S is projective. We have a commutative diagram

$$\begin{matrix} \pi_*(\mathcal{E}) \\ \downarrow \nu(\alpha) \\ (i \circ \pi)_*(\mathcal{E}) \end{matrix} = \begin{pmatrix} 0 \to & J \to & S \to & M \to & 0 \\ & \downarrow i & \downarrow \mu & \downarrow \mathrm{Id} & \\ 0 \to & J \oplus Q_0 \to & L \to & M \to & 0 \end{pmatrix}$$

where $\mu : S \to L$ is the induced map on pushouts. We obtain a commutative diagram for any coefficient module B;

$$\begin{matrix} H^1(M,B) \to & H^1(L,B) \to & H^1(J \oplus Q_0, B) \to & H^2(M,B) \\ \downarrow \mathrm{Id} & \downarrow \mu^* & \downarrow i^* & \downarrow \mathrm{Id} \\ H^1(M,B) \to & H^1(S,B) \to & H^1(J,B) \to & H^2(M,B) \end{matrix}.$$

Clearly $\mathrm{Id} : H^k(M,B) \to H^k(M,B)$ is an isomorphism for $k = 1, 2$, and as Q_0 is projective, $i^* : H^1(J \oplus Q_0, B) \to H^1(J,B)$ is an isomorphism. Thus we see that $\mu^* : H^1(L,B) \to H^1(S,B)$ is surjective. However, $H^1(L,B) = 0$ since L is projective. Hence $H^1(S,B) = 0$ for all coefficient modules B, so that, by (15.7), $S = Q_1/j(Q_0)$ is projective, as desired. $\square$

It is natural to imagine that each $D \in D_1(M)$ is a generalized syzygy; that is, occurs in an exact sequence $(0 \to D \to Q \to M \to 0)$ where Q is projective. This simple view is false in general; however, as we now show, it is correct under the additional hypothesis that M is 1-coprojective:

Proposition 22.2: *Let M be a 1-coprojective Λ-module; then each $D \in D_1(M)$ occurs in an exact sequence $0 \to D \to S \to M \to 0$ where S is projective.*

Proof. We may choose an exact sequence $\mathcal{F} = (0 \to K \xrightarrow{i} F_X \xrightarrow{p} M \to 0)$ where X is a set of generators for M. Since $D \in D_1(M)$ then, by definition, there are projective modules P_1, P_2 such that $D \oplus P_1 \cong K \oplus P_2$. We may now modify $\mathcal{F}$ to get an exact sequence $0 \to K \oplus P_2 \xrightarrow{i'} F_X \oplus P_2 \xrightarrow{p'} M \to 0$ where $i' = i \oplus \mathrm{Id} : K \oplus P_2 \xrightarrow{i'} F_X \oplus P_2$ and where p' is the obvious composite of p with the projection $F_X \oplus P_2 \to F_X$. Now let $\varphi : D \oplus P_1 \cong K \oplus P_2$ be an isomorphism. Putting $j = i' \circ \varphi$ we get an exact sequence $0 \to D \oplus P_1 \xrightarrow{j} F_X \oplus P_2 \xrightarrow{p'} M \to 0$. Evidently j induces an imbedding $j : P_1 \to F_X \oplus P_2$.

Putting $S = (F_X \oplus P_2)/j(P_1)$ and identifying D with $(D \oplus P_1)/P_1$ we get an exact sequence

$$0 \to D \xrightarrow{j_*} S \xrightarrow{\pi} M \to 0$$

where j_*, π are induced by j and p' respectively. Now $H^1(M, \Lambda) = 0$, by hypothesis, so that, by (22.1), S is projective. $\square$

The following statement for n-coprojective modules can be obtained from (22.2) by induction; we leave the details to the reader.

Proposition 22.3: *Let M be a n-coprojective Λ-module where $n \geq 2$; then each $D \in D_n(M)$ occurs in an exact sequence $0 \to D \to S_{n-1} \to \cdots \to S_0 \to M \to 0$ where $S_0, \ldots, S_{n-1}$ are projective.*

Proposition 22.4: *Let $0 \to M' \to S \to M \to 0$ be an exact sequence of Λ-homomorphisms in which S is finitely generated and stably free; then $M' \in \Omega_1(M)$.*

Proof. Choose an integer $n \geq 1$ such that $S \oplus \Lambda^n \cong \Lambda^{m+n}$ and modify the exact sequence as follows:

$$0 \to M' \oplus \Lambda^n \xrightarrow{\left(\begin{smallmatrix} i & 0 \\ 0 & \mathrm{Id} \end{smallmatrix}\right)} S \oplus \Lambda^n \xrightarrow{\pi} M \to 0$$

where $\pi(x, \lambda) = p(x)$. We get an exact sequence $0 \to M' \oplus \Lambda^n \to \Lambda^{m+n} \to M \to 0$ from which we see that $M' \oplus \Lambda^n \in \Omega_1(M)$. From the definition of the stable module $\Omega_1(M)$ it follows, as claimed, that $M' \in \Omega_1(M)$. $\square$

When M is 1-coprojective we obtain the following converse to (22.4):

Proposition 22.5: *Let M be finitely generated and 1-coprojective; if $M' \in \Omega_1(M)$ then there exists an exact sequence of Λ-homomorphisms $0 \to M' \to S \to M \to 0$ in which S is finitely generated and stably free.*

Proof. We repeat the proof of (22.2) with the constraints that $|X| = m$ and $P_i = \Lambda^{n_i}$ for some positive integers m, n_1, n_2. Then the module S occurs in an exact sequence $0 \to \Lambda^{n_1} \to \Lambda^{m+n_2} \to S \to 0$. As S is projective then the sequence splits to give $S \oplus \Lambda^{n_1} \cong \Lambda^{m+n_2}$ and S is stably free as claimed. $\square$

For $n \geq 2$ there is a corresponding statement to (22.3) in which the condition 'stably free' of (22.5) can be improved to 'free' by internal stabilization as follows; we leave the details to the reader.

Proposition 22.6: *Let M be a finitely generated Λ-module and suppose that M is n-coprojective where $n \geq 2$; then each $J \in \Omega_n(M)$ occurs in an exact sequence $0 \to J \to \Lambda^{a_{n-1}} \to \cdots \to \Lambda^{a_0} \to M \to 0$ for some positive integers $a_0, \ldots, a_{n-1}$.*

§23: The dual to Schanuel's Lemma:

There is a dual form of Schanuel's Lemma which holds in restricted circumstances. First consider a pair of exact sequences in $\mathcal{M}\mathrm{od}_\Lambda$

$$\mathcal{E} = (0 \to K \xrightarrow{i} P \xrightarrow{p} M \to 0); \;\; \mathcal{F} = (0 \to K \xrightarrow{j} Q \xrightarrow{q} N \to 0)$$

in which P and Q are projective and form the pushout square

$$\begin{array}{ccc} K & \xrightarrow{\;\;j\;\;} & Q \\ \downarrow i & & \downarrow \eta_Q \\ P & \xrightarrow{\;\;\eta_P\;\;} & \varinjlim(i,j) \end{array}$$

where $\varinjlim(i,j) = (P \oplus Q)/\mathrm{Im}(i \times -j)$. Taking the canonical inclusions and projections

$$i_P : P \to P \oplus Q; \quad i_P(x) = (x, 0) \; \pi_P : P \oplus Q \to P; \quad \pi_P(x,y) = x$$

$$i_Q : Q \to P \oplus Q; \quad i_Q(y) = (0, y) \; \pi_Q : P \oplus Q \to Q; \quad \pi_Q(x,y) = y$$

then put $\eta_P = \mu \circ i_P$; $\eta_Q = \mu \circ i_Q$ where $\mu : P \oplus Q \to (P \oplus Q)/\mathrm{Im}(i \times -j)$ is the identification map. Then there are internal direct sum decompositions

(23.1) $$P \oplus \mathrm{Im}(j) = \mathrm{Im}(i_P) \dot{+} \mathrm{Im}(i \times -j)$$

(23.2) $$\mathrm{Im}(i) \oplus Q = \mathrm{Im}(i_Q) \dot{+} \mathrm{Im}(i \times -j).$$

Note that $p \circ \pi_P : P \oplus Q \to M$ vanishes on $\mathrm{Im}(i \times -j)$ and so induces a homomorphism

$$\xi_P : \varinjlim(i,j) \to M; \quad \xi_P([x,y]) = p(x).$$

Similarly $q \circ \pi_Q : P \oplus Q \to N$ vanishes on $\mathrm{Im}(i \times -j)$ and so induces a homomorphism

$$\xi_Q : \varinjlim(i,j) \to N; \quad \xi_Q([x,y]) = q(y).$$

From the direct sum decompositions (23.1), (23.2) we obtain exact sequences

(23.3) $$0 \to Q \stackrel{\eta_Q}{\to} \varinjlim(i,j) \stackrel{\xi_P}{\to} M \to 0$$

(23.4) $$0 \to P \stackrel{\eta_P}{\to} \varinjlim(i,j) \stackrel{\xi_Q}{\to} N \to 0.$$

We arrive at the following dual form of Schanuel's Lemma:

Theorem 23.5: *Let* $(0 \to K \stackrel{i}{\to} P \stackrel{p}{\to} M \to 0)$ *and* $(0 \to K \stackrel{j}{\to} Q \stackrel{q}{\to} N \to 0)$ *be projective* 0*-complexes in* $\mathcal{M}\mathrm{od}_\Lambda$*. If* M *and* N *are 1-coprojective then*

$$M \oplus Q \cong N \oplus P.$$

Proof. Since $\mathrm{Ext}^1(M,Q) = 0$ then (23.3) splits; hence $\varinjlim(i,j) \cong M \oplus Q$. Similarly, since $\mathrm{Ext}^1(N,P) = 0$ then (23.4) splits showing that $\varinjlim(i,j) \cong N \oplus P$ from which the statement follows. □

§24: Endomorphism rings:

For any Λ-module M we write $\mathrm{End}^0_\Lambda(M) = \mathrm{Hom}^0_\Lambda(M,M)$. Then $\mathrm{End}^0_\Lambda(M)$ is a two sided ideal in $\mathrm{End}_\Lambda(M)$. Moreover, the natural map $\nu : \mathrm{End}_\Lambda(M) \to \mathrm{End}_{\mathcal{D}\mathrm{er}}(M)$ is a ring homomorphism with $\mathrm{Ker}(\nu) = \mathrm{End}^0_\Lambda(M)$.

Let $\alpha = (0 \to J \stackrel{j}{\to} P \stackrel{p}{\to} M \to 0)$ be a projective 0-complex and suppose that $f : M \to M$ a Λ-homomorphism. By the universal property of projective modules there exists a morphism F over f thus:

(24.1) $$\begin{matrix} \alpha \\ \downarrow F \\ \alpha \end{matrix} = \begin{pmatrix} 0 \to & J \stackrel{j}{\to} & P \stackrel{p}{\to} & M \to & 0 \\ & \downarrow f_+ & \downarrow F & \downarrow f & \\ 0 \to & J \stackrel{j}{\to} & P \stackrel{p}{\to} & M \to & 0 \end{pmatrix}.$$

Although the Λ-homomorphism $f_+ : J \to J$ need not be unique, it becomes unique in the category $\mathcal{D}\mathrm{er}(\Lambda)$. To see this, first consider the case of a morphism F over the zero mapping $0 : M \to M$ thus:

$$\begin{array}{c} \alpha \\ \downarrow F \\ \alpha \end{array} = \left(\begin{array}{ccccc} 0 \to & J \stackrel{j}{\to} & P \stackrel{p}{\to} & M \to & 0 \\ & \downarrow f_+ & \downarrow F & \downarrow 0 & \\ 0 \to & J \stackrel{j}{\to} & P \stackrel{p}{\to} & M \to & 0 \end{array} \right).$$

Then $\mathrm{Im}(F) \subset \mathrm{Ker}(p) = \mathrm{Im}(j)$ so that $f_+ = (j^{-1} \circ F) \circ j$ is a factorization of f_+ through the projective P; that is:

(24.2) If $f = 0$ then $f_+ \approx 0$.

Now suppose F' is also a morphism over f

$$\begin{array}{c} \alpha \\ \downarrow F' \\ \alpha \end{array} = \left(\begin{array}{ccccc} 0 \to & J \stackrel{j}{\to} & P \stackrel{p}{\to} & M \to & 0 \\ & \downarrow f'_+ & \downarrow F' & \downarrow f & \\ 0 \to & J \stackrel{j}{\to} & P \stackrel{p}{\to} & M \to & 0 \end{array} \right).$$

Then $F' - F$ is a morphism over zero so that, by (24.2) $f'_+ - f_+ \approx 0$; that is

(24.3) If F, F' are morphisms of α over f then $f'_+ \approx f_+$.

Thus for any projective 0-complex $\alpha = (0 \to J \stackrel{j}{\to} P \stackrel{p}{\to} M \to 0)$ there is a mapping $\widetilde{\rho}_\alpha : \mathrm{End}_\Lambda(M) \to \mathrm{End}_{\mathcal{D}\mathrm{er}}(J)$ determined by the correspondence $\widetilde{\rho}_\alpha(f) = [f_+]$. It is straightforward to see that:

(24.4) $\widetilde{\rho}_\alpha$ is a ring homomorphism.

Now suppose that $f \in \mathrm{End}^0_\Lambda(M)$ and let $f - \xi \circ \eta$ be a factorization of f through a projective module Q thus:

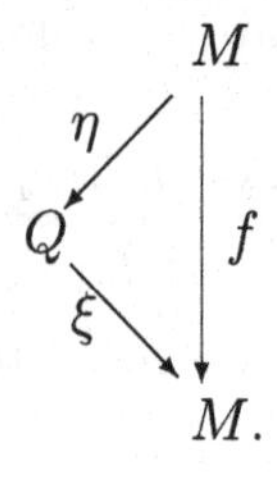

Since p is surjective then by the universal property of the projective Q there exists a homomorphism $\widehat{\xi} : Q \to P$ such that $\xi = p \circ \widehat{\xi}$:

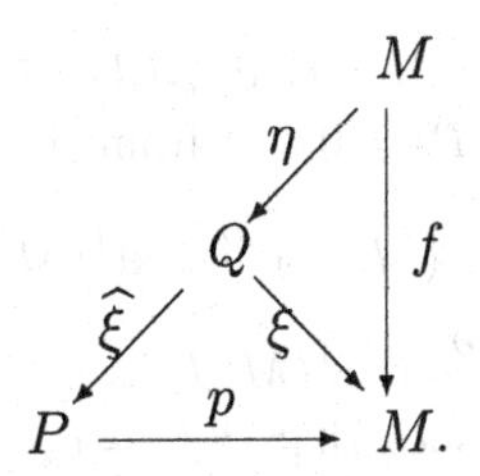

Defining $F = \widehat{\xi} \circ \eta \circ p$ then the following diagram commutes

$$\begin{array}{ccccccc} 0 \to & J \xrightarrow{j} & P \xrightarrow{p} & M \to & 0 \\ & \downarrow 0 & \downarrow F & \downarrow f & \\ 0 \to & J \xrightarrow{j} & P \xrightarrow{p} & M \to & 0 \end{array}$$

and so $\widetilde{\rho}_\alpha(f) = 0$. We have shown:

(24.5) $$\mathrm{End}^0_\Lambda(M) \subset \mathrm{Ker}(\widetilde{\rho}_\alpha).$$

In particular, the correspondence $[f] \mapsto \widetilde{\rho}_\alpha(f)$ defines a ring homomorphism

(24.6) $$\rho_\alpha : \mathrm{End}_{\mathcal{D}\mathrm{er}}(M) \to \mathrm{End}_{\mathcal{D}\mathrm{er}}(J).$$

In general ρ_α need not be injective nor is it clear whether ρ_α is necessarily surjective; however when M is 1-coprojective we obtain:

Theorem 24.7: *Let $\alpha = (0 \to J \xrightarrow{j} P \xrightarrow{p} M \to 0)$ be a projective 0-complex; if M is 1-coprojective then $\rho_\alpha : \mathrm{End}_{\mathcal{D}\mathrm{er}}(M) \to \mathrm{End}_{\mathcal{D}\mathrm{er}}(J)$ is an isomorphism of rings.*

Proof. To show that ρ_α is surjective it suffices to show that

(I): $\widetilde{\rho}_\alpha : \mathrm{End}_\Lambda(M) \to \mathrm{End}_{\mathcal{D}\mathrm{er}}(J)$ is surjective.

As we have already shown that $\mathrm{End}^0_\Lambda(M) \subset \mathrm{Ker}(\widetilde{\rho}_\alpha)$ then to show that ρ_α is injective it suffices to show that

(II): $\mathrm{Ker}(\widetilde{\rho}_\alpha) \subset \mathrm{End}^0_\Lambda(M)$.

To show (I), observe that $\mathrm{Hom}_{\mathcal{D}\mathrm{er}}(P, J) = \mathrm{Ext}^1(P, J) = 0$ so that the exact sequence of (21.11)

$$\mathrm{Hom}_{\mathcal{D}\mathrm{er}}(P, J) \xrightarrow{j^*} \mathrm{Hom}_{\mathcal{D}\mathrm{er}}(J, J) \xrightarrow{\delta_*} \mathrm{Ext}^1(M, J) \xrightarrow{p^*} \mathrm{Ext}^1(P, J)$$

reduces to an isomorphism $\delta_* : \mathrm{End}_{\mathcal{D}\mathrm{er}}(J) \xrightarrow{\simeq} \mathrm{Ext}^1(M, J)$. Since M is coprojective then $\mathrm{Ext}^1(M, P) = 0$ and from the exact sequence

$$\text{(III)}\ \mathrm{Hom}_\Lambda(M, P) \xrightarrow{p_*} \mathrm{Hom}_\Lambda(M, M) \xrightarrow{\partial} \mathrm{Ext}^1(M, J) \xrightarrow{j_*} 0 (= \mathrm{Ext}^1(M, P))$$

it follows that $\mathrm{End}_\Lambda(M) \xrightarrow{\partial} \mathrm{Ext}^1(M, J)$ is surjective. If $[g] \in \mathrm{End}_{\mathcal{D}\mathrm{er}}(J)$ there exists $f \in \mathrm{End}_\Lambda(M)$ such that $\partial(f) = \delta_*([g])$; thus there is a congruence $c : g_*(\alpha) \equiv f^*(\alpha)$

$$\begin{matrix} g_*(\alpha) \\ \downarrow c \\ f^*(\alpha) \end{matrix} = \begin{pmatrix} 0 \longrightarrow & J \longrightarrow & \varinjlim(g, j) \longrightarrow & M \longrightarrow & 0 \\ & \downarrow \mathrm{Id}_J & \downarrow c & \downarrow \mathrm{Id}_M & \\ 0 \longrightarrow & J \longrightarrow & \varprojlim(p, f) \longrightarrow & M \longrightarrow & 0 \end{pmatrix}.$$

In addition, there are natural maps $\nu_1 : \alpha \to g_*(\alpha)$ and $\nu_2 : f^*(\alpha) \to \alpha$ thus

$$\begin{matrix} \alpha \\ \downarrow \nu_1 \\ g_*(\alpha) \end{matrix} = \begin{pmatrix} 0 \longrightarrow & J \xrightarrow{j} & P \xrightarrow{p} & M \longrightarrow & 0 \\ & \downarrow g & \downarrow \nu_1 & \downarrow \mathrm{Id}_M & \\ 0 \longrightarrow & J \longrightarrow & \varinjlim(g, j) \longrightarrow & M \longrightarrow & 0 \end{pmatrix};$$

$$\begin{matrix} f^*(\alpha) \\ \downarrow \nu_2 \\ \alpha \end{matrix} = \begin{pmatrix} 0 \longrightarrow & J \longrightarrow & \varprojlim(p, f) \longrightarrow & M \longrightarrow & 0 \\ & \downarrow \mathrm{Id}_J & \downarrow \nu_2 & \downarrow f & \\ 0 \longrightarrow & J \xrightarrow{j} & P \xrightarrow{p} & M \longrightarrow & 0 \end{pmatrix}.$$

In the composition $\mu = \nu_2 \circ c \circ \nu_1 : \alpha \to \alpha$, the mapping $g : J \to J$ is induced over $f : M \to M$ showing that $\widetilde{\rho}_\alpha(f) = [g]$ as required. This proves (I).

For (II), observe that $\widetilde{\rho}_\alpha = \delta_*^{-1} \circ \partial$ where $\partial : \mathrm{Hom}_\Lambda(M, M) \to \mathrm{Ext}^1(M, J)$ is the boundary map of the exact sequence

$$\cdots \xrightarrow{i_*} \mathrm{Hom}_\Lambda(M, P) \xrightarrow{p_*} \mathrm{Hom}_\Lambda(M, M) \xrightarrow{\partial} \mathrm{Ext}^1(M, J) \xrightarrow{i_*} \mathrm{Ext}^1(M, P) \xrightarrow{p_*} \cdots$$

and δ_* is the isomorphism already noted. Thus $\mathrm{Ker}(\widetilde{\rho}_\alpha) = \mathrm{Ker}(\partial)$. However, from (III), in addition to the surjectivity of ∂, we note that $\mathrm{Ker}(\partial) = \mathrm{Im}(p_*) \subset \mathrm{End}^0_\Lambda(M)$. Hence $\mathrm{Ker}(\widetilde{\rho}_\alpha) \subset \mathrm{End}^0_\Lambda(M)$ as claimed. □

Chapter Four

Extension and restriction of scalars

The rings of most concern to us are group rings $R[G]$ over a variety of coefficient rings R together with their subrings and quotient rings. Primarily we are concerned to study integral group rings $\mathbb{Z}[G]$. However, to analyze these we shall also need to consider some other coefficient rings, notably the rational field $\mathbb{Q}$, the finite fields $\mathbb{F}_p$ and the rings $\widehat{\mathbb{Z}}_p$ of p-adic integers.

In practice the groups G that we encounter are limited to the cyclic groups C_p, C_q and the metacyclic groups $G(p,q)$ where p is an odd prime and q is a divisor of $p-1$. However, within this apparently limited context, we will find it necessary to employ a wide range of techniques from representation theory. These are presented in this chapter.

§25: Extension and restriction of scalars:

Throughout this chapter, Λ and Ω will denote algebras over a commutative ring R. Moreover, we will assume that

(♠) Ω is an R-subalgebra of Λ with the property that Λ *is free as a left* Ω*-module.*

As Λ is a Ω-Λ bimodule then, for any right Ω-module M, the tensor product $M \otimes_\Omega \Lambda$ acquires the structure of a right Λ-module thus:

$$M \otimes_\Omega \Lambda \times \Lambda \to M \otimes_\Omega \Lambda; \quad (m \otimes \lambda, \mu) \mapsto m \otimes (\lambda\mu).$$

Let i denote the inclusion $i : \Omega \hookrightarrow \Lambda$. We obtain a functor $i_* : \mathcal{M}od_\Omega \to \mathcal{M}od_\Lambda$ defined on objects by $i_*(M) = M \otimes_\Omega \Lambda$ and on morphisms as follows; given a morphism $f : M \to N$ in $\mathcal{M}od_\Omega$ we define $i_*(f) : M \otimes_\Omega \Lambda \to N \otimes_\Omega \Lambda$ by

$$i_*(f)(m \otimes \lambda) = f(m) \otimes \lambda.$$

The functor i_* is generally referred to as '*extension of scalars*'. There is a corresponding '*restriction of scalars*' functor $i^* : \mathcal{M}od_\Lambda \to \mathcal{M}od_\Omega$ where for $N \in \mathcal{M}od_\Lambda$ the underlying abelian group of $i^*(N)$ is the same as N and the action Ω action is the restriction to $N \times \Omega$ of the Λ-action on N.

Now suppose that $M \in \mathcal{M}od_\Omega$, $N \in \mathcal{M}od_\Lambda$ and that $g : M \otimes_\Omega \Lambda \to N$ is a Λ-linear mapping; define $\nu(g) : M \to N$ by $\nu(g)(m) = g(m \otimes 1)$. Clearly

$$\begin{aligned}\nu(g)(m\omega) &= g(m\omega \otimes 1)\\ &= g(m \otimes \omega)\\ &= g(m \otimes 1)\omega\\ &= \nu(g)(m)\omega\end{aligned}$$

so that $\nu(g)$ is Ω-linear. The correspondence $g \mapsto \nu(g)$ thus defines a mapping

$$\nu : \mathrm{Hom}_\Lambda(i_*(M), N) \to \mathrm{Hom}_\Omega(M, i^*(N)).$$

As a consequence we see that the functors i_*, i^* are left-right adjoint; that is:

Theorem 25.1: *The mapping* $\nu : \mathrm{Hom}_\Lambda(i_*(M), N) \to \mathrm{Hom}_\Omega(M, i^*(N))$ *is an isomorphism of R-modules.*

Proof. It is straightforward to see that ν is additive so it suffices to show that ν is invertible. Thus suppose that $f : M \to i^*(N)$ is an Ω-linear map and define $\widehat{f} : M \otimes_\Omega \Lambda \to N$ by

$$\widehat{f}\left(\sum_j m_j \otimes \lambda_j\right) = \sum_j f(m_j)\lambda_j.$$

It is straightforward to check that $\widehat{f}$ is Λ-linear. Consequently the correspondence $f \mapsto \widehat{f}$ defines a mapping $\widehat{\ } : \mathrm{Hom}_\Omega(M, i^*(N)) \to \mathrm{Hom}_\Lambda(i_*(M), N)$ and one checks easily that $\widehat{\ }$ is a two-sided inverse to ν. □

With our convention (♠) it is now straightforward to see that:

(25.2) If P is a projective Λ-module then $i^*(P)$ is projective over Ω.

(25.3) If Q is a projective Ω-module then $i_*(Q)$ is projective over Λ.

Proposition 25.4: *The mapping ν restricts to an isomorphism*

$$\nu : \mathrm{Hom}^0_\Lambda(i_*(M), N) \xrightarrow{\simeq} \mathrm{Hom}^0_\Omega(M, i^*(N))$$

Proof. Suppose $g \in \mathrm{Hom}^0_\Lambda(i_*(M), N)$ factors as follows:

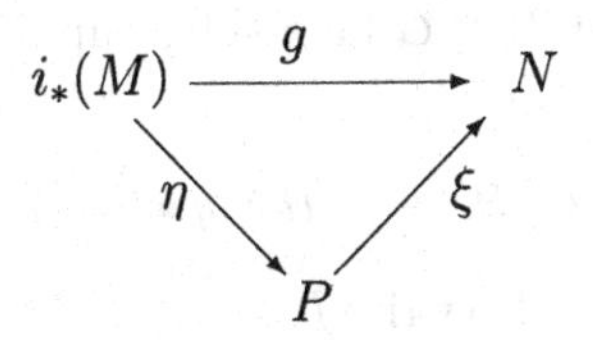

where $P \in \mathcal{P}(\Lambda)$. Then $\nu(g)$ factorizes as

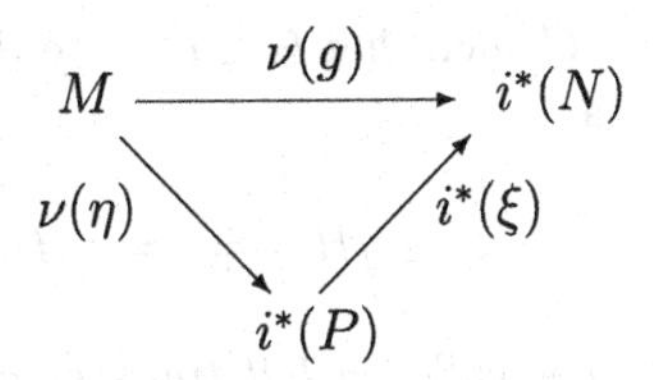

so that $\nu(g) \in \mathrm{Hom}^0_\Omega(M, i^*(N))$ as $i^*(P)$ is projective over Ω. Conversely, suppose $f \in \mathrm{Hom}^0_\Omega(M, i^*(N))$ factors as follows:

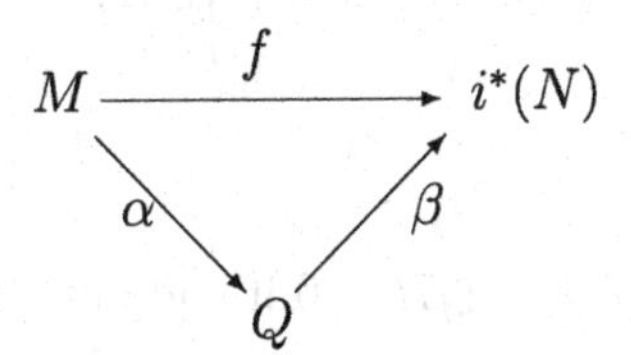

where Q is projective over Ω; then $\widehat{f}$ factorizes as

$$\begin{array}{ccc} i_*(M) & \xrightarrow{\widehat{f}} & N \\ & {\scriptstyle \alpha \otimes Id} \searrow \quad \nearrow {\scriptstyle \widehat{\beta}} & \\ & i_*(Q) & \end{array}$$

so that $\widehat{f} \in \mathrm{Hom}^0_\Lambda(i_*(M), N)$ as $i_*(Q)$ is projective over Λ. The conclusion now follows from (25.1) as ν and $f \mapsto \widehat{f}$ are mutually inverse. □

Corollary 25.5: *ν induces an isomorphism of R-modules*

$$\nu : \mathrm{Hom}_{\mathcal{D}\mathrm{er}(\Lambda)}(i_*(M), N) \to \mathrm{Hom}_{\mathcal{D}\mathrm{er}(\Omega)}(M, i^*(N)).$$

The left-right adjointness of i_*, i^* in (25.1) and (25.5) is completely general provided Λ is free over Ω. In §32 below, we shall encounter a special case in which i_*, i^* are also right-left adjoint.

§26: Transversals and cocycles:

Let G be a group and let $H \subset G$ be a subgroup of finite index. There is a canonical left action of G on G/H

$$\bullet : G \times G/H \to G/H \; ; \; g \bullet (\gamma H) = g\gamma H.$$

When convenient we shall also write $I = G/H$. We obtain a homomorphism

$$s : G \longrightarrow \mathfrak{S}_I \; ; \; \mathbf{s}(g)(\gamma H) = g\gamma H.$$

For each $i \in I$ choose $\xi_i \in G$ such that $\xi_i \in i$. Note the self-indexing nature of the choice. In particular,

(26.1) $$\xi_i \in \gamma H \Longleftrightarrow i = \gamma H.$$

The collection $\xi = (\xi_i)_{i \in I}$ is called a *left transversal to H in G*. We note that

(26.2) $$G = \bigcup_{i \in I} \xi_i H.$$

Moreover

(26.3) $$\xi_i H \cap \xi_j H \neq \emptyset \text{ if and only if } i = j.$$

The left action of $g \in G$ on $I = G/H$ produces a permutation s of the indices by

(26.4) $$g \bullet \xi_i H = \xi_{\mathbf{s}(g)(i)} H.$$

Thus, for all $i \in I$ and each $g \in G$, there exists a unique $h_i(g) \in H$ such that

(26.5) $$g \cdot \xi_i = \xi_{\mathbf{s}(g)(i)} h_i(g).$$

Proposition 26.6: *For all $\gamma, \delta \in G$, $h_i(\gamma\delta) = h_{\mathbf{s}(\delta)(i)}(\gamma) h_i(\delta)$.*

Proof. First note that $\delta \cdot \xi_i = \xi_{\mathbf{s}(\delta)(i)} h_i(\delta)$ so that

$$\begin{aligned}\gamma \cdot (\delta \cdot \xi_i) &= \gamma \cdot (\xi_{\mathbf{s}(\delta)(i)} h_i(\delta)) \\ &= \xi_{\mathbf{s}(\gamma)[\mathbf{s}(\delta)(i)]} h_{\mathbf{s}(\delta)(i)}(\gamma) h_i(\delta) \\ &= \xi_{\mathbf{s}(\gamma\delta)(i)} h_{\mathbf{s}(\delta)(i)}(\gamma) h_i(\delta)\end{aligned}$$

However $$(\gamma\delta)\cdot\xi_i = \xi_{\mathbf{s}((\gamma\delta))(i)}h_i(\gamma\delta)$$

so that the result follows on comparing the coefficients of $\xi_{\mathbf{s}((\gamma\delta))(i)}$. □

$\mathbf{h} = (h_i)_{i\in I}$ is called the *left cocycle associated to the transversal* $\xi = (\xi_i)_{i\in I}$. When $i = H$ it is possible to choose $\xi_i = \mathrm{Id}_H$. In this case the transversal is said to be *normalized.* Without further mention we shall assume that all transversals are normalized. When $i = H$ we have some additional properties:

(26.7) If $\omega \in H$ then $\mathbf{s}(\omega)(H) = H$.

(26.8) If ξ is normalized and $\omega \in H$ then $h_H(\omega) = \omega$.

There are corresponding notions of *right transversal* and *right cocycle.* Consider the right action of G on $H\backslash G$

$$\bullet : H\backslash G \times G \to H\backslash G; (H\gamma)\bullet g = H\gamma g.$$

Put $J = H\backslash G$ and for each $j \in J$, choose $\xi_j \in G$ such that $\xi_j \in j$. Again note the self-indexing nature of the choice. The collection $\xi = (\xi_j)_{j\in H\backslash G}$ is called a *right transversal to H in G* when

(26.9) $$G = \bigcup_{j\in J} H\xi_j$$

and

(26.10) $$H\xi_i \cap H\xi_j \neq \emptyset \text{ if and only if } i = j.$$

By change of emphasis we regard the right action of G on $H\backslash G$ as a permutation $\mathbf{t}$ of the indices $j \in J$ as follows:

(26.11) $$H\xi_j \bullet g = H\xi_{\mathbf{t}(g^{-1})(j)}.$$

It now follows that, for all $j \in J$ and each $g \in G$ there exists a unique $h_j(g) \in H$ with the property that

(26.12) $$\xi_j \cdot g = h_j(g)\xi_{\mathbf{t}(g^{-1})(j)}.$$

In this case

Proposition 26.13: *For all* $\gamma, \delta \in G$, $h_j(\gamma\delta) = h_j(\gamma)h_{\mathbf{t}(\gamma^{-1})(j)}(\delta)$.

The collection $\mathbf{h} = (h_j)_{j\in J}$ is the *right cocycle associated to the right transversal* $\xi = (\xi_j)_{j\in I}$. There is an obvious relationship between right and left transversals and cocycles thus: let $\theta : G/H \to H\backslash G$ be the bijection $\theta(\gamma H) = H\gamma^{-1}$. Then

(26.14) $$\mathbf{t}(g) = \theta \circ \mathbf{s}(g) \circ \theta^{-1}.$$

(26.15) $(\xi_i)_{i\in I}$ is a left transversal $\Longleftrightarrow$ $(\xi_{\theta(i)}^{-1})_{i\in I}$ is a right transversal.

In which case the right cocycle associated to $(\xi_{\theta(i)}^{-1})_{i\in I}$ is $(k_{\theta(i)})_{i\in I}$ where

(26.16) $$k_{\theta(i)}(g) = h_i(g^{-1})^{-1}.$$

§27: Wreath products and group extensions:

Let H be a group; if J is a nonempty set, we denote by H^J the J-fold self-product of H. Formally, H^J consists of functions $\mathbf{h} : J \to H$ and forms a group under pointwise multiplication $(\mathbf{h}\cdot\mathbf{k})(j) = \mathbf{h}(j)\mathbf{k}(j)$. The identity is the constant function $\mathbf{1}(j) = \mathrm{Id}_H$ and inverses are given by $\mathbf{h}^{-1}(j) = \mathbf{h}(j)^{-1}$. On defining

(27.1) $${}^{\sigma}\mathbf{h}(j) = \mathbf{h}(\sigma^{-1}(j))$$

we obtain a left action of the permutation group $\mathfrak{S}_J$ of J on H^J thus;

$$\mathfrak{S}_J \times H^J \to H^J; \quad (\sigma, \mathbf{h}) \mapsto {}^{\sigma}\mathbf{h}.$$

The semi-direct product $H^J \ltimes \mathfrak{S}_J$ is called the *wreath product* of H by J. We represent elements of $H^J \ltimes \mathfrak{S}_J$ in the form $(\mathbf{h};\sigma)$ where $\mathbf{h} \in H^J$ and $\sigma \in \mathfrak{S}_J$. The group law on $H^J \ltimes \mathfrak{S}_J$ is then

$$(\mathbf{h};\sigma)\cdot(\mathbf{k};\tau) = (\mathbf{h}\cdot{}^{\sigma}\mathbf{k};\sigma\tau).$$

If H is a subgroup of G then put $J = H\backslash G$ and let $\bullet : J \times G \to J$ be the canonical right action $(H\gamma)\bullet g = H\gamma g$. We obtain a homomorphism

$$G \longrightarrow \mathfrak{S}_J \ ; \ g \mapsto \mathbf{t}(g)$$

where $\mathbf{t}(g^{-1})(H\gamma) = H\gamma g$. Let $(\xi_j)_{j\in J}$ be a right transversal to H and let $(h_j)_{j\in J}$ be the right cocycle associated with $(\xi_j)_{j\in J}$. Now define a function $\mathbf{h} : G \to H^J$ by $\mathbf{h}(g)(j) = h_j(g)$. It follows easily from (26.13) and (27.1) that:

(27.2) $$\mathbf{h}(\gamma\delta) = \mathbf{h}(\gamma)^{\mathbf{t}(\gamma)}\mathbf{h}(\delta).$$

Finally, define a function $\iota_\xi : G \longrightarrow H^J \rtimes \mathfrak{S}_J$ by $\iota_\xi(g) = (\mathbf{h}(g); \mathbf{t}(g))$.

Theorem 27.3: *$\iota_\xi : G \to H^J \rtimes \mathfrak{S}_J$ is an injective group homomorphism.*

Proof. For γ, $\delta \in G$ we have

$$\begin{aligned}\iota_\xi(\gamma)\iota_\xi(\delta) &= (\mathbf{h}(\gamma)\ ;\ \mathbf{t}(\gamma)) \cdot (\mathbf{h}(\delta); \mathbf{t}(\delta)) \\ &= (\mathbf{h}(\gamma)^{\mathbf{t}(\gamma)}\mathbf{h}(\delta)\ ;\ \mathbf{t}(\gamma)\mathbf{t}(\delta)).\end{aligned}$$

However $\mathbf{t}(\gamma)\mathbf{t}(\delta) = \mathbf{t}(\gamma\delta)$, so that, by (27.2),

$$\iota_\xi(\gamma)\iota_\xi(\delta) = (\mathbf{h}(\gamma\delta)\ ;\ \mathbf{t}(\gamma\delta)) = \iota_\xi(\gamma\delta)$$

and hence ι_ξ is a group homomorphism as claimed. To show that ι_ξ is injective, suppose that $g \in G$ satisfies $\iota_\xi(g) = \mathrm{Id}$. We must show that $g = 1$. As $\mathbf{t}(g) = \mathrm{Id}$ and $h_j(g) = 1$ for all $j \in J$ then the expression $\xi_j \cdot g = h_j(g)\xi_{\mathbf{t}g^{-1}(j)}$ of (26.12) above reduces to $\xi_j \cdot g = \xi_j$. Hence $g = 1$, so completing the proof. □

The imbedding ι_ξ clearly depends upon the choice of transversal $\xi = (\xi_j)_{j\in J}$. Any other right transversal $\eta = (\eta_j)_{j\in J}$ also gives an imbedding $\iota_\eta : G \to H^J \rtimes \mathfrak{S}_J$. We claim that any two such imbeddings are conjugate in the following sense:

Theorem 27.4: *Let $\xi = (\xi_j)_{j\in J}$ and $\eta = (\eta_j)_{j\in J}$ be right transversals to H in G; then there exists $P \in H^J \times \{\mathrm{Id}\} \subset H^J \rtimes \mathfrak{S}_J$ such that, for all $g \in G$*

$$\iota_\eta(g) = P \cdot \iota_\xi(g) \cdot P^{-1}.$$

Proof. The transversals ξ, η determine right cocycles $(h_j)_{j\in J}$, $(k_j)_{j\in J}$ such that

$$\xi_j \cdot g = h_j(g)\xi_{\mathbf{t}(g^{-1})(j)}\ ;\ \eta_j \cdot g = k_j(g)\eta_{\mathbf{t}(g^{-1})(j)} \tag{27.5}$$

so that ι_η is defined by

$$\iota_\eta(g) = (\mathbf{k}(g); \mathbf{t}(g)) \tag{27.6}$$

where $\mathbf{k} = (k_j)_{j\in J}$. Now η_j and ξ_j determine the same element of $J = H\backslash G$. The rule of equality for right cosets implies that for each $j \in J$ and $g \in G$ there exists $p_j \in H$ such that

$$\eta_j = p_j\xi_j;\quad \xi_{\mathbf{t}(g^{-1})(j)} = p^{-1}_{\mathbf{t}(g^{-1})(j)}\eta_{\mathbf{t}(g^{-1})(j)}. \tag{27.7}$$

It follows from (27.2), (27.5) and (27.7) that

(27.8) $$k_j(g) = p_j \cdot h_j(g) \cdot p^{-1}_{\mathbf{t}(g^{-1})(j)}.$$

Now define $\mathbf{p} \in H^J$ by $\mathbf{p}(j) = p_j$ and put $P = (\mathbf{p};\ \mathrm{Id}) \in H^J \rtimes \mathfrak{S}_J$. It follows from (27.8) that for all $g \in G$, $\iota_\eta(g) = P \cdot \iota_\xi(g) \cdot P^{-1}$ as claimed.

□

§28: Induced representations:

If R is a ring we denote by $M_n(R)$ the ring of $n \times n$ matrices over R. If d is a positive integer we recall the 'block decomposition' of matrices in $M_{dn}(R)$. Regard a matrix $A \in M_d(M_n(R))$ as a collection $A = (A^{*,*}_{j,k})_{1 \le j,k \le d}$ where each $A^{*,*}_{j,k} \in M_n(R)$; that is each $A^{*,*}_{j,k}$ is an $n \times n$ matrix over R thus:

$$A^{*,*}_{j,k} = (A^{r,s}_{j,k})^{1 \le r,s \le n}$$

where $A^{r,s}_{j,k} \in R$. Define $\natural : M_d(M_n(R)) \to M_{dn}(R)$ by $\natural(A)_{r+(j-1)n,s+(k-1)n} = A^{r,s}_{j,k}$. It is straightforward to check that $\natural : M_d(M_n(R)) \to M_{dn}(R)$ is a ring isomorphism. Hence we have:

(28.1) $GL_{dn}(R) \cong GL_d(M_n(R))$ for any ring R.

Note that we have a 'diagonal' representation $\Delta : GL_n(R)^{(d)} \to GL_d(M_n(R))$

$$\Delta(A_1, A_2, \ldots, A_d) = \begin{pmatrix} A_1 & & & \\ & A_2 & & \\ & & \ddots & \\ & & & A_d \end{pmatrix}.$$

Now take $\mathfrak{S}_d = \mathfrak{S}_{\{1,\ldots,d\}}$. For any ring S there is a permutation representation $\pi_S : \mathfrak{S}_d \to GL_d(S)$ defined by

$$\pi_S(\sigma)_{j,k} = \begin{cases} 1 & j = \sigma(k) \\ 0 & j \ne \sigma(k) \end{cases}.$$

Otherwise expressed $\pi_S(\sigma)_{j,k} = \delta_{j,\sigma(k)}$ where δ is the Kronecker delta. In the present context we concentrate on the case $S = M_n(R)$ and put $\pi = \pi_{M_n(R)}$; that is, $\pi : \mathfrak{S}_d \to GL_d(M_n(R))$. It is straightforward to see that:

(28.2) $\mathrm{Im}(\pi)$ normalize $\mathrm{Im}(\Delta)$.

By (28.2) the correspondence $(A_1, A_2, \ldots, A_d; \sigma) \mapsto (\Delta(A_1, A_2, \ldots, A_d); \pi(\sigma))$ defines a homomorphism $(\Delta, \pi) : GL_n(R)^{(d)} \rtimes \mathfrak{S}_d \to GL_d(M_n(R))$. As both Δ and π are injective then so is (Δ, π). Composing with $\natural : GL_d(M_n(R)) \to GL_{dn}(R)$ we obtain an imbedding $\natural \circ (\Delta, \pi) : GL_n(R)^{(d)} \rtimes \mathfrak{S}_d \longrightarrow GL_{dn}(R)$. Given a representation $\rho : H \to GL_n(R)$ we denote by

$$\rho_* : H^{(d)} \rtimes \mathfrak{S}_d \to GL_n(R)^{(d)} \rtimes \mathfrak{S}_d$$

the homomorphism $\rho_*(h_1, h_2, \ldots, h_d; \sigma) = (\rho(h_1), \rho(h_2), \ldots, \rho(h_d); \sigma)$. Composition with $\natural \circ (\Delta, \pi)$ now gives

$$\widehat{\rho} = \natural \circ (\Delta, \pi) \circ \rho_* : H^{(d)} \rtimes \mathfrak{S}_d \to GL_{dn}(R).$$

Now suppose given a subgroup $H \subset G$ of finite index d in G and a representation $\rho : H \to GL_n(R)$. By (27.3), a choice of transversal ξ for H gives an imbedding $\iota_\xi : G \to H^{(d)} \rtimes \mathfrak{S}_d$ and composition with $\widehat{\rho}$ gives the *induced representation*

$$\rho_\xi = \widehat{\rho} \circ \iota_\xi : G \to GL_{dn}(R).$$

Up to conjugacy in $GL_{dn}(R)$, the induced representation is independent of the particular transversal chosen; for if η is also a transversal then, by (27.4),

$$\iota_\eta(g) = T \cdot \iota_\xi(g) \cdot T^{-1}$$

for some $T \in H^{(d)} \times \{\mathrm{Id}\}$. Taking $S = \widehat{\rho}(T)$ we obtain $\rho_\eta(g) = S \circ \rho_\xi \circ S^{-1}$.

§29: Lattices and representations:

Let R be a commutative ring and let H be a group. By a *lattice over* $R[H]$ we mean an $R[H]$-module whose underlying R-module is finitely generated and free. Suppose that L is a lattice over $R[H]$ with $\mathrm{rk}_R(L) = n$. Then L determines a representation $\rho_L : H \to GL_n(R)$ as follows: choose an R-basis $\mathcal{E} = (e_i)_{1 \le i \le n}$ for L and define $\rho_{L,\mathcal{E}} : H \to GL_n(R)$ by

(29.1) $$\rho_{\mathcal{E},L}(h) = (a_{ji}(h))_{1 \le i,j \le n} \cdot \iff e_i \cdot h^{-1} = \sum_{j=1}^{n} e_j a_{ji}(h).$$

One checks easily that $\rho_{L,\mathcal{E}} : H \to GL_n(R)$ is a homomorphism. Evidently $\rho_{L,\mathcal{E}}$ depends on the choice of basis $\mathcal{E}$. We next consider the effect of change of basis. If $\mathcal{F} = (\varphi_i)_{1 \le i \le n}$ is also an R-basis for L write

$$\varphi_i \cdot h^{-1} = \sum_{j=1}^{n} \varphi_j b_{ji}(h)$$

so that $\rho_{\mathcal{F},L}(h) = (b_{ji}(h))_{1 \le i,j \le n}$. Let $P = (p_{ji})_{1 \le i,j \le n}$ be the matrix which expresses the basis $\mathcal{E}$ in terms of $\mathcal{F}$ thus: $e_i = \sum_{j=1}^{n} \varphi_j p_{ji}$.

Evidently $P \in GL_n(R)$ and the standard change of basis formula from elementary linear algebra shows that:

(29.2) $$\rho_{\mathcal{F},L}(h) = P \rho_{\mathcal{E},L}(h) P^{-1}.$$

That is, an $R[H]$-lattice L with $\mathrm{rk}_R(L) = n$ gives rise to a class of representations $H \to GL_n(R)$, any two of which are conjugate within $GL_n(R)$. We denote this class by $rep(L)$. It is straightforward to see that if L' is an $R[H]$-lattice isomorphic to L then $rep(L') \equiv rep(L)$. Conversely, given a homomorphism $\rho : H \to GL_n(R)$ where $\rho(h) = (\rho(h)_{ji})_{1 \le i,j \le n}$ we define a lattice $\mathcal{L}(\rho)$ over $R[H]$ as follows; express an arbitrary element $\mathbf{x} \in R^n$ in standard coordinates as

$$\mathbf{x} = \begin{pmatrix} x_1 \\ x_2 \\ \vdots \\ x_n \end{pmatrix}$$

where $x_i \in R$. Then the action $R^n \times H \to R^n$; $\mathbf{x} \cdot h = \rho(h^{-1})(\mathbf{x})$ of H on R^n extends by linearity to a right $R[H]$-action

$$R^n \times R[H] \to R^n; \ \mathbf{x} \cdot \left(\sum_{h \in H} h b_h \right) = \sum_{h \in H} \rho(h^{-1})(\mathbf{x}) b_h$$

and thereby defines the structure of an $R[H]$-lattice, denoted by $\mathcal{L}(\rho)$, on R^n. Now suppose given another represention $\rho' : H \to GL_n(R)$ and that ρ' is conjugate to ρ in $GL_n(R)$; that is, there exists $P \in GL_n(R)$ such that, for all $h \in H$,

$$\rho'(h) = P \rho(h) P^{-1}.$$

Then the linear map $R^n \to R^n$; $\mathbf{x} \mapsto P\mathbf{x}$ defines an $R[H]$-isomorphism

$$P : \mathcal{L}(\rho) \xrightarrow{\simeq} \mathcal{L}(\rho').$$

It is straightforward to see that $\rho \in rep(\mathcal{L}(\rho))$ and that, if $\rho \in rep(L)$ then $\mathcal{L}(\rho) \cong L$. We have shown:

Theorem 29.3: *The correspondence $L \mapsto rep(L)$ defines a bijective mapping*

$$\left\{\begin{array}{c}\text{Isomorphism classes}\\ \text{of } R[H]\text{-lattices } L\\ \text{with } \mathrm{rk}_R(L) = n\end{array}\right\} \xrightarrow{\simeq} \left\{\begin{array}{c}GL_n(R)\text{ conjugacy}\\ \text{classes of representations}\\ \rho : H \to GL_n(R)\end{array}\right\}$$

The inverse mapping is given by $\rho \mapsto \mathcal{L}(\rho)$.

Let G be a finite group with $|G| = n$; then listing the elements of G thus $(e_i)_{1 \leq i \leq n}$ where $e_1 = 1$, the group ring $R[G]$ itself gives rise to the so called *regular representation* $\rho_{\mathrm{reg}} : G \to GL_n(R)$, defined by

(29.4) $$\rho_{\mathrm{reg}}(g)(e_i) = e_i \cdot g^{-1}.$$

§30: Induced modules:

Let Ω be a ring. If $M \in \mathcal{M}od_\Omega$ and $N \in {}_\Omega\mathcal{M}od$ we denote by $M \otimes_\Omega N$ the universal additive abelian group generated by symbols of the form $m \otimes n$ where $m \in M$, $n \in N$, subject to the following relations:

(i) $(m_1 + m_2) \otimes n = m_1 \otimes n + m_2 \otimes n$ $(m_i \in M, \lambda \in \Lambda)$
(ii) $m \otimes (n_1 + n_2) = m \otimes n_1 + m \otimes n_2$ $(m \in M, n_i \in \Lambda)$
(iii) $m\omega \otimes n = m \otimes \omega n$ $(m \in M, \omega \in \Omega, n \in N)$

If N is an Ω-bimodule then $M \otimes_\Omega N$ acquires the structure of right Ω-module under the action

$$M \otimes_\Omega N \times \Omega \to M \otimes_\Omega N; \quad (m \otimes n, \omega) \mapsto m \otimes (n\omega)$$

This is the case if $N = \Omega^{(d)}$ when we obtain:

(30.1) There is an isomorphism of right Ω-modules $M \otimes_\Omega \Omega^{(d)} \cong M^{(d)}$.

Our primary concern is with right modules. We show how to induce a *right* $R[G]$-module from a *right* $R[H]$-module by means of a *left* transversal. Note the difference in laterality. As before, G will denote a group, $H \subset G$ a subgroup of finite index d, $I = G/H$ and $\mathbf{s} : G \to \mathfrak{S}_I$ the homomorphism

$\mathbf{s}(g)(\gamma H) = g\gamma H$. Let M be a right $R[H]$-module. We denote by M^I set of functions $\mathbf{m} : I \to M$. Clearly M^I is also a right $R[H]$-module under the action

$$M^I \times R[H] \to M^I; \quad (\mathbf{m}, \omega) \mapsto \mathbf{m} \cdot \omega$$

where $(\mathbf{m} \cdot \omega)(i) = \mathbf{m}(i)\omega$. It is clear that:

(30.2) If M is a lattice over $R[H]$ with $\mathrm{rk}_R(M) = r$ then M^I is also a lattice over $R[H]$ and $\mathrm{rk}_R(M^I) = dr$.

Now let $\mathbf{h} = (h_i)_{i \in I}$ be the left cocycle associated to a left transversal $\xi = (\xi_i)_{i \in I}$ to H in G. Define $\cdot : M^I \times G \to M^I$ by

(30.3) $$(\mathbf{m} \cdot g)(i) = [\mathbf{m}(\mathbf{s}(g)(i))]h_i(g)$$

Proposition 30.4: *The above defines a right action of G on M^I.*

Proof.

$$\begin{aligned} [(\mathbf{m} \cdot \gamma) \cdot \delta](i) &= [(\mathbf{m} \cdot \gamma)(\mathbf{s}(\delta)(i))]h_i(\delta) \\ &= [\mathbf{m}\{\mathbf{s}(\gamma)(\mathbf{s}(\delta)(i))\}]h_{\mathbf{s}(\delta)(i)}(\gamma)h_i(\delta) \\ &= [\mathbf{m}\{\mathbf{s}(\gamma\delta)(i)\}]h_{\mathbf{s}(\delta(i))}(\gamma)h_i(\delta) \\ \text{whilst} \quad [\mathbf{m} \cdot (\gamma\delta)](i) &= [\mathbf{m}(\mathbf{s}(\gamma\delta)(i))]h_i(\gamma\delta). \end{aligned}$$

However, by (26.6) we have $h_i(\gamma\delta) = h_{\mathbf{s}(\delta)(i)}(\gamma)h_i(\delta)$ so that, as required

$$(\mathbf{m} \cdot \gamma) \cdot \delta = \mathbf{m} \cdot (\gamma\delta).$$

□

As the above action is clearly via R-linear mappings it extends by linearity to a right $R[G]$-module structure on M^I which we denote by $\mathrm{Ind}(M, \xi)$. We note that:

Proposition 30.5: *If ξ, η are left transversals to H in G then*

$$\mathrm{Ind}(M, \xi) \cong_{R[G]} \mathrm{Ind}(M, \eta).$$

Proof. Let $\xi = (\xi_i)_{i \in I}$ and $\eta = (\eta_i)_{i \in I}$ be left transversals to H and let $\mathbf{h} = (h_i)_{i \in I}$. $\mathbf{k} = (k_i)_{i \in I}$ be their associated cocycles. Put $p_i = \xi_i^{-1}\eta_i \in H$. It is straightforward to check that, for each $g \in G$ and each $i \in I$, we have

(*) $$p_{\mathbf{s}(g)(i)}k_i(g) = h_i(g)p_i.$$

Let $\cdot : M^I \times G \to M^I$ be the right action defined by ξ; that is:

$$(\mathbf{m} \cdot g)(i) = \mathbf{m}(\mathbf{s}(g)(i))h_i(g).$$

Similarly let $\diamond : M^I \times G \to M^I$ be the right action defined by η:

$$(\mathbf{m} \diamond g)(i) = \mathbf{m}(\mathbf{s}(g)(i))k_i(g).$$

Define $P : M^I \to M^I$ by $P(\mathbf{m})(i) = \mathbf{m}(i)p_i$. Then P is evidently an R-linear bijection. Moreover, for all $g \in G$ and all $\mathbf{m} \in M^I$ it follows from (*) that,

$$P(\mathbf{m} \cdot g) = P(\mathbf{m}) \diamond g$$

Thus P extends by R-linearity to an $R[G]$-isomorphism

$$P : \mathrm{Ind}(M, \xi) \xrightarrow{\simeq} \mathrm{Ind}(M, \eta)$$ □

Proposition 30.6: *If M is a right lattice over $R[H]$ then $M \otimes_{R[H]} R[G]$ is a right lattice over $R[G]$ with* $\mathrm{rk}_R(M \otimes_{R[H]} R[G]) = d\,\mathrm{rk}_R(M)$.

Proof. Choose a right transversal $\eta = (\eta_j)_{j \in H\backslash G}$. Then $(\eta_j)_{j \in H\backslash G}$ is an $R[H]$-basis for the right $R[H]$-module structure on $R[G]$ so that

$$R[G] \cong_{R[H]} \underbrace{R[H] \oplus \cdots\cdot \oplus R[H]}_{d}.$$

However, as $R[G]$ is an $R[H]$-bimodule then, by (30.1),

$$M \otimes_{R[H]} R[G] \cong_{R[H]} \underbrace{M \oplus \cdots\cdot \oplus M}_{d},$$

from which the conclusion follows immediately. □

Let i denote the inclusion $i : R[H] \hookrightarrow R[G]$. When M is a right $R[H]$-lattice we now exhibit $\mathrm{Ind}(M, \xi)$ is an explicit model for the module $i_*(M)$. To see this, choose a left transversal $\xi = (\xi_i)_{i \in I}$ and let $\mathbf{h} = (h_i)_{i \in I}$ denote the cocycle associated to ξ.

Put $$M_H = \{\mathbf{m} \in M^I : \mathbf{m}(i) = 0 \text{ if } i \neq H\}.$$

If $m \in M$ denote by $\widehat{\mathbf{m}} \in M_H$ the element

$$\widehat{\mathbf{m}}(\gamma H) = \begin{cases} m & i = H \\ 0 & i \neq H. \end{cases}$$

Proposition 30.7: *If $m \in M$ and $\omega \in H$ then* $\widehat{\mathbf{m} \cdot \omega} = \widehat{\mathbf{m}} \cdot \omega$.

Proof. By definition $(\widehat{\mathbf{m}} \cdot \omega)(i) = \widehat{\mathbf{m}}(\mathbf{s}(\omega)(i))h_i(\omega)$. As $\mathbf{s}(\omega)$ is bijective if $\omega \in H$ then by (26.7), it follows that $\mathbf{s}(\omega)(i) = H \iff i = H$. Thus

$$(\widehat{\mathbf{m}} \cdot \omega)(i) = \begin{cases} mh_H(\omega) & i = H \\ 0 & i \neq H. \end{cases}$$

Moreover, by (26.8), $h_H(\omega) = \omega$ so that

$$(\widehat{\mathbf{m}} \cdot \omega)(i) = \begin{cases} m\omega & i = H \\ 0 & i \neq H \end{cases}$$

and so $\widehat{\mathbf{m}} \cdot \omega = \widehat{\mathbf{m} \cdot \omega}$. □

Clearly M_H is an additive subgroup of M^I. Moreover, the mapping $M \to M_H$, $m \mapsto \widehat{\mathbf{m}}$ is evidently bijective. It follows immediately that:

Corollary 30.8: *M_H is an $R[H]$-submodule of M^I and $M_H \cong_{R[H]} M$.*

More generally, for $j \in I$ we define $M_j = \{\mathbf{m} \in M^I : \mathbf{m}(i) = 0 \text{ if } i \neq j\}$. Evidently M_j is an additive subgroup of M^I. Moreover, as $I = G/H$ is finite then M^I is the internal sum $M^I = \sum_{j \in I} M_j$. Taking $j = \gamma H$ it is straightforward to see that the mapping $\widehat{\mathbf{m}} \mapsto \widehat{\mathbf{m}} \cdot \gamma^{-1}$ gives an isomorphism of abelian groups $M_H \xrightarrow{\simeq} M_{\gamma H} = M_j$. It now follows that $M^I = M_H \cdot R[G]$; that is:

Proposition 30.9: *M_H generates $\mathrm{Ind}(M, \xi)$ as a module over $R[G]$.*

Now consider the mapping $\natural : M \times R[G] \to \mathrm{Ind}(M, \xi)$ given by $\natural(m, \lambda) = \widehat{\mathbf{m}} \cdot \lambda$. Then $\natural$ is bi-additive and for $m \in M$, $\omega \in H$, $\lambda \in R[G]$ it follows from (30.7) that

$$\natural(m \cdot \omega, \lambda) = \natural(m, \omega\lambda).$$

By the universal property of $- \otimes_{R[H]} R[G]$ there exists a unique $R[G]$-linear mapping

$$\widetilde{\natural} : M \otimes_{R[H]} R[G] \longrightarrow \mathrm{Ind}(M, \xi)$$

making the following diagram commute:

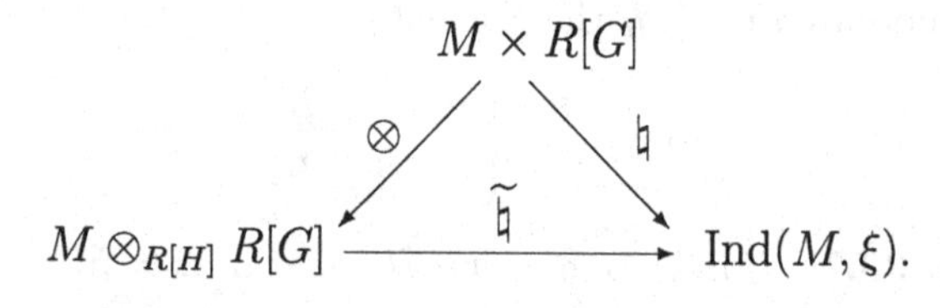

Proposition 30.10: *$\widetilde{\natural} : M \otimes_{R[H]} R[G] \to \mathrm{Ind}(M, \xi)$ is an isomorphism of right $R[G]$-modules.*

Proof. It suffices to prove that $\widetilde{\natural}$ is bijective. It follows from (30.2) and (30.6) that, as abelian groups both $\mathrm{Ind}(M, \xi)$ and $M \otimes_{R[H]} R[G]$ are free of rank $= drk_{\mathbb{Z}}(M)$. However, by (30.9) above, $\widetilde{\natural}$ is surjective. Hence $\widetilde{\natural}$ is bijective as claimed. □

Thus $\mathrm{Ind}(M, \xi)$ is an explicit model for the *induced module* $i_*(M) = M \otimes_{R[H]} R[G]$. There is a completely dual theory which induces *left* $R[G]$-modules from *left* $R[H]$-modules using a *right* transversal. We leave the details to the reader.

§31: Duality:

If M is a module over the ring Λ we define the dual M^* by $M^* = \mathrm{Hom}_\Lambda(M, \Lambda)$. M^* is evidently an abelian group. Moreover, if M is a *right* Λ-module then M^* acquires the structure of a *left* Λ-module via $\Lambda \times \mathrm{Hom}_\Lambda(M, \Lambda) \to \mathrm{Hom}_\Lambda(M, \Lambda)$ on writing

$$(\lambda \cdot \mathbf{a})(m) = \lambda \cdot \mathbf{a}(m).$$

If $f : M \to N$ is homomorphism of right Λ-modules there is an induced homomorphism $f^* : N^* \to M^*$ of left Λ-modules defined by

$$f^*(\mathbf{a}) = \mathbf{a} \circ f.$$

If $g : N \to P$ is also a homomorphism of right Λ-modules then

$$(g \circ f)^* = f^* \circ g^*.$$

Consequently duality defines a contravariant functor $^* : \mathcal{M}od_\Lambda \to {}_\Lambda\mathcal{M}od$. Similarly, if M is *left* Λ-module, $M^* = \mathrm{Hom}_\Lambda(M, \Lambda)$ acquires the structure of a *right* Λ-module $\mathrm{Hom}_\Lambda(M, \Lambda) \times \Lambda \to \mathrm{Hom}_\Lambda(M, \Lambda)$ on writing

$$(\mathbf{a} \cdot \lambda)(m) = \mathbf{a}(m) \cdot \lambda$$

and in this case duality defines a contravariant functor $^* : {}_\Lambda\mathcal{M}od \to \mathcal{M}od_\Lambda$. Our primary interest is in right modules. The change in laterality under duality is an inconvenient fact. Ideally we would like the dual of a right module to be a right module. Under propitious circumstances we now show how to modify the definition to ensure that is the case.

By an *anti-involution* θ on a ring Λ we mean a mapping $\theta : \Lambda \to \Lambda$ satisfying

(i) $\theta(\lambda + \mu) = \theta(\lambda) + \theta(\mu)$;
(ii) $\theta(\lambda \cdot \mu) = \theta(\mu) \cdot \theta(\lambda)$;
(iii) $\theta(1_\Lambda) = 1_\Lambda$;
(iv) $\theta \circ \theta = \mathrm{Id}_\Lambda$.

For the rest of this section we will assume that Λ is a ring with anti-involution θ. Thus denoting the opposite ring to Λ by Λ^{opp} then θ is a self-inverse ring isomorphism $\theta : \Lambda \to \Lambda^{opp}$. When Λ is equipped with such an anti-involution θ we can convert a left module $N = (N, \circ)$ to a right module $\overline{N} = (N, \diamond)$ via θ as follows:

$$\diamond : N \times \Lambda \to N$$

$$n \diamond \lambda = \theta(\lambda) \circ n$$

$\overline{N}$ is called the *conjugate* to N. The *conjugate dual* to M is $\overline{(M^*)}$; that is, the dual module $M^* = \mathrm{Hom}_\Lambda(M, \Lambda)$ after conversion to a right module. It is straightforward to see that $\overline{(M^*)} = (\overline{M})^*$; in practice we denote the conjugate dual by $M^\bullet$. We note that repeating the duality functor is similar to repeating conjugate duality; that is,

(31.1) $$M^{\bullet\bullet} \cong M^{**} \text{ for any module } M \in \mathcal{M}od_\Lambda.$$

For any right Λ-module M there is a homomorphism $\natural : M \to M^{\bullet\bullet}$ defined by

$$\natural(m)(\mathbf{a}) = \mathbf{a}(m)$$

so that the correspondence $M \mapsto M^{\bullet\bullet}$ defines a natural transformation $\natural$: $\mathrm{Id} \to {}^{\bullet\bullet}$. We say that M is *reflexive* when $\natural : M \longrightarrow M^{\bullet\bullet}$ is an isomorphism. In general, modules are far from being reflexive. However, as we shall now see, the modules of primary interest to us are reflexive.

Thus for the rest of this section we will take G to be a finite group, R to be a commutative ring and Λ to be the group algebra $\Lambda = R[G]$. Then Λ admits a canonical anti-involution '$\overline{}$' given by

$$\overline{\sum_{g \in G} a_g \cdot g} = \sum_{g \in G} a_g \cdot g^{-1}.$$

Now suppose that M is a Λ-lattice such that, with respect to an R basis $\{e_i\}_{1 \le i \le n}$, M is described by the representation $\rho : G \to GL_n(R)$. Then

$M^\bullet$ has the dual basis $\{e_i^\bullet\}_{1\leq i\leq n}$, defined by $e_i^\bullet(e_j) = \delta_{ij}$. with respect to which the representation $\rho^\bullet$ is given by conjugate transpose; that is:

(31.2) $$\rho^\bullet(g) = \rho(g^{-1})^t.$$

Evidently $\rho^{\bullet\bullet}(g) = \rho(g)$ so that $\natural : M \longrightarrow M^{\bullet\bullet}$ is an isomorphism and we have:

(31.3) Any $R[G]$-lattice is reflexive.

For any $R[G]$-modules M, N, $\mathrm{Hom}_{R[G]}(M,N)$ is naturally an R-module. If, in addition, M, N are $R[G]$-lattices we have canonical R-isomorphisms as follows:

(31.4) $$\mathrm{Hom}_{R[G]}(N^\bullet, M^\bullet) \cong \mathrm{Hom}_{R[G]}(M,N).$$

(31.5) $$\mathrm{Hom}^0_{R[G]}(N^\bullet, M^\bullet) \cong \mathrm{Hom}^0_{R[G]}(M,N).$$

(31.6) $$\mathrm{Hom}_{\mathcal{D}\mathrm{er}}(N^\bullet, M^\bullet) \cong \mathrm{Hom}_{\mathcal{D}\mathrm{er}}(M,N).$$

(31.5) follows from (31.4) by observing that $f : M \to N$ factors through the projective P if and only if $f^\bullet : N^\bullet \to M^\bullet$ factors through the projective $P^\bullet$. (31.6) follows trivially from (31.4) and (31.5). We note that in the regular representation ρ_{reg} of (29.4), each $\rho_{\mathrm{reg}}(g)$ is a permutation matrix; hence

(31.7) $$\rho_{\mathrm{reg}}(g) = \rho_{\mathrm{reg}}(g^{-1})^t.$$

Thus the group ring $R[G]$ is self-dual; that is;

(31.8) $$R[G]^\bullet \cong R[G].$$

§32: The Eckmann-Shapiro Theorem:

The adjunction theorems of §25 hold in complete generality provided Λ is free over Ω. However in certain circumstances there is a dual form as we now explain. In this context we will take $\Lambda = R[G]$ and $\Omega = R[H]$ where H is a subgroup of finite index d in G.

For the sake of clarity $M^\flat$ will denote the Ω-dual of an Ω-lattice M and $N^\bullet$ will denote the Λ-dual of the Λ-lattice N. We consider the extent to which duality is preserved under restriction and extension of scalars. This is seen most easily by dealing directly with representations. Thus suppose that N is a Λ-lattice and let $\rho : G \to GL_n(R)$ be a representation defining N; then the representation $\rho^\bullet$ defining $N^\bullet$ is given by

$$\rho^\bullet(g) = \rho(g^{-1})^t.$$

Then the restriction of $\rho^\bullet$ to H defines $i^*(N^\bullet)$; that is,

$$\rho^\bullet_{|H}(h) = \rho_{|H}(h^{-1})^t$$

which coincides with the representation defining $i^*(N)^\flat$; that is:

(32.1) If N is a Λ-lattice then $i^*(N^\bullet) \cong i^*(N)^\flat$.

Now suppose that M is an Ω-lattice defined by a representation $\rho : H \to GL_m(R)$. As in §27 let

$$\rho_* : H^{(d)} \rtimes \mathfrak{S}_d \to GL_n(R)^{(d)} \rtimes \mathfrak{S}_d$$

be the homomorphism $\rho_*(h_1, h_2, \ldots, h_d; \sigma) = (\rho(h_1), \rho(h_2), \ldots, \rho(h_d); \sigma)$. The composition is now a homomorphism $\natural \circ (\Delta, \pi) \circ \rho_* : H^{(d)} \rtimes \mathfrak{S}_d \to GL_{dn}(R)$. By (27.3), a choice of right transversal ξ for H gives an imbedding $\iota_\xi : G \to H^{(d)} \rtimes \mathfrak{S}_d$ and, as in §27, the composition

$$\varphi = \natural \circ (\Delta, \pi) \circ \rho_* \circ \iota_\xi : G \to GL_{dn}(R)$$

is the induced representation which defines the Λ-lattice $N = i_*(M)$. The dual representation $\varphi^\bullet$ is then the composition

$$\varphi^\bullet = \natural \circ (\Delta, \pi) \circ \widetilde{\rho}_* \circ \iota_\xi : G \to GL_{dn}(R)$$

where $\widetilde{\rho}_*(h_1, h_2, \ldots, h_d; \sigma) = (\rho(h_1^{-1})^t, \rho(h_2^{-1})^t, \ldots, \rho(h_d^{-1})^t; (\sigma^{-1})^t)$. Observe that $\rho(h^{-1})^t = \rho^\flat(h)$ whilst σ, being a permutation representation, is orthogonal and $(\sigma^{-1})^t = \sigma$. Hence

$$\widetilde{\rho}_*(h_1, h_2, \ldots, h_d; \sigma) = (\rho^\flat(h_1), \rho^\flat(h_2), \ldots, \rho^\flat(h_d); \sigma) = \rho^\flat_*(h_1, h_2, \ldots, h_d; \sigma)$$

and so $\varphi^\bullet = \natural \circ (\Delta, \pi) \circ \rho^\flat_* \circ \iota_\xi : G \to GL_{dn}(R)$
which is a representation defining the induced module $i_*(M^\flat)$. Thus we have:

(32.2) $$i_*(M^\flat) \cong i_*(M)^\bullet.$$

As a consequence we have:

Theorem 32.3: *Let M be an Ω-lattice and N a Λ-lattice; then there is an isomorphism of R-modules*

$$\mathrm{Hom}_\Omega(i^*(N), M) \xrightarrow{\simeq} \mathrm{Hom}_\Lambda(N, i_*(M)).$$

Proof. $M^\flat$ is an Ω-lattice and $N^\bullet$ is a Λ-lattice. Hence ν^{-1} induces an isomorphism of R-modules $\mathrm{Hom}_\Omega(M^\flat, i^*(N^\bullet)) \xrightarrow{\simeq} \mathrm{Hom}_\Lambda(i_*(M^\flat), N^\bullet)$. As $i^*(N^\bullet) \cong i^*(N)^\flat$ and $i_*(M^\flat) \cong i_*(M)^\bullet$ we obtain an isomorphism

$$\mathrm{Hom}_\Omega(M^\flat, i^*(N)^\flat) \xrightarrow{\simeq} \mathrm{Hom}_\Lambda(i_*(M)^\bullet, N^\bullet).$$

However, by §31, duality gives isomorphisms $\mathrm{Hom}_\Omega(M^\flat, i^*(N)^\flat) \cong \mathrm{Hom}_\Omega(i^*(N), M)$ and $\mathrm{Hom}_\Lambda(i_*(M)^\bullet, N^\bullet) \cong \mathrm{Hom}_\Lambda(N, i_*(M))$ so giving the required isomorphism

$$\mathrm{Hom}_\Omega(i^*(N), M) \xrightarrow{\simeq} \mathrm{Hom}_\Lambda(N, i_*(M)) \qquad \square$$

With the above hypotheses, duality also gives R-module isomorphisms

$$\mathrm{Hom}_{\mathcal{D}\mathrm{er}}(M^\flat, i^*(N)^\flat) \cong \mathrm{Hom}_{\mathcal{D}\mathrm{er}}(i^*(N), M);$$

$$\mathrm{Hom}_{\mathcal{D}\mathrm{er}}(i_*(M)^\bullet, N^\bullet) \cong \mathrm{Hom}_{\mathcal{D}\mathrm{er}}(N, i_*(M)).$$

From these we derive the corresponding statement over the derived module categories:

Theorem 32.4: *Let M be an Ω-lattice and N a Λ-lattice; then there is an isomorphism of R-modules*

$$\mathrm{Hom}_{\mathcal{D}\mathrm{er}(\Omega)}(i^*(N), M) \xrightarrow{\simeq} \mathrm{Hom}_{\mathcal{D}\mathrm{er}(\Lambda)}(N, i_*(M)).$$

§33: Syzygies and lattices:

First consider an exact sequence of modules $0 \to J \to R^a \to P \to 0$ over the ring R; if P is projective then $P \oplus J \cong R^a$ so J is also projective. More generally, suppose $0 \to J \to R^{a_n} \to \cdots \to R^{a_0} \to P \to 0$ is an exact sequence of R-modules for integers $a_r \geq 1$. By decomposing the sequence as the Yoneda product of

$$0 \to J \to R^{a_n} \to K \to 0 \quad \text{and} \quad 0 \to K \to R^{a_{n-1}} \to \cdots \to R^{a_0} \to P \to 0$$

we see inductively that K is projective and hence J is projective. With the convention that $a_k = 0$ for $k > n$ and writing $b = \sum_{r\geq 0} a_{2r}$, $c = \sum_{r\geq 0} a_{2r+1}$

we find that

(*) $$P \oplus R^b \cong J \oplus R^c \text{ if } n \text{ is odd;}$$

(**) $$P \oplus J \oplus R^c \cong R^b \text{ if } n \text{ is even.}$$

In either case, if P is stably free then so also is J; that is:

(33.1) Let $0 \to J \to R^{a_n} \to \cdots \to R^{a_0} \to P \to 0$ be an exact sequence of R-modules; if P is stably free then J is also stably free.

For the remainder of this section Λ will denote a group ring $\Lambda = R[G]$ where R is a commutative ring and G is finite; by a *weak lattice* over Λ we mean a Λ-module whose underlying R-module is finitely generated and stably free. Let S be a finitely generated stably free module over Λ and write $S \oplus \Lambda^a \cong \Lambda^b$; then

$$i^*(S) \oplus R^m \cong R^n$$

where $m = a|G|$ and $n = b|G|$ and so $i^*(S)$ is stably free over R; thus:

(33.2) If S is finitely generated and stably free over Λ then S is a weak Λ-lattice.

Proposition 33.3: *Let M be a weak Λ-lattice and suppose that $J \in \Omega_n(M)$ where $n \geq 1$; then J is a weak Λ-lattice.*

Proof. If $n \geq 2$ then, by (22.6), there exists an exact sequence of Λ-modules

$$0 \to J \to \Lambda^{a_{n-1}} \to \cdots \to \Lambda^{a_0} \to M \to 0.$$

Applying i^* gives an exact sequence $0 \to i^*(J) \to R^{b_{n-1}} \to \cdots \to R^{b_0} \to i^*(M) \to 0$ where $b_r = a_r|G|$. As $i^*(M)$ is stably free the conclusion follows from (33.1).

When $n = 1$ then, by (22.5), there exists an exact sequence of Λ-modules

$$0 \to J \to S \to M \to 0$$

where S is finitely generated stably free. Now $0 \to i^*(J) \to i^*(S) \to i^*(M) \to 0$ is exact and, by hypothesis, $i^*(M)$ is stably free, hence projective over R. Thus

$$i^*(J) \oplus i^*(M) \cong_R i^*(S).$$

Again by hypothesis, $i^*(M)$ is stably free. As S is stably free then so also is $i^*(S)$. Hence $i^*(J)$ is stably free and this completes the proof. □

Recall that the ring R is said to have *stably free cancellation* (= property SFC) when every stably free R-module is free. We note that:

(33.4) Let M be a weak Λ-lattice; if R has property SFC then M is a Λ-lattice.

As a consequence of duality we have the following which we note for future reference:

Proposition 33.5: *Let M be a lattice over $\Lambda = R[G]$; if R has property SFC then there exists an exact sequence*

$$0 \to M \to \Lambda^m \to N \to 0$$

where N is also a Λ-lattice.

Proof. The Λ-dual $M^\bullet$ is also a Λ-lattice and so there exists an exact sequence

$$0 \to K \to \Lambda^m \to M^\bullet \to 0;$$

K is a Λ-lattice, hence $K^\bullet$ is also a Λ-lattice and dualization gives an exact sequence

$$0 \to M^{\bullet\bullet} \to (\Lambda^\bullet)^m \to K^\bullet \to 0.$$

As $M^{\bullet\bullet} \cong M$ and $\Lambda^\bullet \cong \Lambda$ the conclusion follows on taking $N = K^\bullet$. □

§34: The Eckmann-Shapiro relations in cohomology:

In this section we take $\Lambda = R[G]$ where G is a finite group and R is a commutative ring with property SFC; then we have the following special case of (33.3):

Proposition 34.1: *Let M be a Λ-lattice and let $J \in \Omega_n(M)$ where $n \geq 1$; then J is a Λ-lattice.*

Proposition 34.2: *If M is a Λ-lattice then M is 1-coprojective.*

Proof. Take a surjective Λ-homomorphism $p : \Lambda^a \twoheadrightarrow M$ and put $J = \mathrm{Ker}(p)$. Then the exact sequence $0 \to J \xrightarrow{j} \Lambda^a \xrightarrow{p} M \to 0$ gives the cohomology sequence

$$\mathrm{Hom}_\Lambda(\Lambda^a, i_*(R)) \xrightarrow{j^*} \mathrm{Hom}_\Lambda(J, i_*(R)) \xrightarrow{\partial} \mathrm{Ext}^1_\Lambda(M, i_*(R)) \xrightarrow{p^*} \mathrm{Ext}^1_\Lambda(\Lambda^a, i_*(R)).$$

Likewise, the exact sequence $0 \to i^*(J) \xrightarrow{j} i^*(\Lambda^a) \xrightarrow{p} i^*(M) \to 0$ gives the cohomology sequence

$$\mathrm{Hom}_R(i^*(\Lambda^a), R) \xrightarrow{j^*} \mathrm{Hom}_R(i^*(J), R) \xrightarrow{\partial} \mathrm{Ext}^1_R(i^*(M), R) \xrightarrow{p^*} \mathrm{Ext}^1_R(i^*(\Lambda^a), R).$$

We now have a commutative diagram

$$\begin{array}{ccccccc}
\mathrm{Hom}_R(i^*(\Lambda^a), R) & \xrightarrow{j^*} & \mathrm{Hom}_R(i^*(J), R) & \xrightarrow{\partial} & \mathrm{Ext}^1_R(i^*(M), R) & \xrightarrow{p^*} & \mathrm{Ext}^1_R(i^*(\Lambda)^a, R) \\
\downarrow & & \downarrow & & & & \\
\mathrm{Hom}_\Lambda(\Lambda^a, i_*(R)) & \xrightarrow{j^*} & \mathrm{Hom}_\Lambda(J, i_*(R)) & \xrightarrow{\partial} & \mathrm{Ext}^1_\Lambda(M, i_*(R)) & \xrightarrow{p^*} & \mathrm{Ext}^1_\Lambda(\Lambda^a, i_*(R))
\end{array}$$

where the downarrows are the natural isomorphisms from (32.3). As $i^*(\Lambda)^a$ is free over R then $\mathrm{Ext}^1_R(i^*(\Lambda)^a, R) = 0$. Clearly also $\mathrm{Ext}^1_\Lambda(\Lambda^a, i_*(R)) = 0$. Consequently we have a commutative diagram with exact rows

$$\begin{array}{ccccccc}
\mathrm{Hom}_R(i^*(\Lambda^a), R) & \xrightarrow{j^*} & \mathrm{Hom}_R(i^*(J), R) & \xrightarrow{\partial} & \mathrm{Ext}^1_R(i^*(M), R) & \to & 0 \\
\downarrow & & \downarrow & & & & \\
\mathrm{Hom}_\Lambda(\Lambda^a, i_*(R)) & \xrightarrow{j^*} & \mathrm{Hom}_\Lambda(J, i_*(R)) & \xrightarrow{\partial} & \mathrm{Ext}^1_\Lambda(M, i_*(R)) & \to & 0
\end{array}$$

As the down arrows are isomorphisms then $\mathrm{Ext}^1_R(i^*(M), R) \cong \mathrm{Ext}^1_\Lambda(M, i_*(R))$. As $i^*(M)$ is projective over R then $\mathrm{Ext}^1_R(i^*(M), R) = 0$. Hence $\mathrm{Ext}^1_\Lambda(M, i_*(R)) = 0$. As $i_*(R) = \Lambda$ it follows that M is 1-coprojective □

Proposition 34.3: *If M is a Λ-lattice then M is n-coprojective for all $n \geq 1$.*

Proof. By (34.2) it suffices to consider the case $n \geq 2$. Thus take $J \in \Omega_{n-1}(M)$. Then J is a Λ-lattice by (34.1) and so $\mathrm{Ext}^1_\Lambda(J, \Lambda) = 0$ by (34.2). However, by dimension shifting, $\mathrm{Ext}^n_\Lambda(M, \Lambda) = \mathrm{Ext}^1_\Lambda(J, \Lambda)$ and so $\mathrm{Ext}^n_\Lambda(M, \Lambda) = 0$. □

Corollary 34.4: *Let M, N be lattices over $R[H]$, $R[G]$ respectively where H is a subgroup of G; then for each $n \geq 1$ there are natural isomorphisms*

$$\mathrm{Ext}^n_{R[G]}(N, i_*(M)) \cong \mathrm{Ext}^n_{R[H]}(i^*(N), M).$$

Proof. Let $0 \to J \to P_{n-1} \to \cdots \to P_0 \to N \to 0$ be an exact sequence of $R[G]$-modules with each P_r projective. Then by (21.13)

$$\mathrm{Hom}_{\mathcal{D}\mathrm{er}(R[G])}(J, i_*(M)). \cong \mathrm{Ext}^n_{R[G]}(N, i_*(M)).$$

As the sequence $0 \to i^*(J) \to i^*(P_{n-1}) \to \cdots \to i^*(P_0) \to i^*(N) \to 0$ is exact and each $i^*(P_r)$ is projective, it follows from (21.13) that

$$\mathrm{Hom}_{\mathcal{D}\mathrm{er}(R[H])}(i^*(J), M) \cong \mathrm{Ext}^n_{R[H]}(i^*(N), M)$$

By (32.4) $\mathrm{Hom}_{\mathcal{D}\mathrm{er}(R[G])}(J, i_*(M)) \cong \mathrm{Hom}_{\mathcal{D}\mathrm{er}(R[H])}(i^*(J), M)$ from which the conclusion follows as stated. □

Corollary (34.4) is the *Eckmann-Shapiro Theorem.* In similar fashion we have its dual, the *adjointness theorem*, which generalizes (25.5).

Corollary 34.5: *Let M, N be lattices over $R[H]$, $R[G]$ respectively; then for each $n \geq 1$ there are natural isomorphisms*

$$\mathrm{Ext}^n_{R[G]}(i_*(M), N) \cong \mathrm{Ext}^n_{R[H]}(M, i^*(N)).$$

Proof. Let $0 \to K \to Q_{n-1} \to \cdots \to Q_0 \to M \to 0$ be an exact sequence of $R[H]$-modules with each P_r projective. Then

$$\mathrm{Ext}^n_{R[H]}(M, i^*(N)) \cong \mathrm{Hom}_{\mathcal{D}\mathrm{er}(R[H])}(K, i^*(N)).$$

Moreover, the sequence $0 \to i_*(K) \to i_*(Q_{n-1}) \to \cdots \to i_*(Q_0) \to i_*(M) \to 0$ of $R[G]$-modules is exact and each $i_*(Q_r)$ is projective. By (21.13) we have

$$\mathrm{Hom}_{\mathcal{D}\mathrm{er}(R[G])}(i_*(K), N) \cong \mathrm{Ext}^n_{R[G]}(i_*(M), N).$$

By (25.5) we have $\mathrm{Hom}_{\mathcal{D}\mathrm{er}(R[H])}(K, i^*(N)) \cong \mathrm{Hom}_{\mathcal{D}\mathrm{er}(R[G])}(i_*(K), N)$ from which the stated conclusion now follows. □

§35: Frobenius reciprocity:

We have the following 'reciprocity' theorem which is conventionally attributed to Frobenius.

Proposition 35.1: *Let L be an $R[H]$ lattice and M an $R[G]$ lattice; then there is an isomorphism of $R[G]$-modules*

$$i_*(L) \otimes M \cong i_*(L \otimes_{R[H]} i^*(M)).$$

Proof. As an $R[H]$-module, $i_*(L) \otimes M \cong L \otimes_{R[H]} R[G] \otimes_R M$ on which G acts by

$$(l \otimes g \otimes m) \cdot \gamma = l \otimes g\gamma \otimes m\gamma.$$

The underlying R-module of $i_*(L \otimes i^*(M))$ is $L \otimes_R M \otimes_{R[H]} R[G]$ with G-action

$$(l \otimes m \otimes g) \cdot \gamma = l \otimes m \otimes g\gamma.$$

Define $\Phi : L \otimes_{R[H]} R[G] \otimes_R M \to L \otimes_R M \otimes_{R[H]} R[G]$ by

$$\Phi(l \otimes g \otimes m) = l \otimes mg^{-1} \otimes g.$$

Then Φ is a bijective R-linear homomorphism with inverse given by

$$\Phi^{-1}(l \otimes m \otimes g) = l \otimes g \otimes mg.$$

One checks easily that $\Phi((l \otimes g \otimes m) \cdot \gamma) = \Phi(l \otimes g \otimes m) \cdot \gamma$ so that Φ is an isomorphism of $R[G]$-modules $\Phi : i_*(L) \otimes M \xrightarrow{\simeq} i_*(L \otimes_R i^*(M))$. □

Now take the special case where H is the trivial subgroup of G so that $R[H] = R$.
Then $i_* : \mathcal{M}od_R \to \mathcal{M}od_{R[G]}$ will denote the 'extension of scalars' functor

$$i_*(M) = M \otimes_R R[G]$$

and $i^* : \mathcal{M}od_{R[G]} \to \mathcal{M}od_R$ will denote the 'restriction of scalars' functor.

Let L, M be lattices over $R[G]$. Then $i^*(L)$ and $i^*(M)$ are finitely generated projective R-modules and hence

(35.2) $i^*(L) \otimes_R i^*(M)$ is finitely generated projective over R.

We define $L \otimes M$ to be the $R[G]$-module whose underlying R module is $i^*(L) \otimes_R i^*(M)$ together with G-action given by $(l \otimes m) \cdot g = lg \otimes mg$ where $l \in L$, $m \in M$ and $g \in G$. It is clear that:

(35.3) If L, M are $R[G]$ lattices then $L \otimes M$ is also an $R[G]$ lattice.

As a corollary of (35.1) we obtain:

Proposition 35.4: *Let M be an $R[G]$-lattice where* $\mathrm{rk}_R(M) = m$; *then*

$$R[G] \otimes M \cong \underbrace{R[G] \oplus \cdots \oplus R[G]}_{m}.$$

Proof. Observe that $i^*(M) \cong R^{(m)}$ and $R[G] = i_*(R)$. Hence

$$\begin{aligned} R[G] \otimes M &\cong i_*(R \otimes_R i^*(M)) \\ &\cong \underbrace{i_*(R) \oplus \cdots\cdots \oplus i_*(R)}_{m} \\ &\cong \underbrace{R[G] \oplus \cdots\cdots \oplus R[G]}_{m}. \end{aligned}$$

□

Proposition 35.5: *Let P be a finitely generated projective R-module; then $i_*(P)$ is a finitely generated projective $R[G]$-module.*

Proof. There exists a projective R-module Q such that $P \oplus Q \cong_R R^{(n)}$ for some integer $n \geq 0$. Thus $i_*(P) \oplus i_*(Q) \cong_{R[G]} R[G]^{(n)}$ and so $i_*(P)$ is finitely generated projective as claimed. □

Moreover, taking $Q = \Omega^{(m)}$ in the above we see that:

Proposition 35.6: *Let P be a finitely generated stably free R-module; then $i_*(P)$ is a finitely generated stably free $R[G]$-module.*

This immediately implies the following result which, though weaker than (35.4), is also useful:

Proposition 35.7: *Let M be a weak $R[G]$ lattice; then $R[G] \otimes M$ is a finitely generated stably free module over $R[G]$.*

The following is a straightforward consequence of (35.4).

Proposition 35.8: *Let M, P be $R[G]$-lattices where P is projective then $P \otimes M$ is a finitely generated projective $R[G]$-module.*

The above consequence of Frobenius reciprocity shows that, for any finite group G the augmentation ideal I_G has a universal status for the study of syzygies of modules over G; thus we have:

Proposition 35.9: *For any finite group G, $\underbrace{I_G \otimes \cdots \otimes I_G}_{n} \in \Omega_n(\mathbb{Z})$.*

Proof. By induction on n. The case $n = 1$ follows tautologically from the augmentation sequence $0 \to I_G \to \Lambda \to \mathbb{Z} \to 0$ where $\Lambda = \mathbb{Z}[G]$. Now put

$J = \underbrace{I_G \otimes \cdots \otimes I_G}_{n-1}$. Assuming the statement is true for $n-1$ then

$$\Omega_1(J) = \Omega_1(\Omega_{n-1}(\mathbb{Z})) = \Omega_n(\mathbb{Z}).$$

Applying $- \otimes J$ to the augmentation sequence gives an exact sequence

$$0 \to I_G \otimes J \to \Lambda \otimes J \to J \to 0$$

which, in view of (35.4) we may write as

$$0 \to \underbrace{I_G \otimes \cdots \otimes I_G}_{n} \longrightarrow \Lambda^{rk(J)} \longrightarrow J \to 0.$$

Thus $\underbrace{I_G \otimes \cdots \otimes I_G}_{n} \in \Omega_1(J) = \Omega_n(\mathbb{Z})$, so completing the proof. □

A slight generalization shows that:

(35.10) Let M be a lattice over $\mathbb{Z}[G]$; then $M \otimes \underbrace{I_G \otimes \cdots \otimes I_G}_{n} \in \Omega_n(M)$.

Although (35.10) is an interesting theoretical curiosity, it is, alas, of limited practical use so far as computations are concerned. Thus for example, in calculating $\Omega_1(M)$ let a be the minimal number of generators of M over Λ and compare the exact sequence $0 \to L \to \Lambda^a \to M \to 0$ with the exact sequence $0 \to I_G \otimes M \to \Lambda^{rk(M)} \to M \to 0$ obtained by tensoring the augmentation sequence with M. Then by Schanuel's Lemma, we have $I_G \otimes M \oplus \Lambda^a \cong L \oplus \Lambda^{rk(M)}$. Thus provided $\mathrm{rk}(M) - a \geq 2$, which is typically the case, we may apply Jacobinski Cancellation to conclude that

$$I_G \otimes M \cong L \oplus \Lambda^{rk(M)-a}.$$

Thus one expects $I_G \otimes M$ to decompose as the direct sum of a minimal element of $\Omega_1(M)$ with a large free summand so that the task of extracting a minimal element of $\Omega_1(M)$ from $I_G \otimes M$ involves an amount of computation which is generally impractical. Evidently the corresponding task for $\Omega_n(M)$ increases with n. However, see the reference to the thesis of J.D.P. Evans in Appendix A.

Chapter Five

Swan homomorphisms

In his fundamental paper on group cohomology [61] R.G. Swan defined a homomorphism $(\mathbf{Z}/|G|)^* \to \widetilde{K_0}(\mathbf{Z}[G])$ for any finite group G which has since become a standard feature in the classification of projective modules. In [34] we gave a reformulation of Swan's original definition so that, over quite general rings Λ and suitable modules J, it takes the form $S : \mathrm{Aut}_{\mathcal{D}\mathrm{er}}(J) \to \widetilde{K_0}(\Lambda)$. In this chapter we give a simplified account of the theory as formalized in [34].

§36: Swan's projectivity criterion:

By a *projective* 0*-complex* we mean an exact sequence of Λ-modules

$$\mathcal{P} = (0 \to J \xrightarrow{j} P \xrightarrow{p} M \to 0)$$

in which P is projective. Given such a projective 0-complex $\mathcal{P}$ and a Λ-homomorphism $f : J \to J$ then, as in §13, may form the *pushout extension* $f_*(\mathcal{P})$ which is related to $\mathcal{P}$ via the commutative diagram

$$\begin{array}{c} \mathcal{P} \\ \downarrow \\ f_*(\mathcal{P}) \end{array} = \left(\begin{array}{ccccccccc} 0 \longrightarrow & J & \xrightarrow{j} & P & \xrightarrow{p} & M & \longrightarrow & 0 \\ & \downarrow f & & \downarrow \nu & & \downarrow \mathrm{Id} & & \\ 0 \longrightarrow & J & \xrightarrow{\bar{j}} & \varinjlim(f,j) & \xrightarrow{\bar{p}} & M & \longrightarrow & 0 \end{array} \right)$$

where $\bar{j}$, $\bar{p}$, ν are the canonical morphisms. We note that when M is 1-coprojective the isomorphism class of $\varinjlim(f,j)$ depends only upon the class $[f]$ of f in $\mathrm{End}_{\mathcal{D}\mathrm{er}}(J)$.

Proposition 36.1: *Let* $\mathcal{P} = (0 \to J \xrightarrow{j} P \xrightarrow{p} M \to 0)$ *be a projective* 0*-complex where* M *is* 1*-coprojective and let* $f, g : J \to J$ *be* Λ*-homomorphisms.*

$$[f] = [g] \in \mathrm{End}_{\mathcal{D}\mathrm{er}}(J) \Longrightarrow \varinjlim(f,j) \cong \varinjlim(g,j).$$

Proof. As M is 1-coprojective we may apply the cohomology sequence from $\mathrm{Hom}_{\mathcal{D}\mathrm{er}}(-, J)$ to obtain the exact sequence

$$\mathrm{Hom}_{\mathcal{D}\mathrm{er}}(P, J) \xrightarrow{j^*} \mathrm{Hom}_{\mathcal{D}\mathrm{er}}(J, J) \xrightarrow{\delta} \mathrm{Ext}^1(M, J) \xrightarrow{p_*} \mathrm{Ext}^1(P, J).$$

As P is projective then $\mathrm{Hom}_{\mathcal{D}\mathrm{er}}(P, J) \cong \mathrm{Ext}^1(P, J) = 0$ giving an isomorphism

$$\delta : \mathrm{Hom}_{\mathcal{D}\mathrm{er}}(J, J) \xrightarrow{\simeq} \mathrm{Ext}^1(M, J).$$

We recall that δ is defined by $\delta[f] = f_*(\mathcal{P})$. As $[f] = [g]$ then $f_*(\mathcal{P})$ is congruent to $g_*(\mathcal{P})$; that is, there is a commutative diagram with exact rows

$$\begin{matrix} f_*(\mathcal{P}) \\ \downarrow \widehat{h} \\ g_*(\mathcal{P}) \end{matrix} = \left(\begin{matrix} 0 \longrightarrow & J \xrightarrow{j} & \varinjlim(f, j) & \xrightarrow{p} & M \longrightarrow & 0 \\ & \downarrow \mathrm{Id} & \downarrow h & & \downarrow \mathrm{Id} & \\ 0 \longrightarrow & J \xrightarrow{\bar{j}} & \varinjlim(g, j) & \xrightarrow{\bar{p}} & M \longrightarrow & 0 \end{matrix} \right)$$

so that $h : \varinjlim(f, j) \xrightarrow{\simeq} \varinjlim(g, j)$ is an isomorphism. □

The following is a reformulation of *Swan's projectivity criterion* ([60] Prop. 7.1).

Theorem 36.2: *Let* $\mathcal{P} = (0 \to J \xrightarrow{j} P \xrightarrow{p} M \to 0)$ *be a projective 0-complex in which* M *is 1-coprojective and let* $f : J \to J$ *be a* Λ*-homomorphism; then*

$$\varinjlim(f, j) \text{ is projective} \iff [f] \in \mathrm{Aut}_{\mathcal{D}\mathrm{er}}(J).$$

Proof. ($\Longrightarrow$) Put $\varinjlim = \varinjlim(f, j)$. As M is 1-coprojective the exact sequence from $\mathrm{Hom}_{\mathcal{D}\mathrm{er}}(-, J)$ gives the following commutative diagram with exact rows:

$$\begin{matrix} \mathrm{Hom}_{\mathcal{D}\mathrm{er}}(\varinjlim, J) \xrightarrow{\bar{j}^*} & \mathrm{Hom}_{\mathcal{D}\mathrm{er}}(J, J) \xrightarrow{\bar{\delta}} & H^1(M, J) \xrightarrow{\bar{p}^*} & H^1(\varinjlim, J) \\ \downarrow \nu^* & \downarrow f^* & \downarrow \mathrm{Id} & \downarrow \nu^* \\ \mathrm{Hom}_{\mathcal{D}\mathrm{er}}(P, J) \xrightarrow{j^*} & \mathrm{Hom}_{\mathcal{D}\mathrm{er}}(J, J) \xrightarrow{\delta} & H^1(M, J) \xrightarrow{p^*} & H^1(P, J). \end{matrix}$$

By hypothesis both $\varinjlim$ and P are projective so that

$$\mathrm{Hom}_{\mathcal{D}\mathrm{er}}(\varinjlim, J) \cong \mathrm{Hom}_{\mathcal{D}\mathrm{er}}(P, J) = 0 \quad \text{and} \quad H^1(\varinjlim, J) \cong H^1(P, J) = 0$$

and the above reduces to the commutative diagram

$$\begin{array}{ccccccc} 0 \to & \mathrm{Hom}_{\mathcal{D}\mathrm{er}}(J,J) & \stackrel{\overline{\delta}}{\to} & H^1(M,J) \to & 0 \\ & \downarrow f^* & & \downarrow \mathrm{Id} & \\ 0 \to & \mathrm{Hom}_{\mathcal{D}\mathrm{er}}(J,J) & \stackrel{\delta}{\to} & H^1(M,J) \to & 0 \end{array}$$

in which $\overline{\delta}$ and δ are isomorphisms by exactness. Hence f^* is bijective, where

$$f^*([\alpha]) = [\alpha \circ f]$$

for $\alpha \in \mathrm{Hom}_\Lambda(J,J)$. We claim that $[f] \in \mathrm{Aut}_{\mathcal{D}\mathrm{er}}(M)$. To see this, observe that, as f^* is surjective there exists $h \in \mathrm{Hom}_\Lambda(J,J)$ such that $f^*([h]) = [\mathrm{Id}]$; that is

$$h \circ f \approx \mathrm{Id}.$$

To show that $[f] \in \mathrm{Aut}_{\mathcal{D}\mathrm{er}}(M)$ it now suffices to show that $f \circ h \approx \mathrm{Id}$. However

$$\begin{aligned} f^*(f \circ h) &\approx (f \circ h) \circ f \\ &\approx f \circ (h \circ f) \\ &\approx f^*(\mathrm{Id}) \end{aligned}$$

As f^* is injective then $f \circ h \approx \mathrm{Id}$ as required, so proving $(\Longrightarrow)$.

$(\Longleftarrow)$. As M is 1-coprojective, for any Λ-module N we may apply the cohomology sequence from $\mathrm{Hom}_{\mathcal{D}\mathrm{er}}(-,N)$ to obtain the following diagram with exact rows.

$$\begin{array}{ccccccccc} \mathrm{Hom}_{\mathcal{D}\mathrm{er}}(J,N) & \stackrel{\overline{\partial}}{\to} & H^1(M,N) & \stackrel{\overline{p}_*}{\to} & H^1(\varinjlim,N) & \stackrel{\overline{j}^*}{\to} & H^1(J,N) & \stackrel{\overline{\partial}}{\to} & H^2(M,N) \\ \downarrow f^* & & \downarrow \mathrm{Id}^* & & \downarrow \nu^* & & \downarrow f^* & & \downarrow \mathrm{Id}^* \\ \mathrm{Hom}_{\mathcal{D}\mathrm{er}}(J,N) & \stackrel{\partial}{\to} & H^1(M,N) & \stackrel{p_*}{\to} & H^1(P,N) & \stackrel{j^*}{\to} & H^1(J,N) & \stackrel{\partial}{\to} & H^2(M,N) \end{array}$$

For fixed N the correspondence $X \mapsto H^k(X,N)$ is functorial on $\mathcal{D}\mathrm{er}(\Lambda)$. By hypothesis, $[f] \in \mathrm{Aut}_{\mathcal{D}\mathrm{er}}(M)$ so that

$$f^* : \mathrm{Hom}_{\mathcal{D}\mathrm{er}}(J,N) \to \mathrm{Hom}_{\mathcal{D}\mathrm{er}}(J,N) \quad \text{and} \quad f^* : H^1(J,N) \to H^1(J,N)$$

are isomorphisms. Likewise $\mathrm{Id}^* : H^k(M,N) \to H^k(M,N)$ is an isomorphism for $k = 1, 2$. It follows from the Five Lemma that $\nu^* : H^1(\varinjlim, N) \to H^1(P,N)$ is an isomorphism. However as P is projective then $H^1(P, N) = 0$. Hence for all modules N we have $H^1(\varinjlim, N) = 0$. Thus $\varinjlim$ is projective by (15.7). □

§37: Tame modules:

Recall that a Λ-module M is *finitely presented* when there is an exact sequence

$$\Lambda^n \xrightarrow{\partial_1} \Lambda^m \xrightarrow{\partial_0} M \to 0.$$

The following is a straightforward consequence of Schanuel's Lemma; we leave the proof to the reader:

Proposition 37.1: *Let* $\mathcal{P} = (0 \to J \xrightarrow{j} P \xrightarrow{p} M \to 0)$ *be an exact sequence of* Λ*-homomorphisms in which* P *is finitely generated and projective; then*

$$J \text{ is finitely generated} \iff M \text{ is finitely presented.}$$

Given an exact sequence of Λ-homomorphisms $\mathcal{P} = (0 \to J \xrightarrow{j} P \xrightarrow{p} M \to 0)$ we shall say that $\mathcal{P}$ an *admissible* 0-*complex for* J when P is a finitely generated projective and M is finitely presented and 1-coprojective. We say that the module J is *tame* when there exists an admissible 0-complex $\mathcal{P} = (0 \to J \xrightarrow{j} P \xrightarrow{p} M \to 0)$. From (37.1) we see that:

(37.2) A tame module is necessarily finitely generated.

The admissible 0-complex $\mathcal{P} = (0 \to J \xrightarrow{j} P \xrightarrow{p} M \to 0)$ is said to be *free* when $P \cong \Lambda^n$ for some positive integer n; more generally, $\mathcal{P}$ is *stably free* when P is stably free. If Q is a finitely generated projective Λ-module we define

$$\Sigma^Q(\mathcal{P}) = (0 \to J \xrightarrow{\begin{pmatrix} j \\ 0 \end{pmatrix}} P \oplus Q \xrightarrow{\begin{pmatrix} p & 0 \\ 0 & Id \end{pmatrix}} M \oplus Q \to 0).$$

Then $\Sigma^Q(\mathcal{P})$ is also an admissible 0-complex; when $Q = \Lambda^n$ we write

$$\Sigma^n(\mathcal{P}) = \Sigma^{\Lambda^n}(\mathcal{P}).$$

In general if $\mathcal{P} = (0 \to J \xrightarrow{j} P \xrightarrow{p} M \to 0)$ is an admissible 0-complex then we denote by $[\mathcal{P}]$ the class in the reduced projective class group determined by $\mathcal{P}$; that is $[\mathcal{P}] = [P] \in \widetilde{K}_0(\Lambda)$. Clearly we have

(37.3) $$[\Sigma^Q(\mathcal{P})] = [\mathcal{P}] + [Q].$$

Hence

(37.4) $$[\Sigma^n(\mathcal{P})] = [\mathcal{P}].$$

Let $\mathcal{P}$ be an admissible 0-complex; choosing a finitely generated projective module Q such that $P \oplus Q \cong \Lambda^n$ then $\Sigma^Q(\mathcal{P})$ is an admissible free 0-complex; hence:

(37.5) If J is a tame module there exists an admissible free 0-complex for J.

Given admissible 0-complexes $\begin{cases} \mathcal{Q} = (0 \to J \xrightarrow{j} Q \xrightarrow{q} M \to 0) \\ \mathcal{Q}' = (0 \to J \xrightarrow{j'} Q' \xrightarrow{q'} M' \to 0) \end{cases}$

we write $\mathcal{Q}\ {}_J\!\cong \mathcal{Q}'$ when there exists a commutative diagram

$$\begin{array}{ccccccccc} 0 & \longrightarrow & J & \xrightarrow{j} & Q & \xrightarrow{q} & M & \longrightarrow & 0 \\ & & \downarrow \mathrm{Id} & & \downarrow \widehat{h} & & \downarrow h & & \\ 0 & \longrightarrow & J & \xrightarrow{j'} & Q' & \xrightarrow{q'} & M' & \longrightarrow & 0 \end{array}$$

in which h and $\widehat{h}$ are isomorphisms. By an *excellent mapping* $\Phi : \mathcal{Q} \to \mathcal{Q}'$ we mean a morphism of exact sequences as follows:

(37.6) $$\begin{array}{c} \mathcal{Q} \\ \Phi \downarrow \\ \mathcal{Q}' \end{array} = \left(\begin{array}{ccccccccc} 0 & \longrightarrow & J & \xrightarrow{j} & Q & \xrightarrow{q} & N & \longrightarrow & 0 \\ & & \downarrow \mathrm{Id} & & \downarrow \varphi & & \downarrow \varphi_+ & & \\ 0 & \longrightarrow & J & \xrightarrow{j'} & Q' & \xrightarrow{q'} & N' & \longrightarrow & 0 \end{array} \right)$$

in which $\varphi : Q \to Q'$ is surjective and $\mathrm{Ker}(\varphi) \cong \Lambda^k$ for some integer $k \geq 0$. In any such diagram we note that φ_+ is necessarily surjective. We show that:

Proposition 37.7: *Let $\mathcal{Q}, \mathcal{Q}'$ be admissible 0-complexes over J. If $\Phi : \mathcal{Q} \to \mathcal{Q}'$ is an excellent mapping then $\mathcal{Q}_J \cong \Sigma^k(\mathcal{Q}')$ for some integer $k \geq 0$.*

Proof. Suppose given an excellent mapping $\Phi : \mathcal{Q} \to \mathcal{Q}'$ as in (37.6). Then, putting $L = \mathrm{Ker}(\varphi_+)$, we obtain the commutative diagram below, in which all rows and columns are exact and j, j' are injective.

(37.8)

$$\left\{\begin{array}{ccccccc} & & 0 & & 0 & & \\ & & \downarrow & & \downarrow & & \\ & & \Lambda^k & \xrightarrow{\widehat{q}} & L & & \\ & & \downarrow i & & \downarrow i_+ & & \\ J & \xrightarrow{j} & Q & \xrightarrow{q} & N & \to & 0 \\ \downarrow \mathrm{Id} & & \downarrow \varphi & & \downarrow \varphi_+ & & \\ J & \xrightarrow{j'} & Q' & \xrightarrow{q'} & N' & \to & 0 \\ & & \downarrow & & \downarrow & & \\ & & 0 & & 0 & & \end{array}\right.$$

A straightforward diagram chase shows that $\widetilde{q} : \Lambda^k \to L$ is an isomorphism. The right hand column now gives an exact sequence $0 \to L \xrightarrow{i_+} N \xrightarrow{\varphi_+} N' \to 0$ determined by an extension class in $\mathrm{Ext}^1(N', L)$. As N' is 1-coprojective and $L \cong \Lambda^k$ then $\mathrm{Ext}^1(N', L) = 0$. Hence the exact sequence splits and we may choose a homomorphism $r_+ : N \to L$ such that $r_+ \circ i_+ = \mathrm{Id}_L$. Define $r = (\widehat{q})^{-1} \circ r_+ \circ q : Q \to \Lambda^k$. Then

$$\begin{aligned} r \circ i &= (\widehat{q})^{-1} \circ r_+ \circ q \circ i \\ &= (\widehat{q})^{-1} \circ r_+ \circ i_+ \circ \widehat{q} \\ &= (\widehat{q})^{-1} \circ \widehat{q} \\ &= \mathrm{Id}_{\Lambda^k} \end{aligned}$$

Define $T : Q \to Q' \oplus \Lambda^k$ and $T_+ : N \to N' \oplus \Lambda^k$ by

$$T = \begin{pmatrix} \varphi \\ r \end{pmatrix}; \quad T_+ = \begin{pmatrix} \varphi_+ \\ (\widehat{q})^{-1} \circ r_+ \end{pmatrix}.$$

Then T and T_+ are isomorphisms and one checks easily that (T, T_+) defines an isomorphism of exact sequences over Id_J thus:

$$\begin{array}{ccccccc} 0 \longrightarrow & J & \xrightarrow{j} & Q & \xrightarrow{q} & N \longrightarrow & 0 \\ & \downarrow \mathrm{Id} & & \downarrow T & & \downarrow T_+ & \\ 0 \longrightarrow & J & \xrightarrow{\begin{pmatrix} j' \\ \mathrm{Id} \end{pmatrix}} & Q' \oplus \Lambda^k & \xrightarrow{\begin{pmatrix} q' & 0 \\ 0 & \mathrm{Id} \end{pmatrix}} & N' \oplus \Lambda^k \longrightarrow & 0 \end{array}$$

Hence, as claimed, $\mathcal{Q}_J \cong \Sigma^k(\mathcal{Q}')$ □

We next show:

Proposition 37.9: *Let $\mathcal{P}$, $\mathcal{P}'$ be admissible 0-complexes over J; if $[\mathcal{P}] = [\mathcal{P}']$ then for some integer $n \geq 0$ there exists an excellent mapping*

$$\Phi : \Sigma^n(\mathcal{P}) \to \mathcal{P}'.$$

Proof. Applying $\mathrm{Hom}_\Lambda(-, P')$ to $\mathcal{P}$ gives the following exact sequence:

$$\mathrm{Hom}_\Lambda(P, P') \xrightarrow{j^*} \mathrm{Hom}_\Lambda(J, P') \xrightarrow{\delta} \mathrm{Ext}^1(M, P').$$

As M is 1-coprojective then $\mathrm{Ext}^1(M, P') = 0$ so $j^* : \mathrm{Hom}_\Lambda(P, P') \longrightarrow \mathrm{Hom}_\Lambda(J, P')$ is surjective. Hence there exists a Λ-homomorphism $f : P \to P'$ making the following diagram commute:

$$\begin{array}{ccc} J & \xrightarrow{j} & P \\ \downarrow \mathrm{Id} & & \downarrow f \\ J & \xrightarrow{\bar{j}} & P'. \end{array}$$

In particular, f induces a homomorphism $f_+ : P/\mathrm{Im}(j) \to P'/\mathrm{Im}(j')$. By exactness, we may identify $M \cong P/\mathrm{Im}(j)$ and $M' \cong P'/\mathrm{Im}(j')$ and so construct a morphism of exact sequences as follows:

$$\begin{array}{c} \mathcal{P} \\ F\downarrow \\ \mathcal{P}' \end{array} = \left(\begin{array}{ccccccccc} 0 & \longrightarrow & J & \xrightarrow{j} & P & \xrightarrow{p} & M & \longrightarrow & 0 \\ & & \downarrow \mathrm{Id} & & \downarrow f & & \downarrow f_+ & & \\ 0 & \longrightarrow & J & \xrightarrow{j'} & P' & \xrightarrow{p'} & M' & \longrightarrow & 0 \end{array}\right).$$

As P' is finitely generated we may choose a surjective homomorphism $\epsilon_n : \Lambda^n \to P'$ for some integer n. Observe that n can be any integer at least equal to the smallest number of generators of P'. Note that $p' \circ \epsilon_n$ is also surjective. Define

$$\varphi(n) = (f, \epsilon_n) : P \oplus \Lambda^n \to P'; \varphi(n)_+ = (f_+, p' \circ \epsilon_n) : M \oplus \Lambda^n \to M'.$$

It is straightforward to check that the following diagram commutes

$$\begin{array}{c} \Sigma^n(\mathcal{P}) \\ \Phi(n)\downarrow \\ \mathcal{P}' \end{array} = \left(\begin{array}{ccccccccc} 0 & \longrightarrow & J & \xrightarrow{j} & P \oplus \Lambda^n & \xrightarrow{p} & M \oplus \Lambda^n & \longrightarrow & 0 \\ & & \downarrow \mathrm{Id} & & \downarrow \varphi(n) & & \downarrow \varphi_+(n) & & \\ 0 & \longrightarrow & J & \xrightarrow{j'} & P' & \xrightarrow{p'} & M' & \longrightarrow & 0 \end{array}\right).$$

As $[P] = [P']$ then $\mathrm{Ker}(\varphi(n))$ is stably free and so is free whenever n is sufficiently large. On choosing n large enough so that $\mathrm{Ker}(\varphi(n))$ is free the proof is completed by putting $\Phi = \Phi(n)$, $\varphi = \varphi(n)$, $\varphi_+ = \varphi(n)_+$. □

It follows immediately from (37.7) and (37.9) that:

(37.10) Let $\mathcal{P}$, $\mathcal{P}'$ be admissible 0-complexes for J such that $[\mathcal{P}] = [\mathcal{P}']$; then $\Sigma^n(\mathcal{P}) \ {}_J\!\cong \Sigma^k(\mathcal{P}')$ for some positive integers n, k.

§38: The Swan homomorphism:

Proposition 38.1: *Let* $\mathcal{P} = (0 \to J \xrightarrow{j} P \xrightarrow{p} M \to 0)$ *be an admissible 0-complex and let* $f : J \to J$ *be a* Λ*-homomorphism such that* $[f] \in \mathrm{Aut}_{\mathcal{D}\mathrm{er}}(J)$; *then* $\varinjlim(f,j)$ *is a finitely generated projective module.*

Proof. The projectivity of $\varinjlim(f,j)$ follows from (36.2). By construction

$$\varinjlim(f,j) = (J \oplus P)/\mathrm{Im}(f \times -j)$$

and as J and P are finitely generated so also is $\varinjlim(f,j)$. □

Given an admissible 0-complex $\mathcal{P} = (0 \to J \xrightarrow{j} P \xrightarrow{p} M \to 0)$ it follows from (36.2) and the above that there is a well defined function

$$S_{\mathcal{P}} : \mathrm{Aut}_{\mathcal{D}\mathrm{er}}(J) \longrightarrow \widetilde{K}_0(\Lambda); \quad S_{\mathcal{P}}([f]) = [\varinjlim(f,j)].$$

$S_{\mathcal{P}}$ is the *Swan mapping* associated to $\mathcal{P}$.

Proposition 38.2: *If* $\mathcal{P}$, $\mathcal{P}'$ *are admissible 0-complexes such that* $\mathcal{P} \ {}_J\!\cong \mathcal{P}'$ *then* $S_{\mathcal{P}} = S_{\mathcal{P}'}$.

Proof. Suppose given an isomorphism over J

$$\begin{array}{ccccccccc} 0 \longrightarrow & J & \xrightarrow{j} & P & \xrightarrow{p} & M & \longrightarrow 0 \\ & \downarrow \mathrm{Id} & & \downarrow \widehat{h} & & \downarrow h & \\ 0 \longrightarrow & J & \xrightarrow{\overline{j}} & P' & \xrightarrow{\overline{p}} & M' & \longrightarrow 0. \end{array}$$

For any Λ homomorphism $f : J \to J$, $\mathrm{Id}_J \oplus \widehat{h} : J \oplus P \to J \oplus P'$ induces a homomorphism of pushouts making the following diagram commute

$$\begin{array}{ccccccc} 0 \longrightarrow & J & \longrightarrow & \varinjlim(f,j) & \longrightarrow & M & \longrightarrow 0 \\ & \downarrow \mathrm{Id} & & \downarrow (\mathrm{Id} \oplus \widehat{h})_* & & \downarrow h & \\ 0 \longrightarrow & J & \longrightarrow & \varinjlim(f,j') & \longrightarrow & M' & \longrightarrow 0 \end{array}$$

As h is an isomorphism then $(\mathrm{Id} \oplus \widehat{h})_* : \varinjlim(f,j) \xrightarrow{\simeq} \varinjlim(f,j')$ is an isomorphism by the Five Lemma. Hence if $f \in \mathrm{Aut}_{\mathcal{D}\mathrm{er}}(J)$ then $S_{\mathcal{P}}(f) = S_{\mathcal{P}'}(f)$. □

Comparing the diagrams

$$\begin{cases} J & \xrightarrow{j} & P \\ \downarrow \alpha & & \downarrow \nu \\ J & \xrightarrow{\bar{j}} & \varinjlim(\alpha, j) \end{cases}$$

and

$$\begin{cases} J & \xrightarrow{\binom{j}{0}} & P \oplus Q \\ \downarrow \alpha & & \downarrow \begin{pmatrix} \nu & 0 \\ 0 & \mathrm{Id} \end{pmatrix} \\ J & \xrightarrow{\binom{\bar{j}}{0}} & \varinjlim(\alpha, j) \oplus Q \end{cases}$$

we see that $\varinjlim(\alpha, \binom{j}{0}) \oplus Q \cong \varinjlim(\alpha, j) \oplus Q$. Consequently, if $f : J \to J$ represents an element of $\mathrm{Aut}_{\mathcal{D}\mathrm{er}}(J)$ then:

(38.3) $$S_{\Sigma^Q(\mathcal{P})}(f) = S_{\mathcal{P}}(f) \oplus [Q].$$

It now follows immediately that:

(38.4) $$S_{\Sigma^n(\mathcal{P})} = S_{\mathcal{P}}.$$

Hence we have:

Proposition 38.5: *Let $\mathcal{P}$, $\mathcal{Q}$ be admissible 0-complexes for J; then*

$$[\mathcal{P}] = [\mathcal{Q}] \Longrightarrow S_{\mathcal{P}} = S_{\mathcal{Q}}.$$

Proof. If $[\mathcal{P}] = [\mathcal{Q}]$ then, by (37.10) above we have $\Sigma^m(\mathcal{P})\ {}_J\!\cong \Sigma^n(\mathcal{Q})$ for some positive integers n, k. By (38.2) we have $S_{\Sigma^n(\mathcal{P})} = S_{\Sigma^k(\mathcal{Q})}$. It now follows from (38.4) that $S_{\mathcal{P}} = S_{\mathcal{Q}}$. □

If J is tame then by (37.5) there exists a stably free admissible 0-complex

$$\mathcal{F} = (0 \to J \xrightarrow{j} F \xrightarrow{p} M \to 0).$$

Indeed, we may take F to be free. If $\mathcal{F}'$ is also a stably free admissible 0-complex for J then $[\mathcal{F}] = [\mathcal{F}'] = 0 \in \widetilde{K}_0(\Lambda)$. Hence it follows from (38.5) that:

(38.6) If $\mathcal{F}$, $\mathcal{F}'$ are stably free admissible 0-complexes for J then $S_{\mathcal{F}} = S_{\mathcal{F}'}$.

If J is a tame Λ-module and $\mathcal{F}$ is a stably free admissible 0-complex for J we define the *Swan mapping* $S_J : \mathrm{Aut}_{\mathcal{D}\mathrm{er}}(J) \to \widetilde{K}_0(\Lambda)$ by

(38.7) $$S_J = S_{\mathcal{F}}.$$

It is clear from (38.5) that this definition is unambiguous. Moreover, given a tame Λ-module J then for $g \in \mathrm{Aut}_{\mathcal{D}\mathrm{er}}(J)$; we have the following useful tautology:

(38.8) $S_J(g) = [g_*(\mathcal{F})]$ if $\mathcal{F}$ is any admissible stably free 0-complex for J.

Now suppose that $\mathcal{P} = (0 \to J \xrightarrow{j} P \xrightarrow{p} N \to 0)$ is an admissible 0-complex for J and let $\mathcal{F} = (0 \to J \xrightarrow{j} \Lambda^n \xrightarrow{p} M \to 0)$ be a free admissible 0-complex. Then

$$\Sigma^P(\mathcal{F}) = (0 \to J \to \Lambda^n \oplus P \to M \oplus P \to 0)$$

and so $[\mathcal{P}] = [\Sigma^P(\mathcal{F})] = [P]$. It follows from (38.5) that $S_{\mathcal{P}} = S_{\Sigma^P(\mathcal{F})}$. However, by (38.3), $S_{\Sigma^P(\mathcal{F})}(g) = S_{\mathcal{F}}(g) + [P] = S_J(g) + [P]$. Thus if $\mathcal{P}$ is an admissible 0-complex for J then for any $g \in \mathrm{Aut}_{\mathcal{D}\mathrm{er}}(J)$ we see that:

(38.9) $$S_{\mathcal{P}}(g) = S_J(g) + [\mathcal{P}].$$

Proposition 38.10: *If $\mathcal{P}$ is an admissible 0-complex over J and $f, g \in \mathrm{Aut}_{\mathcal{D}\mathrm{er}}(J)$ then*

$$S_{\mathcal{P}}(f \circ g) = S_{g_*(\mathcal{P})}(f).$$

Proof. Writing $\mathcal{P} = (0 \to J \xrightarrow{j} F \xrightarrow{p} M \to 0)$ we have a pushout diagram.

$$\begin{matrix} \mathcal{P} \\ \downarrow \\ g_*(\mathcal{P}) \end{matrix} = \begin{pmatrix} 0 \to & J & \xrightarrow{j} & F & \xrightarrow{p} & M & \to 0 \\ & g\downarrow & & \nu\downarrow & & \mathrm{Id}\downarrow & \\ 0 \to & J & \xrightarrow{j'} & \varinjlim(g,j) & \xrightarrow{p'} & M & \to 0 \end{pmatrix}.$$

To compute $S_{g_*(\mathcal{P})}(f)$ we construct a new pushout diagram

$$\begin{array}{c} g_*(\mathcal{P}) \\ \downarrow \\ f_*(g_*(\mathcal{P})) \end{array} = \left(\begin{array}{cccccccc} 0 \to & J & \stackrel{j'}{\to} & \varinjlim(g,j) & \stackrel{p'}{\to} & M & \to 0 \\ & f\downarrow & & \nu'\downarrow & & \mathrm{Id}\downarrow & \\ 0 \to & J & \stackrel{j''}{\to} & \varinjlim(f,j') & \stackrel{p''}{\to} & M & \to 0 \end{array}\right).$$

Concatenating the two we obtain:

$$\begin{array}{c} \mathcal{P} \\ \downarrow \\ g_*(\mathcal{P}) \\ \downarrow \\ f_*(g_*(\mathcal{P})) \end{array} = \left(\begin{array}{cccccccc} 0 \to & J & \stackrel{j}{\to} & F & \stackrel{p}{\to} & M & \to 0 \\ & g\downarrow & & \nu\downarrow & & \mathrm{Id}\downarrow & \\ 0 \to & J & \stackrel{j'}{\to} & \varinjlim(g,j) & \stackrel{p'}{\to} & M & \to 0 \\ & f\downarrow & & \nu'\downarrow & & \mathrm{Id}\downarrow & \\ 0 \to & J & \stackrel{j''}{\to} & \varinjlim(f,j') & \stackrel{p''}{\to} & M & \to 0 \end{array}\right).$$

We may compare this to the diagram for $(f \circ g)_*(\mathcal{P})$

$$\begin{array}{c} \mathcal{P} \\ \downarrow \\ (f\circ g)_*(\mathcal{P})) \end{array} = \left(\begin{array}{cccccccc} 0 \to & J & \stackrel{j}{\to} & F & \stackrel{p}{\to} & M & \to 0 \\ & f\circ g\downarrow & & \mu\downarrow & & \mathrm{Id}\downarrow & \\ 0 \to & J & \stackrel{i}{\to} & \varinjlim(f\circ g,j) & \stackrel{\pi}{\to} & M & \to 0 \end{array}\right).$$

However, $(f \circ g)_*(\mathcal{P})$ is congruent to $f_*(g_*(\mathcal{P}))$ as the canonical map of pushouts $\natural : \varinjlim(f \circ g, j) \longrightarrow \varinjlim(f, j')$ makes the following diagram commute

$$\begin{array}{cccccccc} 0 \to & J & \stackrel{i}{\to} & \varinjlim(f\circ g,j) & \stackrel{\pi}{\to} & M & \to 0 \\ & \mathrm{Id}\downarrow & & \natural\downarrow & & \mathrm{Id}\downarrow & \\ 0 \to & J & \stackrel{j''}{\to} & \varinjlim(f,j') & \stackrel{p''}{\to} & M & \to 0. \end{array}$$

By the Five Lemma $\natural : \varinjlim(f \circ g, j) \xrightarrow{\simeq} \varinjlim(f, j')$ is an isomorphism; hence

$$S_{\mathcal{P}}(f \circ g) = S_{g_*(\mathcal{P})}(f). \qquad \square$$

Taking a free admissible 0-complex $\mathcal{F}$ over J so that $S_J = S_{\mathcal{F}}$ we obtain

$$S_J(f \circ g) = S_{g_*(\mathcal{F})}(f).$$

However, by (38.9) $S_{g_*(\mathcal{F})}(f) = S_J(f) + [g_*(\mathcal{F})]$ and by (38.8)

$$[g_*(\mathcal{F})] = S_J(g).$$

Hence we obtain:

Theorem 38.11: *The Swan mapping* $S_J : \mathrm{Aut}_{\mathcal{D}\mathrm{er}}(J) \to \widetilde{K}_0(\Lambda)$ *is a homomorphism for any tame* Λ*-module* J.

§39: Invariance properties of the Swan homomorphism:

Let $\gamma : J \to J'$ be a Λ-isomorphism where J is a tame module with an admissible free 0-complex $\mathcal{F} = (0 \to J \xrightarrow{i} F \xrightarrow{p} N \to 0)$. Then J' is also tame and

(39.1) $\mathcal{F}' = (0 \to J' \xrightarrow{i\circ\gamma^{-1}} F \xrightarrow{p} N \to 0)$ is an admissible free 0-complex for J'.

We may use $\mathcal{F}$ to compute S_J and $\mathcal{F}'$ to compute $S_{J'}$. In this connection, suppose that $\alpha : J \to J$ is a Λ-homomorphism defining an element $[\alpha] \in \mathrm{Aut}_{\mathcal{D}\mathrm{er}}(J)$ so that $S_J([\alpha]) = [\varinjlim(\alpha, i)]$; then $\gamma\alpha\gamma^{-1} : J' \to J'$ defines an element $[\gamma\alpha\gamma^{-1}] \in \mathrm{Aut}_{\mathcal{D}\mathrm{er}}(J')$ and $S_{J'}([\gamma\alpha\gamma^{-1}]) = [\varinjlim(\gamma\alpha\gamma^{-1}, i\gamma^{-1})]$. Moreover, we have canonical isomorphisms

$$\varinjlim(\alpha, i) \cong \varinjlim(\gamma\alpha, i) \cong \varinjlim(\gamma\alpha\gamma^{-1}, i\gamma^{-1})$$

showing that $S_J([\alpha]) = S_{J'}([\gamma\alpha\gamma^{-1}])$. Thus $\mathrm{Im}(S_J) \subset \mathrm{Im}(S_{J'})$. The opposite inclusion holds by symmetry showing that for tame modules J, J' we have:

(39.2) $$J \cong_\Lambda J' \Longrightarrow \mathrm{Im}(S_J) = \mathrm{Im}(S_{J'}).$$

Now suppose that $0 \to J_i \to F_i \to N_i \to 0$ are admissible free 0-complexes and let $\alpha_i : J_i \to J_i$ be Λ-homomorphisms defining elements $[\alpha_i] \in \mathrm{Aut}_{\mathcal{D}\mathrm{er}}(J_i)$; then

(39.3) $$S_{J_1 \oplus J_2}\begin{bmatrix} \alpha_1 & 0 \\ 0 & \alpha_2 \end{bmatrix} = S_{J_1}[\alpha_1] + S_{J_2}[\alpha_2].$$

Now suppose that J is a tame module, P is finitely generated projective and let $\alpha : J \oplus P \to J \oplus P$ be a Λ-homomorphism. Representing α by a matrix in the obvious way $\alpha = \begin{pmatrix} \alpha_{11} & \alpha_{12} \\ \alpha_{21} & \alpha_{22} \end{pmatrix}$ then

(39.4) $$\alpha \approx \begin{pmatrix} \alpha_{11} & 0 \\ 0 & 0 \end{pmatrix}$$

so that

(39.5) $$[\alpha] \in \mathrm{Aut}_{\mathcal{D}\mathrm{er}}(J \oplus P) \Longleftrightarrow [\alpha_{11}] \in \mathrm{Aut}_{\mathcal{D}\mathrm{er}}(J).$$

For $[\alpha] \in \mathrm{Aut}_{\mathcal{D}\mathrm{er}}(J \oplus P)$ it follows from (39.3) that

(39.6) $$S_{J \oplus P}([\alpha]) = S_J([\alpha_{11}]).$$

If J is a tame module and P is finitely generated projective then:

(39.7) $$\mathrm{Im}(S_{J \oplus P}) = \mathrm{Im}(S_J).$$

On taking J to be the zero module, which is trivially tame, we obtain:

(39.8) $S_P \equiv 0$ for any finitely generated projective module P.

Now suppose that J, J' are tame modules such that $J \cong_{\mathcal{D}\mathrm{er}} J'$; then there are finitely generated projective modules P, P' such that $J \oplus P \cong_\Lambda J' \oplus P'$. By (39.7) $\mathrm{Im}(S_J) = \mathrm{Im}(S_{J \oplus P})$ and $\mathrm{Im}(S_{J' \oplus P'}) = \mathrm{Im}(S_{J'})$ whilst by (39.2) $\mathrm{Im}(S_{J \oplus P}) = \mathrm{Im}(S_{J' \oplus P'})$. We obtain the following strengthening of (39.2), namely that for tame modules J, J':

(39.9) $$J \cong_{\mathcal{D}\mathrm{er}} J' \Longrightarrow \mathrm{Im}(S_J) = \mathrm{Im}(S_{J'}).$$

§40: The relation between consecutive Swan homomorphisms:

The following is straightforward:

Proposition 40.1: *If the commutative diagram below has exact rows*

$$\begin{pmatrix} 0 \longrightarrow & N & \longrightarrow & P_1 & \xrightarrow{\partial} & P_0 & \longrightarrow & M \longrightarrow 0 \\ & \downarrow \mathrm{Id} & & \downarrow f_1 & & \downarrow f_0 & & \downarrow \mathrm{Id} \\ 0 \longrightarrow & N & \longrightarrow & Q_1 & \xrightarrow{\partial'} & Q_0 & \longrightarrow & M \longrightarrow 0 \end{pmatrix}$$

then the sequence $0 \to P_1 \xrightarrow{\begin{pmatrix} \partial \\ -f_1 \end{pmatrix}} P_0 \oplus Q_1 \xrightarrow{(f_0, \partial')} Q_0 \to 0$ *is exact.*

In the following we suppose given admissible 0-complexes $(0 \to J_2 \xrightarrow{j} P \xrightarrow{p} J_1 \to 0)$, $(0 \to J_2 \xrightarrow{j} P' \xrightarrow{p} J_1 \to 0)$ in which J_1 is also tame.

Proposition 40.2: *Let $\gamma : J_2 \to J_2$ be a Λ-homomorphism such that $\gamma \approx 0$; then there exists a Λ-homomorphism $\widehat{\gamma} : P \to P'$ making the following diagram commute*

$$\begin{array}{ccccccc} 0 \to & J_2 & \xrightarrow{j} & P & \xrightarrow{p} & J_1 & \to 0 \\ & \downarrow \gamma & & \downarrow \widehat{\gamma} & & \downarrow 0 & \\ 0 \to & J_2 & \xrightarrow{j'} & P' & \xrightarrow{p'} & J_1 & \to 0. \end{array}$$

Proof. As $\gamma \approx 0$ there exists a factorisation of γ through a projective Q as follows:

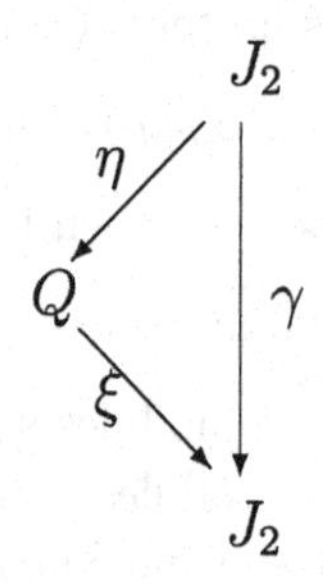

Now form the following commutative diagram

$$\begin{array}{ccccccc} 0 \to & J_2 & \xrightarrow{j} & P & \xrightarrow{p} & J_1 & \to 0 \\ & \eta \downarrow & & \nu \downarrow & & \mathrm{Id} \downarrow & \\ 0 \to & Q & \xrightarrow{\overline{j}} & \underrightarrow{\lim}(\eta, j) & \xrightarrow{\pi} & J_1 & \to 0. \end{array}$$

As Q is projective and J_1 is 1-coprojective then $\mathrm{Ext}^1(J_1, Q) = 0$ so that the lower sequence splits and we may choose a Λ-homomorphism $r : \underrightarrow{\lim}(\eta, j) \to Q$ such that $r \circ \overline{j} = \mathrm{Id}_Q$. Define $\beta : \underrightarrow{\lim}(\eta, j) \to P'$ by $\beta = j' \circ \xi \circ r$. Then the following diagram commutes:

$$\begin{array}{ccccccc} 0 \to & J_2 & \xrightarrow{j} & P & \xrightarrow{p} & J_1 & \to 0 \\ & \eta \downarrow & & \nu \downarrow & & \mathrm{Id} \downarrow & \\ 0 \to & Q & \xrightarrow{\overline{j}} & \underrightarrow{\lim}(\eta, j) & \xrightarrow{\pi} & J_1 & \to 0 \\ & \xi \downarrow & & \beta \downarrow & & 0 \downarrow & \\ 0 \to & J_2 & \xrightarrow{j'} & P' & \xrightarrow{p'} & J_1 & \to 0. \end{array}$$

On defining $\widehat{\gamma} = \beta \circ \nu$ we obtain the commutative diagram

$$\begin{array}{ccccccc} 0 \to & J_2 & \xrightarrow{j} & P & \xrightarrow{p} & J_1 & \to 0 \\ & \downarrow \gamma & & \downarrow \widehat{\gamma} & & \downarrow 0 & \\ 0 \to & J_2 & \xrightarrow{j'} & P' & \xrightarrow{p'} & J_1 & \to 0 \end{array}$$

so completing the proof. □

Now let $\mathcal{F} = (0 \to J_2 \xrightarrow{j} F \xrightarrow{p} J_1 \to 0)$ be a free admissible 0-complex for the tame module J_2. Consider the ring homomorphism $\rho : \mathrm{End}_{\mathcal{D}\mathrm{er}}(J_1) \to \mathrm{End}_{\mathcal{D}\mathrm{er}}(J_2)$ determined, as in Chapter Three, by the relation that $\rho([g_1]) = [g_2]$ whenever there is a commutative diagram of Λ-modules

$$\begin{array}{ccccccc} 0 \longrightarrow & J_2 & \xrightarrow{j} & F_1 & \xrightarrow{p} & J_1 & \longrightarrow 0 \\ & \downarrow g_2 & & \downarrow \mu & & \downarrow g_1 & \\ 0 \longrightarrow & J_2 & \xrightarrow{j} & F_1 & \xrightarrow{p} & J_1 & \longrightarrow 0. \end{array}$$

In what follows we will suppose that J_1 is also tame.

Theorem 40.3: *Let $\mathcal{F} = (0 \to J_2 \xrightarrow{j} F \xrightarrow{p} J_1 \to 0)$ be a free admissible 0-complex in which J_1 is also tame; then $S_{J_1} = -S_{J_2} \circ \rho$.*

Proof. Let $g_1 : J_1 \to J_1$ be a homomorphism representing $[g_1] \in \mathrm{Aut}_{\mathcal{D}\mathrm{er}}(J_1)$ and construct a commutative diagram

$$\begin{array}{ccccccc} 0 \longrightarrow & J_2 & \xrightarrow{j} & F_1 & \xrightarrow{p} & J_1 & \longrightarrow 0 \\ & \downarrow g_2 & & \downarrow \widehat{g} & & \downarrow g_1 & \\ 0 \longrightarrow & J_2 & \xrightarrow{j} & F_1 & \xrightarrow{p} & J_1 & \longrightarrow 0 \end{array}$$

so that, as in §24, $[g_2] = \rho([g_1]) \in \mathrm{Aut}_{\mathcal{D}\mathrm{er}}(J_2)$. Choose a Λ-homomorphism $h : J_2 \to J_2$ such that $[h] = [g_2]^{-1}$ and construct the pushout diagram

$$\begin{array}{ccccccc} 0 \to & J_2 & \xrightarrow{j} & F_1 & \xrightarrow{p} & J_1 & \to 0 \\ & \downarrow h & & \downarrow \nu & & \downarrow \mathrm{Id} & \\ 0 \to & J_2 & \xrightarrow{\overline{j}} & \varinjlim(h, j) & \xrightarrow{\overline{p}} & J_1 & \to 0. \end{array}$$

As $[h] \in \mathrm{Aut}_{\mathcal{D}\mathrm{er}}(J_2)$ then $\varinjlim(h, j)$ is projective and $S_{J_2}([h]) = [\varinjlim(h, j)]$. Composing the two diagrams gives a commutative diagram with exact rows

$$\begin{array}{ccccccc} 0 \to & J_2 & \xrightarrow{j} & F_1 & \xrightarrow{p} & J_1 & \to 0 \\ & \downarrow h \circ g_2 & & \downarrow \nu \circ \widehat{g} & & \downarrow g_1 & \\ 0 \to & J_2 & \xrightarrow{\overline{j}} & \varinjlim(h, j) & \xrightarrow{\overline{p}} & J_1 & \to 0. \end{array}$$

As $[h \circ g_2] = [\mathrm{Id}]$ then $\mathrm{Id}_{J_2} = h \circ g_2 + \gamma$ where $\gamma \approx 0$. As J_1 is coprojective then, by (40.2), there exists a commutative diagram

$$\begin{array}{ccccccc} 0 \to & J_2 & \xrightarrow{j} & F_1 & \xrightarrow{p} & J_1 & \to 0 \\ & \downarrow \gamma & & \downarrow \widehat{\gamma} & & \downarrow 0 & \\ 0 \to & J_2 & \xrightarrow{\overline{j}} & \varinjlim(h, j) & \xrightarrow{\overline{p}} & J_1 & \to 0. \end{array}$$

Consequently, the following diagram commutes:

$$\textbf{(I)} \qquad \begin{array}{ccccccc} 0 \to & J_2 & \xrightarrow{j} & F_1 & \xrightarrow{p} & J_1 & \to 0 \\ & \downarrow \mathrm{Id} & & \downarrow \mu_1 & & \downarrow g_1 & \\ 0 \to & J_2 & \xrightarrow{\overline{j}} & \varinjlim(h, j) & \xrightarrow{\overline{p}} & J_1 & \to 0 \end{array}$$

where $\mu_1 = \nu \circ \widehat{g} + \widehat{\gamma}$. However, as J_1 is also tame there exists a free admissible 0-complex $(0 \to J_1 \xrightarrow{i} F_0 \xrightarrow{\pi} M \to 0)$. Forms the following pushout diagram:

$$\textbf{(II)} \qquad \begin{array}{ccccccc} 0 \to & J_1 & \xrightarrow{i} & F_0 & \xrightarrow{\pi} & M & \to 0 \\ & g_1 \downarrow & & \nu \downarrow & & \mathrm{Id} \downarrow & \\ 0 \to & J_1 & \xrightarrow{\overline{i}} & \varinjlim(g_1, i) & \xrightarrow{\overline{\pi}} & M & \to 0. \end{array}$$

Splicing (I) and (II) together gives a commutative diagram with exact rows

$$\left(\begin{array}{ccccccccc} 0 \longrightarrow & J_2 & \longrightarrow & F_1 & \xrightarrow{\partial} & F_0 & \longrightarrow & M \longrightarrow & 0 \\ & \downarrow \mathrm{Id} & & \downarrow \mu & & \downarrow \mu_0 & & \downarrow \mathrm{Id} & \\ 0 \longrightarrow & J_2 & \longrightarrow & \varinjlim(h, j) & \xrightarrow{\partial'} & \varinjlim(g_1, i) & \longrightarrow & M \longrightarrow & 0 \end{array} \right).$$

where $\partial = i \circ p$ and $\partial' = \bar{i} \circ \bar{p}$. It follows from (40.1) that we have an exact sequence of the form

$$0 \to F_1 \longrightarrow F_0 \oplus \varinjlim(h, j) \longrightarrow \varinjlim(g_1, i) \to 0.$$

As $\varinjlim(g_1, i)$ is projective then the sequence splits to give an isomorphism

$$F_1 \oplus \varinjlim(g_1, i) \cong F_0 \oplus \varinjlim(h, j).$$

As F_1, F_0 are free of finite rank then $[\varinjlim(g_1, i)] = [\varinjlim(h, j)] \in \widetilde{K}_0(\Lambda)$. However,

$$\left[\varinjlim(g_1, i)\right] = S_{J_1}([g_1]) \quad \text{and} \quad \left[\varinjlim(h, j)\right] = S_{J_2}([h]).$$

As S_{J_2} is a homomorphism and $[h] = [g_2]^{-1}$ then $S_{J_2}([h]) = -S_{J_2}([g_2])$. However, as previously observed, $[g_2] = \rho([g_1])$. Hence

$$S_{J_1}([g_1]) = -S_{J_2}(\rho([g_1]). \qquad \square$$

Corollary 40.4: *Let* $0 \to J_2 \to P \to J_1 \to 0$ *be an admissible* 0*-complex in which* J_1 *is also tame; then* $\mathrm{Im}(S_{J_1}) = \mathrm{Im}(S_{J_2})$.

Proof. In the special case where P is free the conclusion follows immediately from (40.3). In general, we may choose a finitely generated projective module Q such that $P \oplus Q \cong \Lambda^a$. Then forming the exact sequence $0 \to J_2 \oplus Q \to P \oplus Q \to J_1 \to 0$ it follows from the above special case that $\mathrm{Im}(S_{J_2 \oplus Q}) = \mathrm{Im}(S_{J_1})$. The conclusion now follows as $\mathrm{Im}(S_{J_2 \oplus Q}) = \mathrm{Im}(S_{J_2})$ by (39.9). $\square$

§41: The dual Swan mapping:

There is a dual construction to the Swan homomorphism which in some sense is more elementary but, in practice, less directly useful. Thus suppose given a projective 0-complex $\mathcal{P} = (0 \to J \xrightarrow{j} P \xrightarrow{p} M \to 0)$. If $f : M \to M$ is a Λ-homomorphism we may form the *pullback extension* $f^*(\mathcal{P})$ related to $\mathcal{P}$ by means of the commutative diagram below:

$$\begin{matrix} f^*(\mathcal{P}) \\ \downarrow \nu \\ \mathcal{P} \end{matrix} = \begin{pmatrix} 0 \longrightarrow & J & \xrightarrow{i} & \varprojlim(p, f) & \xrightarrow{\pi} & M \longrightarrow & 0 \\ & \downarrow \mathrm{Id} & & \downarrow \nu & & \downarrow f & \\ 0 \longrightarrow & J & \xrightarrow{j} & P & \xrightarrow{p} & M \longrightarrow & 0 \end{pmatrix}.$$

Proposition 41.1: *If* $[f] \in \mathrm{Aut}_{\mathcal{D}\mathrm{er}}(M)$ *then* $\varprojlim(p, f)$ *is projective.*

Proof. Consider the natural morphism

$$\begin{matrix} f^*(\mathcal{P}) \\ \downarrow \nu \\ \alpha \end{matrix} = \left(\begin{matrix} 0 \longrightarrow & J & \xrightarrow{i} & \varprojlim(p,f) & \xrightarrow{\pi} & M' & \longrightarrow 0 \\ & \downarrow \mathrm{Id} & & \downarrow \nu & & \downarrow f & \\ 0 \longrightarrow & J & \xrightarrow{j} & P & \xrightarrow{p} & M & \longrightarrow 0 \end{matrix}\right).$$

For brevity we write $\varprojlim = \varprojlim(p,f)$. Given a Λ-module N on applying the cohomology exact sequence extending $\mathrm{Hom}_\Lambda(-,N)$ we obtain a commutative ladder with exact rows

$$\begin{matrix} H^1(M,N) & \xrightarrow{j_*} & H^1(P,N) & \xrightarrow{p_*} & H^1(J,N) & \xrightarrow{\partial} & H^2(M,N) \\ \downarrow f^* & & \downarrow \nu^* & & \downarrow \mathrm{Id} & & \downarrow f^* \\ H^1(M',N) & \xrightarrow{i_*} & H^1(\varprojlim,N) & \xrightarrow{\pi_*} & H^1(J,N) & \xrightarrow{\partial} & H^2(M',N). \end{matrix}$$

Since f^* and Id are isomorphisms it follows that $\nu^* : H^1(P,N) \to H^1(\varprojlim,N)$ is surjective. However, $H^1(P,N) = 0$ since P is projective. Thus $H^1(\varprojlim,N) = 0$ for arbitrary N, and so, by (15.7), $\varprojlim(p,f)$ is projective as claimed. □

We note the above statement does not require any coprojectivity hypothesis on M; the converse statement however does and we have:

Theorem 41.2: *Let $\mathcal{P} = (0 \to J \xrightarrow{j} P \xrightarrow{p} M \to 0)$ be a projective 0-complex in which M is 1-coprojective and let $f : M \to M$ be a Λ-homomorphism; then*

$$\varprojlim(p,f) \text{ is projective} \iff [f] \in \mathrm{Aut}_{\mathcal{D}\mathrm{er}}(M).$$

Proof. By (41.1) it suffices to prove ($\Longrightarrow$). Suppose that $\varprojlim = \varprojlim(p,f)$ is projective and apply the direct sequence from $\mathrm{Hom}_{\mathcal{D}\mathrm{er}}(M,-)$ to obtain the following commutative diagram with exact rows:

$$\begin{matrix} \mathrm{Hom}_{\mathcal{D}\mathrm{er}}(M,\varprojlim) & \xrightarrow{\overline{p}_*} & \mathrm{Hom}_{\mathcal{D}\mathrm{er}}(M,M) & \xrightarrow{\overline{\partial}} & H^1(M,J) & \xrightarrow{\overline{j}_*} & H^1(M,\varprojlim) \\ \downarrow \nu_* & & \downarrow f_* & & \downarrow \mathrm{Id} & & \downarrow \nu_* \\ \mathrm{Hom}_{\mathcal{D}\mathrm{er}}(M,P) & \xrightarrow{p_*} & \mathrm{Hom}_{\mathcal{D}\mathrm{er}}(M,M) & \xrightarrow{\partial} & H^1(M,J) & \xrightarrow{j_*} & H^1(M,P). \end{matrix}$$

As $\varprojlim$ and P are projective then $\mathrm{Hom}_{\mathcal{D}\mathrm{er}}(M,\varprojlim) = \mathrm{Hom}_{\mathcal{D}\mathrm{er}}(M,P) = 0$. As M is 1-coprojective then $H^1(M,\varprojlim) = H^1(M,P) = 0$ so that the above reduces to the following commutative diagram with exact rows

$$\begin{array}{ccccccc} 0 & \to & \mathrm{End}_{\mathcal{D}\mathrm{er}}(M) & \stackrel{\overline{\partial}}{\to} & H^1(M,J) & \to & 0 \\ & & \downarrow f_* & & \downarrow \mathrm{Id} & & \\ 0 & \to & \mathrm{End}_{\mathcal{D}\mathrm{er}}(M) & \stackrel{\partial}{\to} & H^1(M,J) & \to & 0. \end{array}$$

Clearly $f_* : \mathrm{End}_{\mathcal{D}\mathrm{er}}(M) \to \mathrm{End}_{\mathcal{D}\mathrm{er}}(M)$ is bijective. We claim that $[f] \in \mathrm{Aut}_{\mathcal{D}\mathrm{er}}(M)$. To see this note that $f_*([g]) = [f \circ g]$. As f_* is surjective, we may choose $g \in \mathrm{End}_\Lambda(M)$ such that $f_*([g]) = [\mathrm{Id}_M]$; that is, $f \circ g \approx \mathrm{Id}_M$. Observing that

$$f_*(g \circ f) \approx f \circ (g \circ f) \approx (f \circ g) \circ f \approx f \approx f \circ \mathrm{Id}_M = f_*(\mathrm{Id}_M)$$

it follows from the injectivity of f_* that $g \circ f \approx \mathrm{Id}_M$. Thus $f \circ g \approx \mathrm{Id}_M \approx g \circ f$ and hence $f \in \mathrm{Aut}_{\mathcal{D}\mathrm{er}}(M)$. □

The following statement is dual to (36.1):

Proposition 41.3: *Let* $\mathcal{P} = (0 \to J \stackrel{j}{\to} P \stackrel{p}{\to} M \to 0)$ *be a projective* 0*-complex where* M *is* 1*-coprojective and let* $f, g : M \to M$ *be* Λ*-homomorphisms; then*

$$[f] = [g] \in \mathrm{End}_{\mathcal{D}\mathrm{er}}(J) \Longrightarrow \varprojlim(p, f) \cong \varprojlim(p, g).$$

Proof. The exact sequence from $\mathrm{Hom}_{\mathcal{D}\mathrm{er}}(M, -)$ takes the form

$$\mathrm{Hom}_{\mathcal{D}\mathrm{er}}(M, P) \stackrel{p_*}{\to} \mathrm{Hom}_{\mathcal{D}\mathrm{er}}(M, M) \stackrel{\partial}{\to} \mathrm{Ext}^1(M, J) \stackrel{j_*}{\to} \mathrm{Ext}^1(M, P).$$

As P is projective and M is 1-coprojective then $\mathrm{Hom}_{\mathcal{D}\mathrm{er}}(M, P) \cong \mathrm{Ext}^1(M, P) = 0$ giving an isomorphism

$$\delta : \mathrm{Hom}_{\mathcal{D}\mathrm{er}}(M, M) \stackrel{\simeq}{\longrightarrow} \mathrm{Ext}^1(M, J).$$

We recall that ∂ is defined by $\partial[f] = f^*(\mathcal{P})$. As $[f] = [g]$ then $f^*(\mathcal{P})$ is congruent to $g^*(\mathcal{P})$; that is, there is a commutative diagram with exact rows

$$\begin{array}{c} f^*(\mathcal{P}) \\ \downarrow \widehat{h} \\ g^*(\mathcal{P}) \end{array} = \left(\begin{array}{ccccccccc} 0 & \longrightarrow & J & \longrightarrow & \varprojlim(p, f) & \longrightarrow & M & \longrightarrow & 0 \\ & & \downarrow \mathrm{Id} & & \downarrow h & & \downarrow \mathrm{Id} & & \\ 0 & \longrightarrow & J & \longrightarrow & \varinjlim(p, g) & \longrightarrow & M & \longrightarrow & 0 \end{array} \right)$$

so that $h : \varprojlim(p, f) \stackrel{\simeq}{\longrightarrow} \varprojlim(p, g)$ is an isomorphism. □

Now suppose given projective 0-complexes

$$\mathcal{P} = (0 \to J \xrightarrow{j} P \xrightarrow{p} M \to 0); \quad \mathcal{Q} = (0 \to N \xrightarrow{i} Q \xrightarrow{q} J \to 0)$$

and a Λ-homomorphism $f : J \to J$. With this notation we obtain the following relationship between the two constructions:

Theorem 41.4: *If M is 1-coprojective and $[f] \in \mathrm{Aut}_{\mathcal{D}\mathrm{er}}(J)$ then*

$$\varprojlim(q, f) \oplus \varinjlim(f, j) \cong P \oplus Q.$$

Proof. The constructions $f_*(\mathcal{P})$ and $f^*(\mathcal{Q})$ give commutative diagrams with exact rows

$$\begin{array}{ccccccccc} 0 \longrightarrow & N & \xrightarrow{\hat{i}} & \varprojlim(q, f) & \xrightarrow{\hat{q}} & J \longrightarrow & 0 \\ & \downarrow \mathrm{Id} & & \downarrow f_1 & & \downarrow f & & ; \\ 0 \longrightarrow & N & \xrightarrow{i} & Q & \xrightarrow{q} & J \longrightarrow & 0 \end{array}$$

$$\begin{array}{ccccccccc} 0 \longrightarrow & J & \xrightarrow{j} & P & \xrightarrow{p} & M \longrightarrow & 0 \\ & \downarrow f & & \downarrow f_0 & & \downarrow \mathrm{Id} & & . \\ 0 \longrightarrow & J & \xrightarrow{\hat{j}} & \varinjlim(f, j) & \xrightarrow{\hat{p}} & M \longrightarrow & 0 \end{array}$$

Splicing these together gives a commutative diagram with exact rows

$$\begin{array}{ccccccccccc} 0 \longrightarrow & N & \xrightarrow{\hat{i}} & \varprojlim(q, f) & \xrightarrow{\partial} & P & \xrightarrow{p} & M \to & 0 \\ & \downarrow \mathrm{Id} & & \downarrow f_1 & & \downarrow f_0 & & \downarrow \mathrm{Id} & \\ 0 \longrightarrow & N & \xrightarrow{i} & Q & \xrightarrow{\delta} & \varinjlim(f, j) & \xrightarrow{\hat{p}} & M \to & 0 \end{array}$$

where $\partial = j \circ \hat{q}$ and $\delta = \hat{j} \circ q$. As in (40.1), the following sequence is exact

$$0 \to \varprojlim(q, f) \xrightarrow{\binom{\partial}{-f_1}} P \oplus Q \xrightarrow{(f_0, \delta)} \varinjlim(f, j) \to 0.$$

However $\varinjlim(f, j)$ is projective by (36.2). Consequently this sequence splits, whence the conclusion. $\square$

Let J be a finitely generated Λ-module so that, for some positive integer a there exists a free 0-complex

$$0 \to N \xrightarrow{i} \Lambda^a \xrightarrow{q} J \to 0.$$

If $f : J \to J$ defines an automophism in $\mathcal{D}\mathrm{er}(\Lambda)$ then $\varprojlim(q, f)$ is projective. We define a mapping $S^J : \mathrm{Aut}_{\mathcal{D}\mathrm{er}}(J) \to \widetilde{K}_0(\Lambda)$ by means of the correspondence

$$S^J([f]) = [\varprojlim(q, f)].$$

As we are assuming that Λ is Noetherian, it follows that $\varprojlim(q, f)$ is finitely generated. However, this need not be the case over more general rings. In the contexts that primarily concern us, it is convenient to assume that J is tame. Then there is a free admissible 0-complex

$$0 \to J \xrightarrow{j} \Lambda^b \xrightarrow{p} M \to 0.$$

By (41.4) we see that

(41.5) $$\varprojlim(q, f) \oplus \varinjlim(f, j) \cong \Lambda^{a+b}.$$

Moreover, $$[\varprojlim(q, f)] = -[\varinjlim(f, j)] = -S_J([f]).$$

As $S_J([f])$ depends only upon J and $[f]$ the same is true of $[\varprojlim(q, f)]$. If J is a tame module we obtain a mapping $S^J : \mathrm{Aut}_{\mathcal{D}\mathrm{er}}(J) \to \widetilde{K}_0(\Lambda)$ by taking a free 0-complex $0 \to N \xrightarrow{i} \Lambda^a \xrightarrow{q} J \to 0$ and writing

$$S^J([f]) = [\varprojlim(q, f)].$$

S^J is the *dual Swan mapping* associated to J. As the above discussion shows, it is independent of the particular free 0-complex used. Furthermore we have:

(41.6) $$S^J([f]) = -S_J([f]).$$

As S_J is a homomorphism we see also that:

(41.7) S^J is a homomorphism.

As a corollary to (41.6) we have:

(41.8) If J is a tame module then $\mathrm{Im}(S^J) = \mathrm{Im}(S_J)$.

Furthermore, as a consequence of (40.4) and (41.8) we see that:

Corollary 41.9: *Let $0 \to J_2 \to P \to J_1 \to 0$ be an admissible 0-complex in which J_1 is also tame; then $\mathrm{Im}(S^{J_1}) = \mathrm{Im}(S^{J_2})$.*

§42: The stability group of a lattice:

We begin with a general statement:

Proposition 42.1: *Suppose given a commutative diagram of Λ-modules with exact rows as follows:*

$$\begin{matrix} \mathcal{E} \\ \downarrow \\ \mathcal{X} \end{matrix} = \begin{pmatrix} 0 \to & K & \stackrel{i}{\to} & E & \stackrel{p}{\to} & M & \to 0 \\ & f\downarrow & & \widehat{f}\downarrow & & \mathrm{Id}\downarrow & \\ 0 \to & K & \stackrel{j}{\to} & X & \stackrel{q}{\to} & M & \to 0 \end{pmatrix}; \quad \text{then}$$

(i) $f_*(\mathcal{E})$ is congruent to $\mathcal{X}$; in particular
(ii) $X \cong \varinjlim(f, i)$.

Proof. We have a commutative diagram with exact rows

$$\begin{matrix} \mathcal{E} \\ \downarrow \\ f_*(\mathcal{E}) \\ \downarrow \\ \mathcal{X} \end{matrix} = \begin{pmatrix} 0 \to & K & \stackrel{i}{\to} & E & \stackrel{p}{\to} & M & \to 0 \\ & f\downarrow & & \widetilde{f}\downarrow & & \mathrm{Id}\downarrow & \\ 0 \to & K & \stackrel{\widehat{j}}{\to} & \varinjlim(f,i) & \stackrel{q}{\to} & M & \to 0 \\ & \mathrm{Id}\downarrow & & \nu\downarrow & & \mathrm{Id}\downarrow & \\ 0 \to & K & \stackrel{j}{\to} & X & \stackrel{q}{\to} & M & \to 0 \end{pmatrix}$$

where $\widehat{j}$, $\widetilde{f}$ and ν are the canonical maps from the pushout diagram

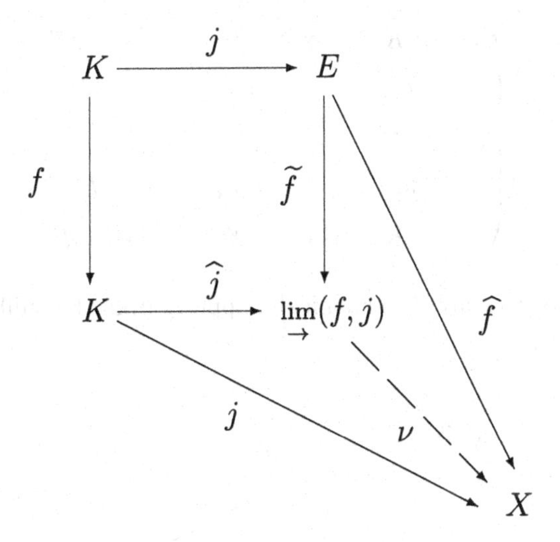

In particular, we have a congruence

$$\begin{matrix} f_*(\mathcal{E}) \\ \downarrow \\ \mathcal{X} \end{matrix} = \begin{pmatrix} 0 \to & K & \stackrel{\widehat{j}}{\to} & \varinjlim(f,i) & \stackrel{q}{\to} & M & \to 0 \\ & \mathrm{Id}\downarrow & & \nu\downarrow & & \mathrm{Id}\downarrow & \\ 0 \to & K & \stackrel{j}{\to} & X & \stackrel{q}{\to} & M & \to 0 \end{pmatrix}$$

so that, by the Five Lemma, we have an isomorphism $X \cong \varinjlim(f,i)$. $\square$

We now suppose given exact sequences $\mathcal{P} = (0 \to K \stackrel{i}{\to} P \stackrel{p}{\to} M' \to 0)$ and $\mathcal{Q} = (0 \to K \stackrel{j}{\to} Q \stackrel{q}{\to} M'' \to 0)$ over Λ in which P is projective and let $h : M' \to M''$ be a Λ-homomorphism. By the universal property of projective modules we can form a commutative diagram

$$\begin{matrix} \mathcal{P} \\ \downarrow \\ \mathcal{Q} \end{matrix} = \begin{pmatrix} 0 \to & K & \stackrel{i}{\to} & P & \stackrel{p}{\to} & M' & \to 0 \\ & \widehat{h}\downarrow & & \widetilde{h}\downarrow & & h\downarrow & \\ 0 \to & K & \stackrel{j}{\to} & Q & \stackrel{q}{\to} & M'' & \to 0 \end{pmatrix};$$

then with the above notation:

Proposition 42.2: $\varprojlim(q,h) \cong \varinjlim(\widehat{h},i)$.

Proof. We can form the following commutative diagram with exact rows:

$$\begin{matrix}\mathcal{P}\\ \downarrow \\ h^*(\mathcal{Q}) \\ \downarrow \\ \mathcal{Q}\end{matrix} = \left(\begin{array}{ccccccc} 0\to & K & \xrightarrow{i} & P & \xrightarrow{p} & M' & \to 0 \\ & \widehat{h}\downarrow & & \nu\downarrow & & \mathrm{Id}\downarrow & \\ 0\to & K & \xrightarrow{\widehat{j}} & \varprojlim(q,h) & \xrightarrow{\pi_{M'}} & M' & \to 0\\ & \mathrm{Id}\downarrow & & \pi_Q\downarrow & & h\downarrow & \\ 0\to & K & \xrightarrow{j} & X & \xrightarrow{q} & M'' & \to 0\end{array}\right)$$

where ν, $\widehat{j}$, π_Q, $\pi_{M'}$ are the canonical mappings from the pullback diagram

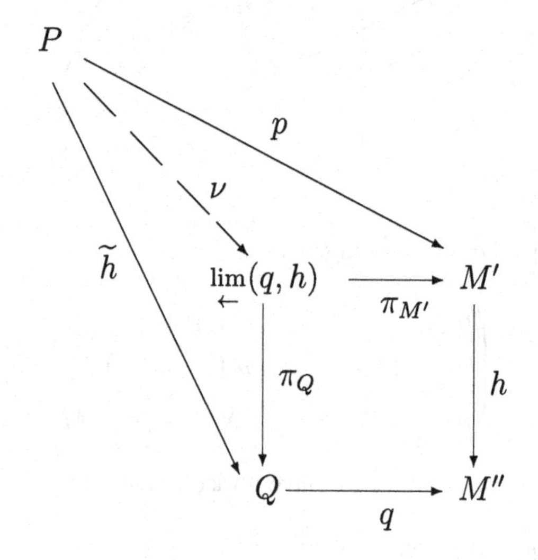

In particular, applying (42.1) to the diagram

$$\begin{array}{ccccccc} 0\to & K & \xrightarrow{i} & P & \xrightarrow{p} & M' & \to 0 \\ & \widehat{h}\downarrow & & \nu\downarrow & & \mathrm{Id}\downarrow & \\ 0\to & K & \xrightarrow{\widehat{j}} & \varprojlim(q,h) & \xrightarrow{\pi_{M'}} & M' & \to 0\end{array}$$

we see that, as claimed, $\varprojlim(q,h) \cong \varinjlim(\widehat{h}, i)$. $\square$

Now suppose that J is a finitely presented 1-coprojective Λ-module; thus we are able to specify an exact sequence

$$\mathcal{E} = (0 \to K \xrightarrow{i} \Lambda^a \xrightarrow{p} J \to 0)$$

where a is a positive integer. In particular, $\mathcal{E}$ is an admissible free 0-complex for the finitely generated Λ-module K. With this notation understood we prove:

Theorem 42.3: *Let J be a finitely presented, tame, 1-coprojective Λ-module and let P be finitely generated projective; then*

$$J \oplus P \sim J \iff [P] \in \mathrm{Im}(S_J).$$

Proof. ($\Longrightarrow$) First consider the special case where $J \oplus P \cong J \oplus \Lambda^b$ for some positive integer b and let $h : J \oplus \Lambda^b \xrightarrow{\simeq} J \oplus P$ be an isomorphism. Modify $\mathcal{E}$, first to the exact sequence

$$\mathcal{F} = (0 \to K \xrightarrow{\widehat{i}} \Lambda^a \oplus \Lambda^b \xrightarrow{\widehat{p}} J \oplus \Lambda^b \to 0)$$

where $\widehat{i}(x) = \begin{pmatrix} i(x) \\ 0 \end{pmatrix}$ and $\widehat{p}\begin{pmatrix} \alpha \\ \beta \end{pmatrix} = \begin{pmatrix} p(\alpha) \\ \beta \end{pmatrix}$ and then also to the exact sequence

$$\mathcal{Q} = (0 \to K \xrightarrow{j} \Lambda^a \oplus P \xrightarrow{q} J \oplus P \to 0)$$

in which j (resp. q) has the same defining formula as $\widehat{i}$ (resp. $\widehat{p}$) though with different domain and codomain. Now, appealing to the universal property of projective modules to lift h, we construct a commutative diagram

$$\begin{matrix} \mathcal{F} \\ \downarrow \\ \mathcal{Q} \end{matrix} = \begin{pmatrix} 0 \to & K & \xrightarrow{\widehat{i}} & \Lambda^a \oplus \Lambda^b & \xrightarrow{\widehat{p}} & J \oplus \Lambda^b & \to 0 \\ & \widehat{h} \downarrow & & \widetilde{h} \downarrow & & h \downarrow & \\ 0 \to & K & \xrightarrow{j} & \Lambda^a \oplus P & \xrightarrow{q} & J \oplus P & \to 0 \end{pmatrix}.$$

By (42.2) above, $\underleftarrow{\lim}(q, h) \cong \underrightarrow{\lim}(\widehat{h}, \widehat{i})$. As h is a Λ-isomorphism it follows from (24.7) that $\widehat{h} \in \mathrm{Aut}_{\mathcal{D}\mathrm{er}}(K)$. Hence $[\underleftarrow{\lim}(q, h)] \in \mathrm{Im}(S_K)$. However, there is now a commutative diagram with exact rows

$$\begin{matrix} \mathcal{F} \\ \downarrow \\ h^*(\mathcal{Q}) \\ \downarrow \\ \mathcal{Q} \end{matrix} = \begin{pmatrix} 0 \to & K & \xrightarrow{i} & \Lambda^a \oplus \Lambda^b & \xrightarrow{p} & J \oplus \Lambda^b & \to 0 \\ & \widehat{h} \downarrow & & \mu \downarrow & & \mathrm{Id} \downarrow & \\ 0 \to & K & \xrightarrow{\widehat{j}} & \underleftarrow{\lim}(q, h) & \xrightarrow{\pi_2} & J \oplus \Lambda^b & \to 0 \\ & \mathrm{Id} \downarrow & & \pi_1 \downarrow & & h \downarrow & \\ 0 \to & K & \xrightarrow{j} & \Lambda^a \oplus P & \xrightarrow{q} & J \oplus P & \to 0 \end{pmatrix}$$

where μ, π_1, π_2 are the canonical homomorphisms coming from the pullback diagram

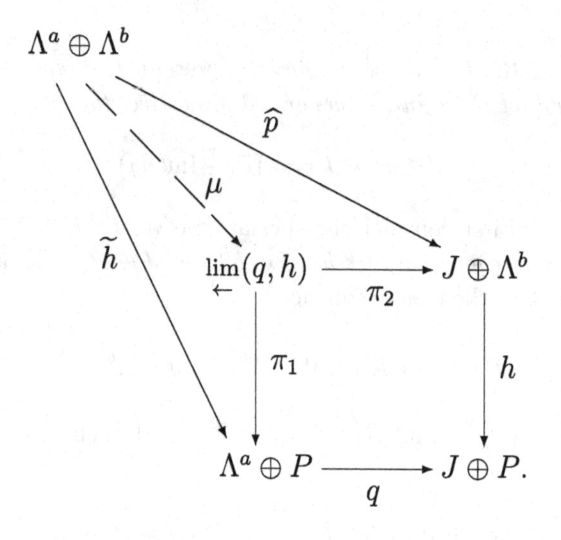

As $h : J \oplus \Lambda^b \to J \oplus P$ is an isomorphism then, applying the Five Lemma to the bottom two rows

$$\begin{array}{ccccccc} 0 \to & K & \stackrel{\widehat{j}}{\to} & \lim\limits_{\longleftarrow}(q,h) & \stackrel{\pi_2}{\to} & J \oplus \Lambda^b & \to 0 \\ & \mathrm{Id}\downarrow & & \pi_1 \downarrow & & h\downarrow & \\ 0 \to & K & \stackrel{j}{\to} & \Lambda^a \oplus P & \stackrel{q}{\to} & J \oplus P & \to 0 \end{array}$$

then π_1 an isomorphism. As noted above, $[\lim\limits_{\longleftarrow}(q,h)] = [\lim\limits_{\longrightarrow}(\widehat{h},\widehat{\imath})] \in \mathrm{Im}(S_K)$ so that $[P \oplus \Lambda^a] = [\lim\limits_{\longrightarrow}(\widehat{h},\widehat{\imath})] \in \mathrm{Im}(S_K)$.

In general, there are positive integers c, d such that $J \oplus P \oplus \Lambda^c \cong J \oplus \Lambda^d$. By the above special case, $[P \oplus \Lambda^c] \in \mathrm{Im}(S_K)$. From (40.3) and the existence of the above exact sequence $\mathcal{E}$ it follows that $\mathrm{Im}(S_K) = \mathrm{Im}(S_J)$. As, $[P] = [P \oplus \Lambda^c]$ it follows, as claimed, that $[P] \in \mathrm{Im}(S_J)$. This proves ($\Longrightarrow$).

($\Longleftarrow$) As J is tame we first choose an admissible free 0-complex for J thus:

$$0 \to J \stackrel{i}{\to} \Lambda^a \stackrel{p}{\to} M \to 0.$$

Suppose that $[P] \in \mathrm{Im}(S_J)$. Then there exists a Λ-homomorphism $h : J \to J$ such that $[h] \in \mathrm{Aut}_{\mathcal{D}\mathrm{er}}(J)$ and a commutative diagram with exact rows

$$\begin{array}{ccccccccc} 0 \to & J & \stackrel{i}{\to} & \Lambda^a & \stackrel{p}{\to} & M & \to 0 \\ & h\downarrow & & \widetilde{h}\downarrow & & \mathrm{Id}\downarrow & \\ 0 \to & J & \stackrel{\widehat{j}}{\to} & \varinjlim(h,i) & \stackrel{\pi_2}{\to} & M & \to 0 \end{array}$$

for which $[P] = [\varinjlim(h,i)] \in \widetilde{K}_0(\Lambda)$. Comparing the rows by means of Schanuel's Lemma we have $J \oplus \varinjlim(h,i) \cong J \oplus \Lambda^a$. Hence $J \oplus P \sim J \oplus \varinjlim(h,i) \cong J \oplus \Lambda^a \sim J$ so completing the proof. □

§43: A criterion for monogenicity:

In this section, S will denote a commutative ring and Λ an S-algebra, free of finite rank as a module over S. Moreover, we assume:

(*) Any stably free Λ-module is free.

We shall say that a finitely generated Λ-module M is a *quasi-free* Λ-lattice when its underlying S-module is free of finite S-rank and for some integer $m \geq 1$.

$$\mathrm{rk}_S(M) = m \cdot \mathrm{rk}_S(\Lambda).$$

The integer m is then the Λ-rank of M, written $\mathrm{rk}_\Lambda(M) = m$.

Proposition 43.1: *Let $\mathcal{P} = (0 \to R \to P \to K \to 0)$ be an exact sequence of Λ-lattices where P is quasi-free projective with $\mathrm{rk}_\Lambda(P) = 1$ and K is 1-coprojective; then there exists an exact sequence $0 \to R \to \Lambda \to K \to 0 \Longleftrightarrow [P] \in \mathrm{Im}(S^K)$.*

Proof. $(\Longleftarrow)$ As K is a Λ-lattice, it is finitely generated over Λ. Hence there is an exact sequence $\mathcal{X}$ of the form

(43.2) $$\mathcal{X} = (0 \to J \to \Lambda^N \stackrel{p}{\to} K \to 0).$$

Clearly we have $N \cdot \mathrm{rk}_S(\Lambda) = \mathrm{rk}_S(J) + \mathrm{rk}_S(K)$. As $[P] \in \mathrm{Im}(S^K)$ there exists $f \in \mathrm{Aut}_{\mathcal{D}\mathrm{er}}(K)$ such that $[P] = [\varprojlim(p,f)] \in \widetilde{K}_0(\Lambda)$. Put

$Q = \varprojlim(p, f)$. As $f \in \mathrm{Aut}_{\mathcal{D}\mathrm{er}}(K)$ then Q is projective by (41.1). Moreover, from the extension $f^*(\mathcal{X}) = (0 \to J \to Q \to K \to 0)$ we see that $\mathrm{rk}_S(Q) = \mathrm{rk}_S(J) + \mathrm{rk}_S(K)$, so that

$$\mathrm{rk}_S(Q) = N \cdot \mathrm{rk}_S(\Lambda).$$

Thus Q is quasi-free of rank N over Λ. Comparing $f^*(\mathcal{X})$ with $\mathcal{P}$ by means of Schanuel's Lemma, we see that

(43.3) $$R \oplus Q \cong J \oplus P.$$

We modify (43.2) to an exact sequence $0 \to J \oplus P \to \Lambda^N \oplus P \to K \to 0$. Substitution of $J \oplus P$ gives an exact sequence $0 \to R \oplus Q \xrightarrow{j} \Lambda^N \oplus P \to K \to 0$ and hence an exact sequence

(43.4) $$0 \to R \xrightarrow{j} T \to K \to 0$$

where $T = (\Lambda^N \oplus P)/Q$. As K is 1-coprojective then T is projective by the 'de-stabilization lemma' (22.1). From the exact sequence $(0 \to Q \to \Lambda^N \oplus P \to T \to 0)$ it follows that $T \oplus Q \cong \Lambda^N \oplus P$. As $[Q] = [P] \in \widetilde{K}_0(\Lambda)$ then there exist positive integers a, b such that $Q \oplus \Lambda^a \cong P \oplus \Lambda^b$ and hence

$$T \oplus \Lambda^b \oplus Q \cong \Lambda^{N+a} \oplus Q.$$

Choosing a projective module Q' such that $Q \oplus Q' \cong \Lambda^c$ then T is stably free as $T \oplus \Lambda^{b+c} \cong \Lambda^{N+a+c}$; that is, by assumption on Λ it follows that $T \cong \Lambda^m$ where $m = \mathrm{rk}_\Lambda(T)$. Then $\mathrm{rk}_S(T) = \mathrm{rk}_S(R) + \mathrm{rk}_S(K)$ from the exact sequence (43.4). From the exact sequence $\mathcal{P}$ we see also that $\mathrm{rk}_S(P) = \mathrm{rk}_S(R) + \mathrm{rk}_S(K)$. Thus $\mathrm{rk}_S(T) = \mathrm{rk}_S(P)$. Hence $\mathrm{rk}_\Lambda(T) = 1$ and so $T \cong \Lambda$. Substitution in (43.4) gives an exact sequence $0 \to R \to \Lambda \to K \to 0$ so proving $(\Longleftarrow)$.

$(\Longrightarrow)$ Suppose there is an exact sequence of the form

$$\mathcal{E} = (0 \to R \xrightarrow{i} \Lambda \xrightarrow{p} K \to 0).$$

As K is 1-coprojective we have an exact sequence

$$\mathrm{Hom}_{\mathcal{D}\mathrm{er}}(K, \Lambda) \xrightarrow{p^*} \mathrm{Hom}_{\mathcal{D}\mathrm{er}}(K, K) \xrightarrow{\partial} \mathrm{Ext}^1(K, R) \xrightarrow{i^*} \mathrm{Ext}^1(K, \Lambda)$$

where $\partial(f) = f^*(\mathcal{E})$. However, $\mathrm{Hom}_{\mathcal{D}\mathrm{er}}(K, \Lambda) = 0$. As K is 1-coprojective, we also have $\mathrm{Ext}^1(K, \Lambda) = 0$. Thus $\partial : \mathrm{Hom}_{\mathcal{D}\mathrm{er}}(K, K) \xrightarrow{\simeq} \mathrm{Ext}^1(K, R)$ is an isomorphism. Consequently, the extension $\mathcal{P}$ is parametrized as

$$\mathcal{P} \equiv f^*(\mathcal{E})$$

for some $f \in \mathrm{End}_\Lambda(K)$; that is, we have a commutative diagram

$$\begin{matrix} \mathcal{P} \\ \downarrow \\ \mathcal{F} \end{matrix} = \begin{pmatrix} 0 \longrightarrow & R & \longrightarrow & P \longrightarrow & K \longrightarrow & 0 \\ & \downarrow \mathrm{Id} & & \downarrow \widehat{\nu} & \downarrow f & \\ 0 \longrightarrow & R & \longrightarrow & \Lambda \longrightarrow & K \longrightarrow & 0 \end{pmatrix}$$

in which $P \cong \varprojlim(p, f^*)$. Moreover as P is projective then f defines an element $[f] \in \mathrm{Aut}_{\mathcal{D}\mathrm{er}}(K)$. Thus $[P] = S^K([f])$, so completing the proof. □

§44: Swan homomorphisms in their original context:

The sense in which we have used the term 'Swan homomorphism' is a generalization of that introduced by Swan in [61]. We conclude this chapter by re-interpreting the original version in our notation.

Let $\Lambda = \mathbb{Z}[G]$ be the integral group ring of a finite group G and let

$$\mathcal{E} = (0 \to I \overset{i}{\hookrightarrow} \Lambda \overset{\epsilon}{\to} \mathbb{Z} \to 0)$$

be the augmentation exact sequence. If $\mathbf{n} \in \mathbb{Z}$ we consider the pullback construction

$$\begin{matrix} 0 \longrightarrow & I & \xrightarrow{j} & \varprojlim(\epsilon, \mathbf{n}) & \xrightarrow{\pi} & \mathbb{Z} \longrightarrow & 0 \\ & \downarrow \mathrm{Id} & & \downarrow \natural & & \downarrow \times\mathbf{n} & \\ 0 \longrightarrow & I & \xrightarrow{i} & \Lambda & \xrightarrow{\epsilon} & \mathbb{Z} \longrightarrow & 0. \end{matrix}$$

In Swan's notation one writes $(I, \mathbf{n}) = \varprojlim(\epsilon, \mathbf{n})$. These are the original Swan modules. Swan showed that

(44.1) $\qquad (I, \mathbf{n})$ is projective $\Longleftrightarrow [\mathbf{n}] \in (\mathbb{Z}/|G|)^*$.

In the present context $(\mathbb{Z}/|G|)^*$ corresponds to $\mathrm{Aut}_{\mathcal{D}\mathrm{er}}(\mathbb{Z})$ and the mapping

$$(\mathbb{Z}/|G|)^* \to \widetilde{K}_0(\Lambda); \quad [\mathbf{n}] \mapsto [(I, \mathbf{n})]$$

corresponds to the 'dual Swan homomorphism' $S^{\mathbb{Z}} : \mathrm{Aut}_{\mathcal{D}\mathrm{er}}(\mathbb{Z}) \to \widetilde{K}_0(\Lambda)$.

Next consider the alternative construction via endomorphisms of I; thus given a Λ-homomorphism $\alpha : I \to I$ we perform the pushout construction

$$\begin{array}{ccccccccc} 0 \longrightarrow I & \xrightarrow{i} & \Lambda & \xrightarrow{\epsilon} & \mathbb{Z} \longrightarrow 0 \\ \downarrow \alpha & & \downarrow \natural & & \downarrow \mathrm{Id} \\ 0 \longrightarrow I & \xrightarrow{j} & \varinjlim(\alpha, i) & \xrightarrow{\pi} & \mathbb{Z} \longrightarrow 0. \end{array}$$

By (34.2), $\mathbb{Z}$ is 1-coprojective over $\Lambda = \mathbb{Z}[G]$. Hence from (24.7) and the exact sequence $0 \to I \xrightarrow{i} \Lambda \xrightarrow{\epsilon} \mathbb{Z} \to 0$ there is a ring isomorphism

$$\rho^{-1} : \mathrm{End}_{\mathcal{D}\mathrm{er}}(I) \xrightarrow{\simeq} \mathrm{End}_{\mathcal{D}\mathrm{er}}(\mathbb{Z})(\cong \mathbb{Z}/|G|)$$

so that, in the above diagram, $[\alpha] \in \mathrm{Aut}_{\mathcal{D}\mathrm{er}}(I) \iff \rho^{-1}([\alpha]) \in (\mathbb{Z}/|G|)^*$. Hence from (36.2) we see that:

(44.2) $\qquad \varinjlim(\alpha, i)$ is projective $\iff \rho^{-1}(\alpha) \in (\mathbb{Z}/|G|)^*$.

When $[\mathbf{n}] \in (\mathbb{Z}/|G|)^*$, Swan also gave a criterion for $(I, \mathbf{n})$ to be free, which in the above notation can be re-phrased as follows:

(44.3) $(I, \mathbf{n})$ is free $\iff [\mathbf{n}] = \rho^{-1}(\alpha)$ for some Λ-isomorphism $\alpha : I \to I$.

When $\Lambda = \mathbb{Z}[G]$ satisfies the Eichler condition then every stably free Λ-module is free and hence:

Theorem 44.4: *If $\Lambda = \mathbb{Z}[G]$ satisfies the Eichler condition then the following conditions are equivalent:*

(i) *each projective Swan module over Λ is free;*
(ii) *the canonical homomorphism $\nu : \mathrm{Aut}_\Lambda(I) \to \mathrm{Aut}_{\mathcal{D}\mathrm{er}}(I)$ is surjective;*
(iii) *the dual Swan homomorphism $S^{\mathbb{Z}} : (\mathbb{Z}/|G|)^* \to \widetilde{K}_0(\Lambda)$ is identically zero;*
(iv) *the Swan homomorphism $S_{\mathbb{Z}} : (\mathbb{Z}/|G|)^* \to \widetilde{K}_0(\Lambda)$ is identically zero.*

§45: The Swan homomorphism for cyclic groups:

We describe the cyclic group C_m of order m via the presentation

$$C_m = \langle y | y^m = 1 \rangle$$

and put $\Sigma(m) = \sum_{r=0}^{m-1} y^r \in \mathbb{Z}[C_m]$. We now suppose given integers r, k such that

(45.1) $\qquad 2 \le r \le m - 1$ and $r(k + 1) = 1 + m$.

With these conditions the following identity holds in $\mathbb{Z}[C_m]$.

(45.2) $$(1 + y + \cdots + y^{r-1})\left(\sum_{j=0}^{k} y^{jr}\right) = 1 + \Sigma(m)$$

To elaborate on this, we shall consider positive integers N, π and write

$$C_{\pi N} = \langle T \mid T^{\pi N} = 1\rangle \; ; \; C_N = \langle t \mid t^N = 1\rangle$$

so that $$\Sigma(\pi N) = \sum_{j=0}^{\pi N-1} T^j \; ; \; \Sigma(N) = \sum_{j=0}^{N-1} t^j.$$

Then the correspondence $T \mapsto t$ induces a surjective ring homomorphism

$$p : \mathbb{Z}[C_{\pi N}] \longrightarrow \mathbb{Z}[C_N]$$

under which we have:

(45.3) $$p(\Sigma(\pi N)) = \pi\Sigma(N).$$

The following observation is due to Swan (cf. Corollary (6.1) of [61]):

Theorem 45.4: *Let N, r be coprime integers satisfying $2 \le r \le N - 1$; then there exists $v(t) \in \mathbb{Z}[C_N]$ such that, for some integer $\pi \ge 1$,*

$$(1 + t + \cdots + t^{r-1})v(t) = 1 + \pi\Sigma(N).$$

Proof. Let a denote the order of $[r]$ in the unit group $(\mathbb{Z}/N)^*$ so that,

$$r^a = 1 + bN$$

for some integer $b \ge 1$. In particular, r divides $1 + bN$. Let π denote the least positive integer for which r divides $1 + \pi N$ and write $r(k + 1) = 1 + \pi N$.

If $\pi = 1$ the conclusion follows from (45.2) on putting $v(t) = \left(\sum_{j=0}^{k} t^{jr}\right)$.

If $\pi \ge 2$ then from (45.2), on substituting T for y and putting $m = \pi N$ we observe that

$$(1 + T + \cdots + T^{r-1})\left(\sum_{j=0}^{k} T^{jr}\right) = 1 + \Sigma(\pi N).$$

Moreover, $p(1 + T + \cdots + T^{r-1}) = 1 + t + \cdots + t^{r-1}$ so that, putting

$$v(t) = p(1 + T + \cdots + T^{r-1})$$

we see that $(1 + t + \cdots + t^{r-1})v(t) = 1 + p(\Sigma(\pi N))$. The conclusion now follows as, by (45.3), $p(\Sigma(\pi N)) = \pi\Sigma(N)$. □

We now restrict attention to the cyclic group $C_q = \langle t \mid t^q = 1 \rangle$ and put

$$\Sigma = \Sigma(q) = \sum_{j=0}^{q-1} t^j \in \mathbb{Z}[C_q].$$

We note the following where $I_{C_q} = \mathrm{Ker}(\epsilon : \mathbb{Z}[C_q] \to \mathbb{Z})$ is the augmentation ideal.

(45.5) There is an isomorphism of $\mathbb{Z}[C_q]$-modules $(\mathbb{Z}[C_q]/\Sigma(q)) \xrightarrow{\simeq} I_{C_q}$.

Theorem 45.6: *The canonical mapping* $\natural : \mathrm{Aut}_{\mathbb{Z}[C_q]}(I_{C_q}) \longrightarrow \mathrm{Aut}_{\mathcal{D}\mathrm{er}}(I_{C_q})$ *is surjective for any integer* $q \geq 2$.

Proof. In view of (45.5) it suffices to show that the canonical mapping

$$\natural : \mathrm{Aut}_{\mathbb{Z}[C_q]}(\mathbb{Z}[C_q]/(\Sigma)) \longrightarrow \mathrm{Aut}_{\mathcal{D}\mathrm{er}}(\mathbb{Z}[C_q]/(\Sigma))$$

is surjective. On making the identification $\mathrm{End}_{\mathcal{D}\mathrm{er}}(\mathbb{Z}[C_q]/(\Sigma)) \cong \mathbb{Z}/q$ we obtain a commutative diagram of surjective ring homomorphisms

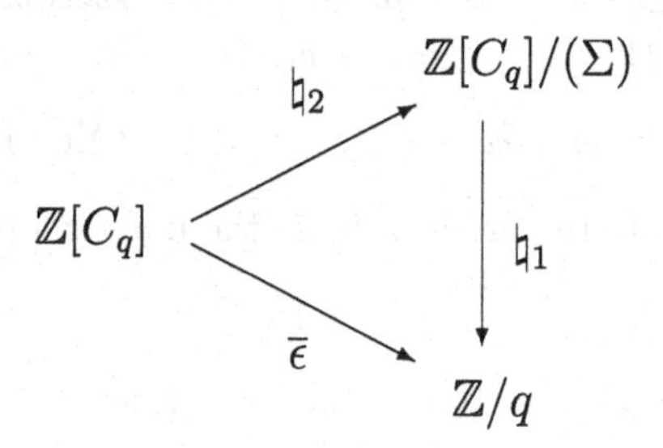

where $\natural_1$, $\natural_2$ are the canonical mappings and $\overline{\epsilon}$ is the mod q reduction of the augmentation homomorphism. Now suppose $u \in (\mathbb{Z}/q)^*$ is a unit mod q. We claim there exists a unit $\widehat{u} \in (\mathbb{Z}[C_q]/(\Sigma))^*$ such that $\natural_1(\widehat{u}) = u$. If $u = 1$ we simply take $\widehat{u} = 1$ so we may suppose that $u \neq 1$. In this case we represent u in the form $u = [r]$ where r is an integer coprime to q in the range $2 \leq r \leq q-1$ and define

$$u(t) = \sum_{j=0}^{r-1} t^j \in \mathbb{Z}[C_q].$$

Clearly $\overline{\epsilon}(u(t)) = u$ Putting $\widehat{u} = \natural_2(u(t))$ we see that $\natural_1(\widehat{u}) = u$. It suffices to show that there exists $\widehat{v} \in \mathbb{Z}[C_q]/(\Sigma)$ such that $\widehat{u} \cdot \widehat{v} = 1$. However, from (45.4) we see there exists $v(t) \in \mathbb{Z}[C_q]$ such that $u(t) \cdot v(t) = 1 + \pi\Sigma$ for some integer π. Applying $\natural_2$ we see that $\widehat{u} \cdot \widehat{v} = 1$ where $\widehat{v} = \natural_2(v(t))$. The stated claim now follows on making the identifications $\mathrm{End}_{\mathbb{Z}[C_q]}(\mathbb{Z}[C_q]/(\Sigma)) \cong \mathbb{Z}[C_q]/(\Sigma)$ and $\mathrm{End}_{\mathcal{D}\mathrm{er}(C_q)}(\mathbb{Z}[C_q]/(\Sigma)) \cong \mathbb{Z}/q$. □

As C_q satisfies the Eichler condition it now follows from (44.4) that:

(45.7) Every projective Swan module over $\mathbb{Z}[C_q]$ is free.

Theorem 45.8: *For any finite cyclic group C_q, the Swan mapping*

$$S_{I_{C_q}} : \mathrm{Aut}_{\mathcal{D}\mathrm{er}}(I_{C_q}) \longrightarrow \widetilde{K}_0(\mathbb{Z}[C_q])$$

is identically zero.

§46: Cancellation over $\mathbb{Z}[C_m]$:

We conclude this chapter with some cancellation theorems for modules over $\mathbb{Z}[C_m]$ for any integer $m \geq 2$. We observed in (4.14) that $\mathcal{P}(\mathbb{Z}[C_m])$ is a cancellation semigroup, by virtue of the fact that $\mathbb{Z}[C_m]$ satisfies the Eichler condition; hence:

(46.1) Let Q be a module over $\mathbb{Z}[C_m]$; then for any positive integer a,

$$Q \oplus \mathbb{Z}[C_m] \cong \mathbb{Z}[C_m]^{(a+1)} \Longrightarrow Q \cong \mathbb{Z}[C_m]^{(a)}.$$

The cancellation statements in the rest of this section are also well known. However, as they are slightly more involved we include proofs for the sake of completeness. Throughout we shall make a number of appeals to the augmentation sequence

$$\mathcal{E} = (0 \to I_C \xrightarrow{i} \mathbb{Z}[C_m] \xrightarrow{\epsilon} \mathbb{Z} \to 0)$$

and its dual $$\mathcal{E}^* = (0 \to \mathbb{Z} \xrightarrow{\epsilon^*} \mathbb{Z}[C_m] \xrightarrow{i^*} I_C^* \to 0).$$

In this context we again recall:

(46.2) $$I_C^* \cong_{\mathbb{Z}[C_m]} I_C.$$

Theorem 46.3: *Let $0 \to I_C \to S \to \mathbb{Z} \to 0$ be an exact sequence of modules over $\mathbb{Z}[C_m]$; if S is projective then $S \cong \mathbb{Z}[C_m]$.*

Proof. Put $\mathcal{S} = (0 \to I_C \to S \to \mathbb{Z} \to 0)$. By the Corepresentation Theorem, $\mathrm{Ext}^1(\mathbb{Z}, I_C) \cong \mathrm{End}_{\mathcal{D}\mathrm{er}}(I_C)$ from which it follows that, for some $\mathbb{Z}[C_m]$-endomorphism $\alpha : I_C \to I_C$, $\mathcal{S}$ can be constructed as a pushout extension $\mathcal{S} \equiv \alpha_*(\mathcal{E})$ thus

$$\textbf{(46.4)} \quad \begin{matrix} \mathcal{E} \\ \downarrow \natural \\ \alpha_*(\mathcal{E}) \end{matrix} = \begin{pmatrix} 0 \to & I_C & \xrightarrow{i} & \mathbb{Z}[C_m] & \xrightarrow{p} & \mathbb{Z} & \to 0 \\ & \downarrow \alpha & & \downarrow \natural & & \downarrow \mathrm{Id} & \\ 0 \to & I_C & \xrightarrow{i} & \varinjlim(\alpha, i) & \xrightarrow{p} & \mathbb{Z} & \to 0 \end{pmatrix}.$$

In particular, $S \cong \varinjlim(\alpha, i)$. Denote by $[\alpha]$ the class of α in the derived module category. As S is projective it follows from (36.2) that $[\alpha] \in \mathrm{Aut}_{\mathcal{D}\mathrm{er}}(I_C)$. However the canonical mapping $\mathrm{Aut}_{\mathbb{Z}[C_m]}(I_C) \to \mathrm{Aut}_{\mathcal{D}\mathrm{er}}(I_C)$ is surjective by (45.6). Thus in (46.4) we may suppose that $\alpha : I_C \to I_C$ is an isomorphism over $\mathbb{Z}[C_m]$. It follows from the Five Lemma that $\natural : \mathbb{Z}[C_m] \to \varinjlim(\alpha, i)$ is an isomorphism. The stated conclusion follows as $\varinjlim(\alpha, i) \cong S$. □

We also have a dual version:

Corollary 46.5: *Let* $0 \to \mathbb{Z} \to S \to I_C^* \to 0$ *be an exact sequence of modules over* $\mathbb{Z}[C_m]$*; if* S *is projective then* $S \cong \mathbb{Z}[C_m]$.

Theorem 46.6: *Let* Q *be a projective module over* $\mathbb{Z}[C_m]$*; then*

$$\mathbb{Z} \oplus Q \cong \mathbb{Z} \oplus \mathbb{Z}[C_m]^{(a)} \Longrightarrow Q \cong \mathbb{Z}[C_m]^{(a)}.$$

Proof. First modify the dual augmentation sequence $(\mathcal{E}^*)$ to an exact sequence

$$\textbf{(46.7)} \qquad 0 \to \mathbb{Z} \oplus \mathbb{Z}[C_m]^{(a)} \to \mathbb{Z}[C_m]^{(a+1)} \to I_C^* \to 0.$$

Now use the isomorphism $\mathbb{Z} \oplus Q \cong \mathbb{Z} \oplus \mathbb{Z}[C_m]^{(a)}$ to modify (46.7) to an exact sequence

$$0 \to \mathbb{Z} \oplus Q \xrightarrow{j} \mathbb{Z}[C_m]^{(a+1)} \longrightarrow I_C^* \to 0.$$

By the de-stabilization lemma (22.1), we obtain an exact sequence

$$0 \to \mathbb{Z} \to S \to I_C^* \to 0$$

where $S = \mathbb{Z}[C_m]^{(a+1)}/j(Q)$ is projective. From (46.5) it follows that $S \cong \mathbb{Z}[C_m]$ and so we have an exact sequence $0 \to Q \to \mathbb{Z}[C_m]^{(a+1)} \to \mathbb{Z}[C_m] \to 0$. Clearly this sequence splits to give an isomorphism $Q \oplus \mathbb{Z}[C_m] \cong \mathbb{Z}[C_m]^{(a+1)}$. The isomorphism $Q \cong \mathbb{Z}[C_m]^{(a)}$ now follows from (46.1). □

Theorem 46.8: *Let* $0 \to I_C \to Q \to \mathbb{Z}[C_m]^{(a)} \to I_C \to 0$ *be an exact sequence of modules over* $\mathbb{Z}[C_m]$*; if* Q *is projective then* $Q \cong \mathbb{Z}[C_m]^{(a)}$.

Proof. First split the exact sequence $0 \to I_C \to Q \to \mathbb{Z}[C_m]^{(a)} \to I_C \to 0$ thus:

(46.9) $$0 \to K \to \mathbb{Z}[C_m]^{(a)} \to I_C \to 0;$$

(46.10) $$0 \to I_C \to Q \to K \to 0.$$

Now modify (46.10) to an exact sequence

(46.11) $$0 \to I_C \to Q \oplus \mathbb{Z}[C_m] \to K \oplus \mathbb{Z}[C_m] \to 0.$$

As $I_C^* \cong I_C$ we may compare (46.9) and $(\mathcal{E}^*)$ via Schanuel's Lemma to conclude:

$$K \oplus \mathbb{Z}[C_m] \cong \mathbb{Z} \oplus \mathbb{Z}[C_m]^{(a)}.$$

Hence we may modify (46.11) to obtain an exact sequence

(46.12) $$0 \to I_C \to Q \oplus \mathbb{Z}[C_m] \to \mathbb{Z} \oplus \mathbb{Z}[C_m]^{(a)} \to 0.$$

Comparing (46.12) with $(\mathcal{E})$ via the dual to Schanuel's Lemma we obtain:

$$\mathbb{Z} \oplus Q \oplus \mathbb{Z}[C_m] \cong \mathbb{Z} \oplus \mathbb{Z}[C_m]^{(a+1)}.$$

Then $Q \oplus \mathbb{Z}[C_m] \cong \mathbb{Z}[C_m]^{(a+1)}$ by (46.6) and $Q \cong \mathbb{Z}[C_m]^{(a)}$ by (46.1). □

Finally we show that:

Theorem 46.13: *If* P *is a projective module over* $\mathbb{Z}[C_m]$ *then*

$$I_C \oplus P \cong I_C \oplus \mathbb{Z}[C_m]^{(a)} \Longrightarrow P \cong \mathbb{Z}[C_m]^{(a)}.$$

Proof. Stabilize the augmentation sequence $\mathcal{E}$ by P to get

$$\Sigma^P(\mathcal{E}) = (0 \to I_C \oplus P \xrightarrow{i'} \mathbb{Z}[C_m] \oplus P \xrightarrow{\eta} \mathbb{Z} \to 0)$$

where $i' = i \oplus \mathrm{Id}$ and $\eta = \epsilon \circ \pi_P$. Let $h : I_C \oplus \mathbb{Z}[C_m]^{(a)} \xrightarrow{\simeq} I_C \oplus P$ be an isomorphism, put $j = i' \circ h$ and consider the extension

$$\mathcal{E}' = (0 \to I_C \oplus \mathbb{Z}[C_m]^{(a)} \xrightarrow{j} \mathbb{Z}[C_m] \oplus P \xrightarrow{\eta} \mathbb{Z} \to 0).$$

Then $\mathcal{E}'$ is classified up to congruence by $c \in \mathrm{Ext}^1(\mathbb{Z}, I_C \oplus \mathbb{Z}[C_m]^{(a)}) \cong \mathbb{Z}/m$. Moreover, as $\mathbb{Z}[C_m] \oplus P$ is projective then $c \in (\mathbb{Z}/m)^*$. Now consider the extension

$$\mathcal{S} = (0 \to I_C \to S \to \mathbb{Z} \to 0)$$

classified up to congruence also by $c \in \mathrm{Ext}^1(\mathbb{Z}, I_C) \cong \mathbb{Z}/m$. As $c \in (\mathbb{Z}/m)^*$, then S is projective and so, by (46.3), $S \cong \mathbb{Z}[C_m]$. Now stabilize $\mathcal{S}$ by $\mathbb{Z}[C_m]^{(a)}$ to get

$$\Sigma^a(\mathcal{S}) = (0 \to I_C \oplus \mathbb{Z}[C_m]^{(a)} \to S \oplus \mathbb{Z}[C_m]^{(a)} \to \mathbb{Z} \to 0).$$

Then $\Sigma^a(\mathcal{S})$ is also classified up to congruence by $c \in \mathrm{Ext}^1(\mathbb{Z}, I_C \oplus \mathbb{Z}[C_m]^{(a)})$. Hence $\Sigma^a(\mathcal{S})$ is congruent to $\mathcal{E}'$ and so $\mathbb{Z}[C_m] \oplus P \cong S \oplus \mathbb{Z}[C_m]^{(a)}$. As $S \cong \mathbb{Z}[C_m]$ then $\mathbb{Z}[C_m] \oplus P \cong \mathbb{Z}[C_m]^{(a+1)}$ and the conclusion follows by (46.1). $\square$

Chapter Six

Modules over quasi-triangular algebras

Let A denote a Dedekind domain and π a principal maximal ideal of A so that π represents the trivial class in the class group $Cl(A)$. We shall denote by $\mathcal{T}_q(A,\pi)$ the following subring of $M_q(A)$ consisting of *quasi-triangular matrices*

$$\mathcal{T}_q(A,\pi) = \{X = (x_{rs})_{1\leq r,s\leq q} \in M_q(A) | x_{rs} \in \pi \text{ if } r > s\}.$$

We shall show that $\mathcal{P}(\mathcal{T}_q(A,\pi))$ is a cancellation semigroup; in fact, on putting $\mathbb{N}^{(q)} = \underbrace{\mathbb{N} \times \cdots \times \mathbb{N}}_{q}$ and $\mathbb{N}_+^{(q)} = \mathbb{N}^{(q)} - \{(0,0,\ldots,0)\}$ then

$$\mathcal{P}(\mathcal{T}_q(A,\pi)) \cong \mathbb{N}_+^{(q)} \oplus Cl(A).$$

§47: Modules over $\mathcal{T}_q(\mathbb{F})$:

Let $\mathcal{T}_q(\mathbb{F})$ denote the ring of upper triangular $q \times q$ matrices over the field $\mathbb{F}$:

$$\mathcal{T}_q(\mathbb{F}) = \{X = (X_{ij}) \in M_q(\mathbb{F}) : X_{ij} = 0 \text{ for } i > j\}.$$

Then $\mathcal{T}_q(\mathbb{F})$ is an Artinian ring with nilpotent radical $\mathfrak{rad}$ given by

$$\mathfrak{rad} = \{X \in M_q(\mathbb{F}) : X_{ij} = 0 \text{ for } i \geq j\}.$$

Defining $\nu(X) = (X_{11}, X_{22}, \ldots, X_{qq})$, we obtain a surjective ring homomorphism

$$\nu : \mathcal{T}_q(\mathbb{F}) \to \mathbb{F}^{(q)} = \underbrace{\mathbb{F} \times \mathbb{F} \times \cdots \times \mathbb{F}}_{q}$$

which in turn induces a homomorphism of semigroups

$$\nu : \mathcal{P}(\mathcal{T}_q(\mathbb{F})) \to \mathcal{P}(\mathbb{F}^{(q)}); \quad \nu(P) = P \otimes_{\mathcal{T}_q(\mathbb{F})} \mathbb{F}^{(q)}.$$

Moreover, $\mathrm{Ker}(\nu) = \mathfrak{rad}$ so that $\mathcal{T}_q(\mathbb{F})/\mathfrak{rad} \cong \mathbb{F}^{(q)}$. By Nakayama's Lemma (cf [43], pp. 181–182)

(47.1) $\nu : \mathcal{P}(\mathcal{T}_q(\mathbb{F})) \to \mathcal{P}(\mathbb{F}^{(q)})$ is injective.

We will show that ν is also surjective. To see this first define

$$\overline{R}(i) = \{(x_1, x_2, \ldots, x_q) \in \mathbb{F}^{(q)} | x_k = 0 \text{ if } k \neq i\}$$

giving a direct sum decomposition $\mathbb{F}^{(q)} = \overline{R}(1) \dot{+} \overline{R}(2) \dot{+} \cdots \dot{+} \overline{R}(q)$. Elementary Wedderburn theory shows that, as modules over $\mathbb{F}^{(q)}$,

(47.2) $\overline{R}(i) \cong \overline{R}(j) \Longleftrightarrow i \neq j.$

Consider the subsemigroup $\mathbb{N}_+^{(q)} = \{\mathbf{n} = (n_1, n_2, \ldots, n_q) \in \mathbb{N}^{(q)} | \sum_{i=1}^q n_i > 0\}$ of $\mathbb{N}^{(q)}$. Writing $\overline{R}(\mathbf{n}) = \overline{R}(1)^{(n_1)} \dot{+} \overline{R}(2)^{(n_2)} \dot{+} \cdots \dot{+} \overline{R}(q)^{(n_q)}$ we see that:

(47.3) $\overline{R}(\mathbf{m}) \cong \overline{R}(\mathbf{n}) \Longleftrightarrow \mathbf{m} = \mathbf{n}.$

In particular, there is an isomorphism of semigroups $\mathcal{P}(\mathbb{F}^{(q)}) \cong \mathbb{N}_+^{(q)}$ induced by the correspondence $\overline{R}(\mathbf{a}) \mapsto \mathbf{a}$. Now define $\breve{R}(i) = \{X \in \mathcal{T}_q(\mathbb{F}) | X_{kj} = 0 \text{ if } k \neq i\}$. Then $\mathcal{T}_q(\mathbb{F})$ decomposes as an internal direct sum of right submodules:

$$\mathcal{T}_q(\mathbb{F}) = \breve{R}(1) \dot{+} \breve{R}(2) \dot{+} \ldots \dot{+} \breve{R}(q)$$

so that each $\breve{R}(i)$ is projective over $\mathcal{T}_q(\mathbb{F})$. Moreover, $\nu(\breve{R}(i)) = \overline{R}(i)$. Thus ν is surjective as claimed. Hence $\nu : \mathcal{P}(\mathcal{T}_q(\mathbb{F})) \to \mathcal{P}(\mathbb{F}^{(q)})$ is an isomorphism of semigroups. On writing $\breve{R}(\mathbf{n}) = \breve{R}(1)^{(n_1)} \dot{+} \breve{R}(2)^{(n_2)} \dot{+} \cdots \dot{+} \breve{R}(q)^{(n_q)}$ we see that

$$\breve{R}(\mathbf{m}) \cong \breve{R}(\mathbf{n}) \Longleftrightarrow \mathbf{m} = \mathbf{n}$$

and the correspondence $\mathbf{n} \mapsto \breve{R}(\mathbf{n})$ gives a semigroup isomorphism $\mathbb{N}_+^{(q)} \cong \mathcal{P}(\mathcal{T}_q(\mathbb{F}))$. In consequence we see that:

(47.4) $K_0(\mathcal{T}_q(\mathbb{F})) \cong \mathbb{Z}^{(q)}$ generated by the classes of $\breve{R}(1), \breve{R}(2), \ldots, \breve{R}(q)$.

§48: Stable classification of projective modules over $\mathcal{T}_q(A, I)$

In what follows, A will denote a commutative ring. We fix $q \geq 2$ and let $\mathfrak{R}$ and $\mathfrak{C}$ denote respectively the set of $1 \times q$ and $q \times 1$ matrices over A;

$$\mathfrak{R} = \{(a_1, a_2, \ldots, a_q);\ a_i \in A\}; \quad \mathfrak{C} = \left\{ \begin{pmatrix} a_1 \\ a_2 \\ \vdots \\ a_q \end{pmatrix};\ a_i \in A \right\}.$$

Then $\mathfrak{R}$ is a A-$M_q(A)$ bimodule $\mathfrak{C}$ is a $M_q(A)$-A bimodule, and we have additive functors

$$\Psi : \mathcal{M}\mathrm{od}_A \to \mathcal{M}\mathrm{od}_{M_q(A)}; \qquad \Psi(M) = M \otimes_A \mathfrak{R}$$
$$\Phi : \mathcal{M}\mathrm{od}_{M_q(A)} \to \mathcal{M}\mathrm{od}_A; \quad \Phi(N) = N \otimes_{M_q(A)} \mathfrak{C}.$$

We note Morita's Theorem (cf [43], p. 166), namely that:

(48.1) Ψ and Φ are mutually inverse functors up to natural equivalence.

We denote by $\mathcal{T}_q(A)$ the subring of $M_q(A)$ consisting of upper triangular matrices:

$$\mathcal{T}_q(A) = \{X = (x_{rs})_{1 \leq r,s \leq q} \in M_q(A) | x_{rs} = 0 \text{ if } r > s\}.$$

More generally, if I is an ideal in A we denote by $\mathcal{T}_q(A, I)$ the set

$$\mathcal{T}_q(A, I) = \{X = (x_{rs})_{1 \leq r,s \leq q} \in M_q(A) | x_{rs} \in I \text{ if } r > s\}.$$

It is straightforward to see that:

(48.2) $\qquad \mathcal{T}_q(A, I)$ is a subring of $M_q(A)$.

We refer to elements of $\mathcal{T}_q(A, I)$ as *quasi-triangular* matrices. In the special case where $I = (\pi)$ is principal we write $\mathcal{T}_q(A, I) = \mathcal{T}_q(A, \pi)$. When $\pi = 0$, we simply have $\mathcal{T}_q(A, 0) = \mathcal{T}_q(A)$ whilst the case $\pi = 1$ corresponds to $M_q(A)$. Now let $\natural : M_q(A) \to M_q(A/I)$ be the surjective ring homomorphism induced from the canonical mapping $A \to A/I$. Clearly we have:

(48.3) $\qquad \mathcal{T}_q(A, I) = \natural^{-1}(\mathcal{T}_q(A/I)).$

We note that $\mathcal{T}_q(A, I)$ can be described as a fibre square

(48.4)
$$\begin{cases} \mathcal{T}_q(A,I) & \overset{i}{\hookrightarrow} & M_q(A) \\ \natural\downarrow & & \natural\downarrow \\ \mathcal{T}_q(A/I) & \overset{j}{\hookrightarrow} & M_q(A/I). \end{cases}$$

In each case, $\natural$ denotes the obvious ring homomorphism induced from the canonical mapping $A \to A/I$. We note that the Milnor condition is satisfied for (48.4) as $\natural : M_q(A) \longrightarrow M_q(A/I)$ is surjective. Now suppose that $I = (\pi)$ is a *maximal principal ideal* in A with quotient field $\mathbb{F} = A/\pi$.

Proposition 48.5: *The double coset* $GL_n(M_q(A))\backslash GL_n(M_q(\mathbb{F}))/GL_n(\mathcal{T}_q(\mathbb{F}))$ *consists of a single point for each* $n \geq 1$.

Proof. As $\mathbb{F} = A/\pi$ is a field $GL_q(M_q(\mathbb{F})) = GL_{nq}(\mathbb{F}) = E_{nq}(\mathbb{F})\cdot\Delta$ where $E_{nq}(\mathbb{F})$ is the subgroup of $GL_{nq}(\mathbb{F})$ generated by elementary transvections and Δ is the subgroup consisting of diagonal matrices of the form

$$\Delta(\delta) = \underbrace{\begin{pmatrix} \delta & & & \\ & 1 & & \\ & & \ddots & \\ & & & 1 \end{pmatrix}}_{nq} \quad \text{where } \delta \in \mathbb{F}^*.$$

For $\delta \in \mathbb{F}^*$ let $\Gamma(\delta) \in \mathcal{T}_q(\mathbb{F})$ be the element

$$\Gamma(\delta) = \underbrace{\begin{pmatrix} \delta & & & \\ & 1 & & \\ & & \ddots & \\ & & & 1 \end{pmatrix}}_{q}$$

and let $\Gamma \subset GL_n(\mathcal{T}_q(\mathbb{F}))$ be the subgroup consisting of the diagonal matrices

$$\underbrace{\begin{pmatrix} \Gamma(\delta) & & & \\ & I_q & & \\ & & \ddots & \\ & & & I_q \end{pmatrix}}_{n}$$

Then $GL_{nq}(\mathbb{F})/\Gamma = E_{nq}(\mathbb{F}) \cdot \Delta/\Gamma \xleftrightarrow{\simeq} E_{nq}(\mathbb{F})$. Consequently

$$E_{nq}(A)\backslash GL_{nq}(\mathbb{F})/\Gamma = E_{nq}(A)\backslash E_{nq}(\mathbb{F}) \cdot \Delta/\Gamma \xleftrightarrow{\simeq} E_{nq}(A)\backslash E_{nq}(\mathbb{F}).$$

As the canonical mapping $A \twoheadrightarrow \mathbb{F}$ induces a surjection $E_{nq}(A) \twoheadrightarrow E_{nq}(\mathbb{F})$ then $E_{nq}(A)\backslash E_{nq}(\mathbb{F})$, and hence also $E_{nq}(A)\backslash GL_{nq}(\mathbb{F})/\Gamma$, consists of a single point. The conclusion follows as $GL_n(M_q(A))\backslash GL_n(M_q(\mathbb{F}))/GL_n(\mathcal{T}_q(\mathbb{F}))$ is the image of $E_{nq}(A)\backslash GL_{nq}(\mathbb{F})/\Gamma$ under the obvious mapping. □

Theorem 48.6: *There is an exact sequence*

$$0 \to K_0(\mathcal{T}_q(A, I)) \to K_0(\mathcal{T}_q(\mathbb{F})) \oplus K_0(M_q(A)) \to K_0(M_q(\mathbb{F})) \to 0.$$

Proof. The Milnor exact sequence of (48.4) contains the following segments

(I) $K_1(M_q(\mathbb{F})) \xrightarrow{\partial} K_0(\mathcal{T}_q(A, I)) \to K_0(\mathcal{T}_q(\mathbb{F})) \oplus K_0(M_q(A)) \to K_0(M_q(\mathbb{F})) \to 0$

(II) $K_1(\mathcal{T}_q(\mathbb{F})) \oplus K_1(M_q(A)) \xrightarrow{(j,\natural_2)} K_1(M_q(\mathbb{F})) \xrightarrow{\partial} K_0(\mathcal{T}_q(A, I))$.

In (II), $\mathrm{Coker}(j, \natural_2) = \varinjlim GL_n(M_q(A))\backslash GL_n(M_q(\mathbb{F}))/GL_n(\mathcal{T}_q(\mathbb{F}))$. It follows from (48.5) that $\mathrm{Coker}(j, \natural_2) = 0$. However, by exactness $\mathrm{Im}(\partial) = \mathrm{Coker}(j, \natural_2)$ so that $\mathrm{Im}(\partial) = 0$ and I) reduces to the exact sequence as advertised. □

There is a commutative diagram with exact rows

$$\begin{array}{ccccccccc}
0 & \to & \mathcal{I}_1 & \to & K_0(A) & \xrightarrow{\natural} & K_0(\mathbb{F}) & \to & 0 \\
 & & \downarrow \Psi_0 & & \downarrow \Psi_1 & & \downarrow \Psi_2 & & \\
0 & \to & \mathcal{I}_2 & \to & K_0(M_q(A)) & \xrightarrow{\natural} & K_0(M_q(\mathbb{F})) & \to & 0
\end{array}$$

where Ψ_1, Ψ_2 are the isomorphisms coming from Morita's Theorem. Consequently $\mathcal{I}_2 \cong \mathcal{I}_1$. However, one sees easily that $\mathcal{I}_1 \cong \widetilde{K}_0(A)$. We obtain an exact sequence $0 \to \widetilde{K}_0(A) \to K_0(M_q(A)) \xrightarrow{\natural} K_0(M_q(\mathbb{F})) \to 0$. This sequence splits as $K_0(M_q(\mathbb{F}))$ is a free abelian group so that

$$K_0(M_q(A)) \cong K_0(M_q(\mathbb{F})) \oplus \widetilde{K}_0(A).$$

It follows from (48.6) that $K_0(\mathcal{T}_q(A, I)) \cong K_0(\mathcal{T}_q(\mathbb{F})) \oplus \widetilde{K}_0(A)$. Furthermore, we saw in §47 that $K_0(\mathcal{T}_q(\mathbb{F})) \cong \mathbb{Z}^{(q)}$. We obtain:

Theorem 48.7: *Let I be a maximal ideal in a commutative ring A; then*

$$K_0(\mathcal{T}_q(A, I)) \cong \mathbb{Z}^{(q)} \oplus \widetilde{K}_0(A).$$

§49: A lifting theorem:

In the remainder of this chapter we will establish a precise unstable form of (48.7). Thus without further comment, A will denote a Dedekind domain and π will denote a *principal maximal ideal* of A with quotient field $\mathbb{F} = A/\pi$; In particular, π represents the trivial class in $Cl(A)$. In consequence, any nontrivial class in $Cl(A)$ can be represented by an ideal which is coprime to π. We will show that there is an isomorphism of additive semigroups

$$\mathcal{P}(\mathcal{T}_q(A, \pi)) \cong \mathbb{N}_+^{(q)} \oplus Cl(A)$$

where $Cl(A)$ is the ideal class group of A. In what follows, $\mathfrak{a}$ will denote a nonzero ideal in A which is coprime to π, so that $\mathfrak{a} \otimes_A \mathbb{F} \cong \mathbb{F}$. We regard $\mathbb{F}$ as an A-module via the canonical homomorphism $\natural : A \to \mathbb{F}$. For any $m \geq 2$, $\natural : A \to \mathbb{F}$ induces a homomorphism of groups $\natural_* : GL_m(A) \to GL_m(\mathbb{F})$. Although $\natural_*$ is not surjective in general, nevertheless, on restricting to $SL_n(A)$, we do have :

Proposition 49.1: $\natural_* : SL_m(A) \to SL_m(\mathbb{F})$ *is surjective for all* $m \geq 2$.

We may regard $\alpha \in GL_m(A)$ as acting on the left of $\beta \in GL_m(\mathbb{F})$ via

$$\alpha \bullet \beta = \natural_*(\alpha) \cdot \beta.$$

As $GL_m(\mathbb{F}) \cong SL_m(\mathbb{F}) \rtimes \mathbb{F}^*$ we see that:

Proposition 49.2: $GL_m(A) \backslash GL_m(\mathbb{F})$ *is naturally a quotient of* $\mathbb{F}^*$.

On making the identification $A \otimes_A \mathbb{F} = \mathbb{F}$ then, for each $m \geq 1$, the correspondence $f \mapsto f \otimes 1$ induces a homomorphism $\natural_* : \mathrm{Hom}_A(A^{(m)}, \mathfrak{a}) \to \mathrm{Hom}_{\mathbb{F}}(\mathbb{F}^{(m)}, \mathfrak{a} \otimes_A \mathbb{F})$.

(49.3) $\natural_* : \mathrm{Hom}_A(A^{(m)}, \mathfrak{a}) \to \mathrm{Hom}_{\mathbb{F}}(\mathbb{F}^{(m)}, \mathfrak{a} \otimes_A \mathbb{F})$ is surjective for any $m \geq 1$.

There is then a homomorphism

$$\begin{array}{rcl} \natural_* : \mathrm{Hom}_A(\mathfrak{a}, A) & \longrightarrow & \mathrm{Hom}_{\mathbb{F}}(\mathfrak{a} \otimes_A \mathbb{F}, \mathbb{F}) \\ \natural_*(f)(a \otimes 1) & = & \natural(f(a)) \end{array}$$

Proposition 49.4: *If $\mathfrak{a}$ is coprime to π then $\natural_*$ is surjective.*

Proof. As $Cl(A)$ is finite there exists $e \geq 2$ such that the ideal $\mathfrak{a}^e$ is principal. Put $\mathfrak{b} = \mathfrak{a}^{e-1}$. As $\mathfrak{a}\mathfrak{b}$ is principal there exists an isomorphism of A-modules $\varphi : \mathfrak{a}\mathfrak{b} \xrightarrow{\simeq} A$. Let $\nu : \mathfrak{b} \to \mathrm{Hom}_A(\mathfrak{a}, A)$ be the mapping

$$\nu(b)(a) = \varphi(ab) \ (a \in \mathfrak{a}, b \in \mathfrak{b})$$

and let $\nu_* : \mathfrak{b} \otimes_A \mathbb{F} \to: \mathrm{Hom}_{\mathbb{F}}(\mathfrak{a} \otimes_A \mathbb{F}, \mathbb{F})$ be the mapping

$$\nu_*(b \otimes 1)(a \otimes 1) = \natural(\varphi(ab)).$$

Defining $\natural_{\mathfrak{b}} : \mathfrak{b} \to \mathfrak{b} \otimes_A \mathbb{F}$ by $\natural_{\mathfrak{b}}(b) = b \otimes 1$ then the following diagram commutes:

$$\begin{array}{ccc} \mathfrak{b} & \xrightarrow{\nu} & \mathrm{Hom}_A(\mathfrak{a}, A) \\ \natural_{\mathfrak{b}} \downarrow & & \downarrow \natural_* \\ \mathfrak{b} \otimes_A \mathbb{F} & \xrightarrow{\nu_*} & \mathrm{Hom}_{\mathbb{F}}(\mathfrak{a} \otimes_A \mathbb{F}, \mathbb{F}). \end{array}$$

As $\varphi : \mathfrak{a}\mathfrak{b} \xrightarrow{\simeq} A$ is an isomorphism there exist $a_1, \ldots, a_m \in \mathfrak{a}, b_1, \ldots, b_m \in \mathfrak{b}$ such that $\varphi(\sum_{i=1}^m a_i b_i) = 1$. Hence we have

$$\sum_{i=1}^{m} \varphi(a_i b_i) = 1.$$

We claim that $\natural(\varphi(a_i b_i)) \neq 0$ for some i. Otherwise each $\varphi(a_i b_i) \in \pi$ and, as $1 = \sum_{i=1}^m \varphi(a_i b_i)$, this implies that $A \subset \pi$, contradiction. Consequently, choosing i such that $\natural(\varphi(a_i b_i)) \neq 0$ we see that

$$\nu_*(b_i \otimes 1)(a_i \otimes 1) = \natural(\varphi(a_i b_i)) \neq 0$$

and so $\nu_* \neq 0$. However $\mathfrak{a}$ is coprime to π so that $\mathfrak{a} \otimes_A \mathbb{F} \cong \mathbb{F}$ and hence $\mathrm{Hom}_{\mathbb{F}}(\mathfrak{a} \otimes_A \mathbb{F}, \mathbb{F}) \cong \mathbb{F}$. As $\nu_* \neq 0$ it follows that ν_* is surjective. As $\natural_{\mathfrak{b}}$ is surjective then $\nu_* \circ \natural_{\mathfrak{b}}$ is surjective. As $\natural_* \circ \nu = \nu_* \circ \natural_{\mathfrak{b}}$ then $\natural_*$ is also surjective. □

In similar fashion to (49.3), for each $m \geq 1$ the correspondence $f \mapsto f \otimes 1$ induces a homomorphism $\natural_* : \mathrm{Hom}_A(\mathfrak{a}, A^{(m)}) \to \mathrm{Hom}_{\mathbb{F}}(\mathfrak{a} \otimes_A \mathbb{F}, \mathbb{F}^{(m)})$. It is now a straightforward consequence of (49.4) that:

(49.5) If $\mathfrak{a}$ is coprime to π then $\natural_* : \mathrm{Hom}_A(\mathfrak{a}, A^{(m)}) \to \mathrm{Hom}_{\mathbb{F}}(\mathfrak{a} \otimes_A \mathbb{F}, \mathbb{F}^{(m)})$ is surjective for any $m \geq 1$.

By (5.15) a general finitely generated projective A-module has the form

$$Q(n-1,\mathfrak{a}) = A^{(n-1)} \oplus \mathfrak{a}.$$

From the isomorphism $\mathfrak{a} \oplus \mathfrak{b} \cong A \oplus \mathfrak{ab}$ it follows that

$$\begin{aligned}(A^{(m-1)} \oplus \mathfrak{a}) \oplus (A^{(n-1)} \oplus \mathfrak{b}) &\cong A^{(m+n-2)} \oplus A \oplus \mathfrak{ab} \\ &\cong A^{(m+n-1)} \oplus \mathfrak{ab}.\end{aligned}$$

We have the following, which we note for future reference:

(49.6) $$Q(m-1,\mathfrak{a}) \oplus Q(n-1,\mathfrak{b}) \cong Q(m+n-1,\mathfrak{ab}).$$

When $\mathfrak{a}$ defines the trivial class in $Cl(A)$ then $\mathfrak{a} \cong A$ in which case we simply write:

$$Q(n) = A^{(n)}.$$

We define $$Q_0(n-1,\mathfrak{a}) = \mathbb{F}^{(n-1)} \oplus \mathfrak{a} \otimes \mathbb{F}.$$

Although $\mathfrak{a} \otimes \mathbb{F} \cong \mathbb{F}$ we shall find it temporarily convenient to maintain a formal distinction between them. Evidently

(49.7) $$Q_0(n-1,\mathfrak{a}) = Q(n-1,\mathfrak{a}) \otimes_A \mathbb{F}.$$

We represent $\mathrm{End}_{\mathbb{F}}(Q_0(n-1,\mathfrak{a}))$ by matrices of the form

$$\begin{pmatrix} W & \vdots & X \\ \cdots & \cdots & \cdots \\ Y & \vdots & Z \end{pmatrix}$$

where $$\begin{cases} W \in \mathrm{End}_{\mathbb{F}}(\mathbb{F}^{(n-1)}); & X \in \mathrm{Hom}_{\mathbb{F}}(\mathfrak{a} \otimes \mathbb{F}, \mathbb{F}^{(n-1)}) \\ Y \in \mathrm{Hom}_{\mathbb{F}}(A^{(n-1)}, \mathfrak{a} \otimes \mathbb{F}); & Z \in \mathrm{End}_{\mathbb{F}}(\mathfrak{a} \otimes \mathbb{F}). \end{cases}$$

The isomorphism $\mathfrak{a} \otimes \mathbb{F} \cong \mathbb{F}$ induces an obvious isomorphism

(49.8) $$\mathrm{Aut}_{\mathbb{F}}(Q_0(n-1,\mathfrak{a})) \cong GL_n(\mathbb{F}).$$

Let $SL(Q_0(n-1,\mathfrak{a}))$ denote the subgroup of $\mathrm{Aut}_{\mathbb{F}}(Q_0(n-1,\mathfrak{a}))$ which corresponds to $SL_n(\mathbb{F})$ under the isomorphism (49.8). Then

$SL(Q_0(n-1,\mathfrak{a}))$ is generated by matrices of the following types $(\mathcal{W}), (\mathcal{X}), (\mathcal{Y})$

$(\mathcal{W})$ $$\begin{pmatrix} W & \vdots & 0 \\ \cdots & \cdots & \cdots \\ 0 & \vdots & 1 \end{pmatrix} \qquad \text{where } W \in \mathrm{SL}_{n-1}(\mathbb{F})$$

$$(\mathcal{X}) \qquad \begin{pmatrix} I_{n-1} & \vdots & X \\ \cdots & \cdots & \cdots \\ 0 & \vdots & 1 \end{pmatrix} \qquad \text{where } X \in \mathrm{Hom}_{\mathbb{F}}(\mathfrak{a} \otimes \mathbb{F}, \mathbb{F}^{(n-1)})$$

$$(\mathcal{Y}) \qquad \begin{pmatrix} I_{n-1} & \vdots & 0 \\ \cdots & \cdots & \cdots \\ Y & \vdots & 1 \end{pmatrix} \qquad \text{where } Y \in \mathrm{Hom}_{\mathbb{F}}(\mathbb{F}^{(n-1)}, \mathfrak{a} \otimes \mathbb{F})$$

Likewise we shall represent $\mathrm{End}_A(Q(n-1, \mathfrak{a}))$ by matrices of the form

$$\begin{pmatrix} \widetilde{W} & \vdots & \widetilde{X} \\ \cdots & \cdots & \cdots \\ \widetilde{Y} & \vdots & \widetilde{Z} \end{pmatrix}$$

where $\begin{cases} \widetilde{W} \in \mathrm{End}_A(A^{(n-1)}); & \widetilde{X} \in \mathrm{Hom}_A(\mathfrak{a}, A^{(n-1)}) \\ \widetilde{Y} \in \mathrm{Hom}_A(A^{(n-1)}, \mathfrak{a}); & \widetilde{Z} \in \mathrm{End}_A(\mathfrak{a}) \end{cases}$

so that $\mathrm{Aut}_A(Q(n-1, \mathfrak{a}))$ contains elements of the following three types $(\widetilde{\mathcal{W}}), (\widetilde{\mathcal{X}}), (\widetilde{\mathcal{Y}})$

$$(\widetilde{\mathcal{W}}) \qquad \begin{pmatrix} \widetilde{W} & \vdots & 0 \\ \cdots & \cdots & \cdots \\ 0 & \vdots & 1 \end{pmatrix} \qquad \text{where } \widetilde{W} \in \mathrm{SL}_{n-1}(A)$$

$$(\widetilde{\mathcal{X}}) \qquad \begin{pmatrix} I_{n-1} & \vdots & X \\ \cdots & \cdots & \cdots \\ 0 & \vdots & 1 \end{pmatrix} \qquad \text{where } \widetilde{X} \in \mathrm{Hom}_A(\mathfrak{a}, A^{(n-1)})$$

$$(\widetilde{\mathcal{Y}}) \qquad \begin{pmatrix} I_{n-1} & \vdots & 0 \\ \cdots & \cdots & \cdots \\ Y & \vdots & 1 \end{pmatrix} \qquad \text{where } \widetilde{Y} \in \mathrm{Hom}_A(A^{(n-1)}, \mathfrak{a})$$

Consider the homomorphism $\natural_* : \mathrm{Aut}_A(Q(n-1, \mathfrak{a})) \to \mathrm{Aut}_{\mathbb{F}}(Q_0(n-1, \mathfrak{a}))$. By (49.1) a matrix of type $\mathcal{W}$ lifts through $\natural_*$ to a matrix of type $\widetilde{\mathcal{W}}$; likewise, by (49.4) and (49.3), a matrix of type $\mathcal{X}$ (resp. $\mathcal{Y}$) lifts through $\natural_*$

to a matrix of type $\widetilde{\mathcal{X}}$ (resp. $\widetilde{\mathcal{Y}}$). Let $\Gamma \subset \mathrm{Aut}_A(Q(n-1,\mathfrak{a}))$ be the subgroup generated by all matrices of type $(\widetilde{\mathcal{W}})$, $(\widetilde{\mathcal{X}})$ or $(\widetilde{\mathcal{Y}})$. As $SL(Q_0(n-1,\mathfrak{a}))$ is generated by matrices of the types $(\mathcal{W}),(\mathcal{X}),(\mathcal{Y})$ it follows that $\natural_* : \Gamma \longrightarrow SL(Q_0(n-1,\mathfrak{a}))$ is surjective. We now dispense with the formal distinction between $\mathbb{F}$ and $\mathfrak{a}\otimes\mathbb{F}$ and write $Q_0(n-1,\mathfrak{a}) = Q_0(n)$ so that $SL(Q_0(n-1,\mathfrak{a})) = SL(Q_0(n))$. We state our conclusion as follows:

(49.9) There is a subgroup $\Gamma \subset \mathrm{Aut}_A(Q(n-1,\mathfrak{a}))$ which maps surjectively onto $SL(Q_0(n))$ under $\natural_*$.

As $\mathrm{Aut}_{\mathbb{F}}(Q_0(n)) \cong GL_n(\mathbb{F}) \cong SL(Q_0(n)) \rtimes \mathbb{F}^*$ we have the following analogue of (49.2).

(49.10) $\mathrm{Aut}_A(Q(n\text{-}1,\mathfrak{a}))\backslash\mathrm{Aut}_{\mathbb{F}}(Q_0(n))$ is naturally a quotient of $\mathbb{F}^*$.

§50: Modules over $M_q(A)$:

We maintain our previous conventions concerning A, π and $\mathfrak{a}$. It follows from Morita's Theorem that:

(50.1) Ψ induces an equivalence of additive categories

$$\Psi : \mathcal{P}(A) \xrightarrow{\approx} \mathcal{P}(M_q(A)).$$

We put $\mathfrak{R}(\mathfrak{a}) = \mathfrak{a}\otimes_A\mathfrak{R}$. For $n \geq 1$ we define[8] $P_-(n-1,\mathfrak{a}) = \mathfrak{R}^{(n-1)}\oplus\mathfrak{R}(\mathfrak{a})$. As $P_-(n-1,\mathfrak{a}) = Q(n-1,\mathfrak{a})\otimes_A\mathfrak{R}$ we have the following consequence of (49.6):

(50.2) $$P_-(m-1,\mathfrak{a}) \oplus P_-(n-1,\mathfrak{b}) \cong P_-(m+n-1,\mathfrak{ab}).$$

As we observed in §5, $Q(n-1,\mathfrak{a})$ describes the form of a general finitely generated projective module over A. On applying Ψ and appealing to (50.1) we see that:

(50.3) A finitely generated projective module over $M_q(A)$ is of the form $P_-(n-1,\mathfrak{a})$ for some integer $n \geq 1$ and some nonzero ideal $\mathfrak{a}$.

As before, we denote by $\mathbb{F}$ the quotient field $\mathbb{F} = A/\pi$. Since $\mathbb{F}$ is trivially a Dedekind domain the above considerations all apply. In this case, however,

---

(8) The function of the negative subscript in $P_-(n-1,\mathfrak{a})$ will become apparent in the next section.

we write

$$\overline{\mathfrak{R}} = \{(\lambda_1, \lambda_2, \dots, \lambda_q) \, ; \lambda_i \in \mathbb{F}\} \, ; \quad \overline{\mathfrak{C}} = \left\{ \begin{pmatrix} \lambda_1 \\ \lambda_2 \\ \vdots \\ \lambda_q \end{pmatrix} ; \lambda_i \in \mathbb{F} \right\}$$

$$\overline{\Psi} : \mathcal{M}od_{\mathbb{F}} \to \mathcal{M}od_{M_q(\mathbb{F})}; \qquad \overline{\Psi}(M) = M \otimes_{\mathbb{F}} \overline{\mathfrak{R}}$$

$$\overline{\Phi} : \mathcal{M}od_{M_q(\mathbb{F})} \to \mathcal{M}od_{\mathbb{F}}; \qquad \overline{\Phi}(N) = N \otimes_{M_q(\mathbb{F})} \overline{\mathfrak{C}}.$$

We regard $\overline{\mathfrak{R}}$ as left A-module via the canonical map $A \twoheadrightarrow \mathbb{F}$ and define

$$P_0(n) = \overline{\mathfrak{R}}^{(n)}.$$

A general finitely generated module over $M_q(\mathbb{F})$ then has the form:

$$P_0(n) = Q_0(n) \otimes_{\mathbb{F}} \overline{\mathfrak{R}}.$$

To improve readability, we omit subscripts as follows by writing

$$\mathrm{Aut}(P_-(n-1, \mathfrak{a})) = \mathrm{Aut}_{M_q(A)}(P_-(n-1, \mathfrak{a}));$$

$$\mathrm{Aut}(P_0(n)) = \mathrm{Aut}_{M_q(\mathbb{F})}(P_0(n)).$$

Morita's Theorem now shows that $\mathrm{Aut}(P_-(n-1, \mathfrak{a})) \cong \mathrm{Aut}_A(Q(n-1, \mathfrak{a}))$ and $\mathrm{Aut}(P_0(n)) \cong \mathrm{Aut}_{\mathbb{F}}(Q_0(n))$. Consequently,

$$\mathrm{Aut}(P_-(n-1, \mathfrak{a})) \backslash \mathrm{Aut}(P_0(n)) \cong \mathrm{Aut}_A(Q(n-1, \mathfrak{a})) \backslash \mathrm{Aut}_{\mathbb{F}}(Q_0(n)).$$

Still assuming that $\mathfrak{a}$ is coprime to π, it follows from (49.10) that:

(50.4) $\mathrm{Aut}(P_-(n-1, \mathfrak{a})) \backslash \mathrm{Aut}(P_0(n))$ is naturally a quotient of $\mathbb{F}^*$.

§51: Classification of projective modules over $\mathcal{T}_q(A, \pi)$:

We recall Milnor's classification of projective modules over a ring Λ described as a fibre product of ring homomorphisms

$$(\mathfrak{F}) \qquad \begin{cases} \Lambda & \stackrel{\eta_-}{\to} & \Lambda_- \\ \downarrow \eta_+ & & \downarrow \varphi_- \\ \Lambda_+ & \stackrel{\varphi_+}{\to} & \Lambda_0 \end{cases}$$

in which $\varphi_- : \Lambda_- \to \Lambda_0$ is surjective. Let $P_+ \in \mathcal{P}(\Lambda_+)$ and $P_- \in \mathcal{P}(\Lambda_-)$; we shall say that (P_+, P_-); are *compatible* when $\varphi_+(P_+) \cong_{\Lambda_0} \varphi_-(P_-)$.

Suppose that (P_+, P_-) are compatible; then writing $P_0 = \varphi_-(P_-)$ we see that:

(51.1) Isomorphisms $\alpha : \varphi_+(P_+) \xrightarrow{\simeq} \varphi_-(P_-)$ are in 1-1 correspondence with elements of $\mathrm{Aut}_{\Lambda_0}(P_0)$.

If $P \in \mathcal{P}(\Lambda)$ then by commutativity of $(\mathfrak{F})$ we see that $(\eta_+(P), \eta_-(P))$ are compatible. Given a compatible pair (P_+, P_-) we say that $P \in \mathcal{P}(\Lambda)$ is of *local type* (P_+, P_-) *with respect to* $(\mathfrak{F})$ when $\eta_+(P) \cong P_+$ and $\eta_-(P) \cong P_-$. Milnor's Theorem can then be stated as follows:

(51.2) If (P_+, P_-) are compatible there exists $P \in \mathcal{P}(\Lambda)$ of local type (P_+, P_-).

Moreover:

(51.3) The isomorphism classes of projective modules of a given local type (P_+, P_-) are in 1-1 correspondence with $\mathrm{Aut}(P_-)\backslash\mathrm{Aut}(P_0)/\mathrm{Aut}(P_+)$.

We now turn our attention to the fibre square

$$(\mathfrak{S}) \qquad \begin{cases} \mathcal{T}_q(A,\pi) & \stackrel{i}{\hookrightarrow} & M_q(A) \\ \natural\downarrow & & \natural\downarrow \\ \mathcal{T}_q(\mathbb{F}) & \stackrel{j}{\hookrightarrow} & M_q(\mathbb{F}) \end{cases}$$

where, maintaining our established conventions, A is a Dedekind domain and $\pi \in A$ is prime with quotient field $\mathbb{F} = A/\pi$. By (50.3), the correspondence

$$\textbf{(51.4)} \qquad (N, [\mathfrak{a}]) \mapsto P_-(N-1, \mathfrak{a}) = \mathfrak{R}^{(N-1)} \oplus \mathfrak{R}(\mathfrak{a})$$

is a faithful parametrization of isomorphism classes of projective modules P_- over $M_q(A)$ by pairs $(N, [\mathfrak{a}])$ where N is a positive integer and $[\mathfrak{a}] \in Cl(A)$ is an ideal class whose representative $\mathfrak{a}$ can be chosen to be a nonzero ideal in A coprime to π. In particular, $\natural_*(P_-(N-1, \mathfrak{a})) \cong \overline{\mathfrak{R}}^{(N)}$. By §47, projective modules P_+over $\mathcal{T}_q(\mathbb{F})$ are parametrized faithfully by elements $\mathbf{n} = (n_1, \ldots, n_q) \in \mathbb{N}_-^{(q)}$ thus:

$$\textbf{(51.5)} \qquad P_+(\mathbf{n}) = \check{R}(1)^{(n_1)} \oplus \cdots \oplus \check{R}(q)^{(n_q)}.$$

Observe that $j_*(\breve{R}(i)) \cong \overline{\mathfrak{R}}$ for each i. Thus we have $j_*(P_+(\mathbf{n})) \cong \overline{\mathfrak{R}}^{||n||}$ where $||n|| = \sum_{r=1}^{q} n_r$. It follows directly that:

(51.6) $P_+(\mathbf{n})$ is compatible with $P_-(N-1, \mathfrak{a}) \Longleftrightarrow N = ||\mathbf{n}||$.

As a consequence we see that:

(51.7) Compatible pairs over $(\mathfrak{S})$ are, up to isomorphism, in 1-1-correspondence with elements of $\mathbb{N}_+^{(q)} \oplus Cl(A)$.

We now turn to the multiplicity statement (51.3) of Milnor's Theorem. Given a compatible pair $(P_+(\mathbf{n}), P_-(||\mathbf{n}||-1, \mathfrak{a}))$ the number of isomorphism classes of projective modules with this local type is in 1-1-correspondence with the set

$$\mathrm{Aut}(P_-(||\mathbf{n}||-1, \mathfrak{a}) \backslash \mathrm{Aut}(P_0(||\mathbf{n}||))/\mathrm{Aut}(P_+(\mathbf{n})).$$

By (50.4), $\mathrm{Aut}(P_-(||\mathbf{n}||-1, \mathfrak{a}) \backslash \mathrm{Aut}(P_0(||\mathbf{n}||))$ is a natural quotient of $\mathbb{F}^*$. It is easy to check that, for $1 \le i \le q$, $\mathrm{Aut}(\breve{R}(i))$ contains a copy of $\mathbb{F}^*$ which injects into $\mathrm{Aut}(\overline{\mathfrak{R}})$ via the inclusion $j : \mathcal{T}_q(\mathbb{F}) \hookrightarrow M_q(\mathbb{F})$. Hence $\mathrm{Aut}(P_+(\mathbf{n}))$ contains a copy of $\mathbb{F}^*$ which injects into $\mathrm{Aut}(P_0(||\mathbf{n}||))$. Hence:

(51.8) $\mathrm{Aut}(P_-(||\mathbf{n}||-1, \mathfrak{a}) \backslash \mathrm{Aut}(P_0(||\mathbf{n}||))/\mathrm{Aut}(P_+(\mathbf{n}))$ has only one element.

It follows that, for each $\mathbf{n} \in \mathbb{N}_+^{(q)}$ and each $[\mathfrak{a}] \in Cl(A)$, there exists a unique projective module $P \in \mathcal{P}(\mathcal{T}_q(A, \pi))$ with local type $(P_+(\mathbf{n}), P_-(||\mathbf{n}||-1, \mathfrak{a}))$. We shall refer to this module as $P(\mathbf{n}, [\mathfrak{a}])$.

Theorem 51.9: *The correspondence*

$$\begin{array}{ccc} \mathbb{N}_+^{(q)} \oplus Cl(A) & \xrightarrow{P} & \mathcal{P}(\mathcal{T}_q(A, \pi)) \\ (\mathbf{n}, [\mathfrak{a}]) & \mapsto & P(\mathbf{n}, [\mathfrak{a}]) \end{array}$$

is an isomorphism of semigroups.

Proof. We have shown that P is bijective. It remains to show that P preserves the semigroup operation. Now

$$P(\mathbf{m}+\mathbf{n}, \mathfrak{a}\mathfrak{b}) \longleftrightarrow (P_+(\mathbf{m}+\mathbf{n}), P_-(||\mathbf{m}|| + ||\mathbf{n}|| - 1, \mathfrak{a}\mathfrak{b}))$$

However, $P_+(\mathbf{m}+\mathbf{n}) \cong P_+(\mathbf{m}) \oplus P_+(\mathbf{n})$ and, by (50.2),

$$P_-(||\mathbf{m}|| + ||\mathbf{n}|| - 1, \mathfrak{a}\mathfrak{b}) \cong P_-(||\mathbf{m}|| - 1, \mathfrak{a}) \oplus P_-(||\mathbf{n}|| - 1, \mathfrak{b}).$$

Consequently,

$$P(\mathbf{m}+\mathbf{n}, \mathfrak{ab}) \longleftrightarrow (P_+(\mathbf{m}), P_-(||\mathbf{m}||-1, \mathfrak{a})) \oplus (P_+(\mathbf{n}), P_-(||\mathbf{n}||-1, \mathfrak{b})).$$

Hence, as required, $P(\mathbf{m}+\mathbf{n}, \mathfrak{ab}) \cong P(\mathbf{m}, \mathfrak{a}) \oplus P(\mathbf{n}, \mathfrak{b})$. □

Corollary 51.10: *$\mathcal{P}(\mathcal{T}_q(A, \pi))$ is a cancellation semigroup.*

We may regard (51.10) as an explicit form of Rosen's Theorem ([57]).

§52: Properties of the row modules $R(i)$:

We denote by $R(i)$ the ith row submodule of $\mathcal{T}_q(A, \pi)$; that is:

$$R(i) = \{X = (x_{rs}) \in \mathcal{T}_q(A, \pi) | x_{rs} = 0 \text{ if } r \neq i\}.$$

Then $\mathcal{T}_q(A, \pi)$ decomposes as direct sum of right Λ-modules thus

(52.1) $$\mathcal{T}_q(A, \pi) \cong R(1) \oplus R(2) \oplus \cdots \oplus R(q).$$

We note that:

(52.2) Each $R(i)$ is free over A with $\mathrm{rk}_A(R(i)) = q$.

From the decomposition (52.1) each $R(i)$ is projective over $\mathcal{T}_q(A, \pi)$. In the description of $\mathcal{P}(\mathcal{T}_q(A, \pi))$ given by (51.9) $R(i)$ is described as the fibre product

(52.3) $$\begin{cases} R(i) & \overset{i}{\hookrightarrow} & \mathfrak{R} \\ \natural_1 \downarrow & & \natural_2 \downarrow \\ \breve{R}(i) & \overset{j}{\hookrightarrow} & \overline{\mathfrak{R}}. \end{cases}$$

More generally, if $\mathbf{n} = (n_1, \ldots, n_q) \in \mathbb{N}_+^{(q)}$ we define

$$R(\mathbf{n}) = R(1)^{n_1} \oplus R(2)^{n_2} \oplus \cdots \oplus R(q)^{n_q}.$$

Under the canonical isomorphism

(52.4) $$\mathcal{T}_q(A, \pi) \otimes_A \mathbb{F} \cong \mathcal{T}_q(\mathbb{F})$$

the module $R(\mathbf{n})$ descends to $\breve{R}(\mathbf{n})$. As we saw in §47, the modules $\breve{R}(\mathbf{n})$ are pairwise isomorphically distinct over $\mathcal{T}_q(\mathbb{F})$. Hence:

(52.5) $$R(\mathbf{m}) \cong R(\mathbf{n}) \Longleftrightarrow \mathbf{m} = \mathbf{n}.$$

From (51.9) we obtain the following elaboration of the description in (52.3):

Proposition 52.6: *If $P \in \mathcal{P}(\mathcal{T}_q(A,\pi))$ occurs as the fibre product in the diagram*

$$\begin{cases} P & \overset{i}{\hookrightarrow} & \mathfrak{R}^{\|\mathbf{n}\|} \\ \natural_1 \downarrow & & \natural_2 \downarrow \\ \breve{R}(\mathbf{n}) & \overset{j}{\hookrightarrow} & \overline{\mathfrak{R}}^{\|\mathbf{n}\|} \end{cases}$$

then $P \cong R(\mathbf{n})$.

If $\mathfrak{a}$ is an ideal in A coprime to π then $\mathfrak{R}(\mathfrak{a}) = \mathfrak{a} \otimes_A \mathfrak{R} = \{(a_1, \dots, a_q) | a_i \in \mathfrak{a}\}$. By Milnor's classification, a typical projective module $P \in \mathcal{P}(\mathcal{T}_q(A,\pi))$ is represented as a fibre product

$$\begin{cases} P & \hookrightarrow & P_- \\ \downarrow & & \downarrow \\ \breve{R}(\mathbf{n}) & \hookrightarrow & \overline{\mathfrak{R}}^{\|\mathbf{n}\|} \end{cases}$$

where $P_- = \bigoplus_{i=1}^{\|\mathbf{n}\|} \mathfrak{R}(\mathfrak{a}_i)$ for some ideals $(\mathfrak{a}_i)_{1 \le i \le \|\mathbf{n}\|}$. Hence for any $\nu \ge 1$, $P^{(\nu)}$ is represented as a fibre product

$$\begin{cases} P^{(\nu)} & \hookrightarrow & P_-^{(\nu)} \\ \downarrow & & \downarrow \\ \breve{R}(\nu \cdot \mathbf{n}) & \hookrightarrow & \overline{\mathfrak{R}}^{\nu\|\mathbf{n}\|} \end{cases}$$

where $P_-^{(\nu)} = \bigoplus_{i=1}^{\|\mathbf{n}\|} \mathfrak{R}(\mathfrak{a}_i)^{(\nu)}$. Now take $\nu = |Cl(A)|$. Then $\mathfrak{a}_i^{(\nu)} \cong A^{(\nu)}$ for all i so that $P_-^{(\nu)} \cong \mathfrak{R}^{\nu\|\mathbf{n}\|)}$. Taking $\mathbf{m} = \nu \cdot \mathbf{n}$ then $P^{(\nu)}$ is represented as a fibre product

$$\begin{cases} P^{(\nu)} & \hookrightarrow & \mathfrak{R}^{\|\mathbf{m}\|} \\ \downarrow & & \downarrow \\ \breve{R}(\mathbf{m}) & \hookrightarrow & \overline{\mathfrak{R}}^{\|\mathbf{m}\|} \end{cases}.$$

Again taking $\nu = |Cl(A)|$ it follows from (52.6) that

(52.7) If $P \in \mathcal{P}(\mathcal{T}_q(A,\pi))$ then $P^{(\nu)} \cong R(\mathbf{m})$ for some $\mathbf{m} \in \mathbb{N}_+^{(q)}$.

In particular:

(52.8) Let $P \in \mathcal{P}(\mathcal{T}_q(A,\pi))$; if $|Cl(A)| = 1$ then $P \cong R(\mathbf{m})$ for some $\mathbf{m} \in \mathbb{N}_+^{(q)}$.

The condition that $|Cl(A)| = 1$ is simply the requirement that A be a principal ideal domain. In particular, the hypotheses of (52.8) are satisfied when A is a local Dedekind domain with unique prime π; in this case we get:

Corollary 52.9: *Let A be local Dedekind ring with prime π; if $P \in \mathcal{P}(\mathcal{T}_q(A,\pi))$ then $P \cong R(\mathbf{m})$ for some $\mathbf{m} \in \mathbb{N}_+^{(q)}$.*

§53: Duality properties of the modules $R(i)$:

We conclude this chapter by studying the duality properties of the $R(i)$. Fix the following notation

$$\mathcal{T}_q = \mathcal{T}_q(A,\pi); \quad R(i) = i^{th} \text{ row of } \mathcal{T}_q; \quad C(j) = j^{th} \text{ column of } \mathcal{T}_q.$$

$R(i)$, $C(j)$ are respectively right and left ideals in $\mathcal{T}_q$. Define $Q = (q_{ij}) \in M_q(A)$ by

$$q_{ij} = \begin{cases} 1 & i+j = q+1 \\ 0 & \text{otherwise} \end{cases}$$

Clearly $Q = Q^t = Q^{-1}$. Define $\theta : \mathcal{T}_q \to \mathcal{T}_q$ by $\theta(A) = QA^tQ$. Then θ is an anti-involution on $\mathcal{T}_q$ which takes a left ideal J to a right ideal $\theta(J)$; in particular:

(53.1) $$\theta(C(k)) = R(q+1-k).$$

If M is a right $\mathcal{T}_q$-module then $\mathrm{Hom}_{\mathcal{T}_q}(M, \mathcal{T}_q)$ is a left $\mathcal{T}_q$-module. In particular:

(53.2) $$\mathrm{Hom}_{\mathcal{T}_q}(R(k), \mathcal{T}_q) \cong C(k).$$

We use θ to convert a left $\mathcal{T}_q$-module M to a right $\mathcal{T}_q$-module ${}^\theta M$ by means of

$$m * \alpha = \theta(\alpha)m$$

where $m \in M$ and $\alpha \in \mathcal{T}_q$. Note that if J is a left ideal in $\mathcal{T}_q$ then $\theta(J)$ is a right ideal in $\mathcal{T}_q$; moreover, we see that θ induces an isomorphism of right

$\mathcal{T}_q$-modules

$$\theta : {}^{\theta}J \xrightarrow{\simeq} \theta(J).$$

If M is a right $\mathcal{T}_q$ module then, properly speaking, the *dual module* of M is the left $\mathcal{T}_q$ module $\mathrm{Hom}_{\mathcal{T}_q}(M, \mathcal{T}_q)$ whilst the right module ${}^{\theta}\mathrm{Hom}_{\mathcal{T}_q}(M, \mathcal{T}_q)$ is the *conjugate dual.* However when the context is clear we refer to ${}^{\theta}\mathrm{Hom}_{\mathcal{T}_q}(M, \mathcal{T}_q)$ simply as the dual and write $M^{\bullet} = {}^{\theta}\mathrm{Hom}_{\mathcal{T}_q}(M, \mathcal{T}_q)$. It follows from (53.1) and (53.2) that:

(53.3) $$R(k)^{\bullet} \cong R(q + 1 - k).$$

Chapter Seven

A fibre product decomposition

Let p be an odd prime and q a divisor of $p-1$. Then the cyclic group C_q acts on C_p via the natural imbedding $C_q \hookrightarrow \mathrm{Aut}(C_p)$. The metacyclic group $G(p,q)$ is the semidirect product

$$G(p,q) = C_p \rtimes C_q$$

where C_q acts on C_p via the natural imbedding $C_q \hookrightarrow \mathrm{Aut}(C_p)$. We denote by A the fixed point ring $A = \mathbb{Z}(\zeta)^{C_q}$ where, as before, $\zeta = \exp(2\pi i/p)$. Then A is a Dedekind domain and contains a unique principal prime ideal (π) which ramifies completely over p. In this chapter, we begin the study of the integral representation theory of $G(p,q)$ by decomposing the integral group ring $\mathbb{Z}[G(p,q)]$ as a fibre product

$$\begin{array}{ccc} \mathbb{Z}[G(p,q)] & \rightarrow & \mathcal{T}_q(A,\pi) \\ \downarrow & & \downarrow \\ \mathbb{Z}[C_q] & \rightarrow & \mathbb{F}_p[C_q] \end{array}$$

where $\mathbb{F}_p$ is the field with p elements.

§54: A fibre product decomposition:

Let $q \geq 2$ be an integer which divides $p-1$. Making a specific choice $\theta \in \mathrm{Aut}(C_p)$ of order q there is then a unique integer a in the range $2 \leq a \leq p-1$ such that $\theta(x) = x^a$. Clearly the residue class $[a] \in \mathbb{F}_p^*$ then has order q. In everything that follows θ and a will have this fixed meaning and we define

$$G(p,q) = G(p,q;a) = \langle x, y \mid x^p = y^{p-1} = 1; \quad yx = x^a y \rangle.$$

To describe the integral group ring $\mathbb{Z}[G(p,q)]$ we first recall the *cyclic algebra construction* (cf. [50], p. 277). Suppose S is a commutative ring and

$\theta : S \to S$ a ring automorphism satisfying $\theta^n = \mathrm{Id}$. The *cyclic ring* $\mathcal{C}_n(S, \theta)$ is the (two-sided) S-module

$$\mathcal{C}_n(S,\theta) = S\mathbf{1} \dot{+} S\mathbf{y} + \cdots \dot{+} S\mathbf{y}^{n-1}$$

which is free of rank n over S with basis $\{\mathbf{1}, \mathbf{y}, \ldots \mathbf{y}^{n-1}\}$ and with multiplication determined by the relations

$$\mathbf{y}^n = \mathbf{1}; \quad \mathbf{y}\xi = \theta(\xi)\mathbf{y}(\xi \in S).$$

So defined, $\mathcal{C}_n(S,\theta)$ is an extension ring of S.

As we saw in (9.9), the integral group ring $\mathbb{Z}[C_p]$ has a canonical fibre product decomposition

(54.1)
$$\begin{array}{ccc} \mathbb{Z}[C_p] & \to & I_C^* \\ \epsilon \downarrow & & \downarrow \\ \mathbb{Z} & \to & \mathbb{F}_p \end{array}$$

where $\mathbb{F}_p$ is the field with p-elements, $\epsilon : \mathbb{Z}[C_p] \to \mathbb{Z}$ is the augmentation map and I_C^* is the integral dual of the augmentation ideal $I_C = \mathrm{Ker}(\epsilon)$. We make the identification $I_C^* = \mathbb{Z}[C_p]/(\Sigma_x)$ where $\Sigma_x = 1 + x + \cdots + x^{p-1}$. Then θ induces a ring automorphism of order q on $\mathbb{Z}[C_p]$. As Σ_x is fixed by θ we obtain an induced ring automorphism on the quotient $I_C^* = \mathbb{Z}[C_p]/(\Sigma_x)$. Likewise the augmentation ideal I_C is stable under θ so that θ induces a ring automorphism on the quotient $\mathbb{Z} = \mathbb{Z}[C_p]/I_C$. We may apply the cyclic algebra construction $\mathcal{C}_q(-,\theta)$ to (54.1) above to obtain a fibre product decomposition

(54.2)
$$\begin{array}{ccc} \mathcal{C}_q(\mathbb{Z}[C_p]) & \to & \mathcal{C}_q(I_C^*,\theta) \\ \downarrow & & \downarrow \\ \mathcal{C}_q(\mathbb{Z}) & \to & \mathcal{C}_q(\mathbb{F}_p). \end{array}$$

We identify $\mathcal{C}_q(\mathbb{Z}[C_p])$ with $\mathbb{Z}[G(p,q)]$. Furthermore, as θ acts trivially on $\mathbb{Z}$ and $\mathbb{F}_p$ we may make the identifications $\mathcal{C}_q(\mathbb{Z}) = \mathbb{Z}[C_q], \mathcal{C}_q(\mathbb{F}_p) = \mathbb{F}_p[C_q]$ and re-interpret the fibre square (54.2) thus:

(54.3)
$$\begin{array}{ccc} \mathbb{Z}[G(p,q)] & \to & \mathcal{C}_q(I_C^*,\theta) \\ \downarrow & & \downarrow \\ \mathbb{Z}[C_q] & \to & \mathbb{F}_p[C_q]. \end{array}$$

Again, let $\Sigma_x = 1 + x + \cdots + x^{p-1} \in \mathbb{Z}[C_p]$ and let $\langle \Sigma \rangle$ denote the two sided ideal in $\mathbb{Z}[G(p,q)]$ generated by Σ_x. It is straightforward to see that

(54.4) $$\langle \Sigma \rangle = \mathrm{span}_{\mathbb{Z}}\{\Sigma_x, \Sigma_x y, \ldots, \Sigma_x y^{p-1}\}.$$

Make the identification $\mathcal{C}_q(I^*, \theta) \otimes \mathbb{Q} \cong \mathcal{C}_q(\mathbb{Q}(\zeta), \theta)$. Applying $- \otimes \mathbb{Q}$ to the fibre square (54.3) and observing that $\mathbb{F}_p[C_q] \otimes \mathbb{Q} = 0$ we obtain a direct product

(54.5) $$\mathbb{Q}[G(p,q)] \cong \mathbb{Q}[C_q] \times \mathcal{C}_q(\mathbb{Q}(\zeta), \theta).$$

Thus $\mathcal{C}_q(\mathbb{Q}(\zeta), \theta)$ is semisimple $\mathbb{Q}$-algebra. It is straightforward to see that the centre $\mathcal{Z}(\mathcal{C}_q(\mathbb{Q}(\zeta), \theta))$ is field, namely the subfield $\mathbb{Q}(\zeta)^\theta \cong R \otimes \mathbb{Q}$ of $\mathbb{Q}(\zeta)$ fixed by θ; hence:

(54.6) $\mathcal{C}_q(\mathbb{Q}(\zeta), \theta)$ is a simple $\mathbb{Q}$-algebra.

§55: The discriminant of an associative algebra:

In what follows A will denote a commutative ring and B an algebra which is free of finite rank over A. Considering B as a right A-module, there is a homomorphism of A-algebras, the *adjoint representation*, $\mathrm{Ad} : B \to \mathrm{End}_A(B)$ given by

$$\mathrm{Ad}(x)(z) = xz.$$

Any A-linear endomorphism $X : B \to B$ has a well defined trace $Tr_A(X) \in A$. We obtain a symmetric bilinear form β on B by

$$\beta : B \times B \to A; \quad \beta(x,y) = \mathrm{Tr}_A(\mathrm{Ad}_A(x)\mathrm{Ad}_A(y)).$$

If $\mathcal{E} = (e_i)_{1 \le i \le n}$ is an A-basis for B we obtain a symmetric $n \times n$ matrix $\beta^{\mathcal{E}}$ by taking

$$(\beta^{\mathcal{E}})_{ij} = \beta(e_i, e_j).$$

We define the *discriminant of A relative to $\mathcal{E}$* by $\mathcal{D}\mathrm{isc}_A(B, \mathcal{E}) = \det(\beta^{\mathcal{E}})$; that is:

(55.1) $$\mathcal{D}\mathrm{isc}_A(B, \mathcal{E}) = \det\left[\mathrm{Tr}_A(\mathrm{Ad}(e_i)\mathrm{Ad}(e_j))_{1 \le i,j \le n}\right].$$

Clearly $\mathcal{D}\mathrm{isc}_A(B, \mathcal{E}) \in A$. To consider the effect of varying the basis $\mathcal{E}$ suppose $\Phi = (\varphi_j)_{1 \le j \le n}$ is also an A-basis for B and denote by $P = (p_{kj})$ the matrix which expresses Φ in terms of $\mathcal{E}$ thus:

$$\varphi_j = \sum_{k=1}^{n} e_k p_{kj}.$$

A straightforward calculation then shows that $\beta^{\Phi} = P^t \beta^{\mathcal{E}} P$ and hence

(55.2)
$$\mathcal{D}\mathrm{isc}_A(B, \Phi) = \det(P)^2 \mathcal{D}\mathrm{isc}_A(B, \mathcal{E}).$$

Evidently $\det(P) \in A^*$, and the *absolute discriminant* $\mathcal{D}\mathrm{isc}_A(B)$ is defined as the class of $\mathcal{D}\mathrm{isc}_A(B, \mathcal{E})$ in the (multiplicative) quotient monoid $A/(A^*)^2$.

The above formalism has a relative interpretation. Suppose $B' \subset B$ is a subalgebra which is also free of finite rank n over A and that Φ is an A-basis for B'. We may still express Φ in terms of $\mathcal{E}$

$$\varphi_j = \sum_{k=1}^{n} e_k p_{kj}.$$

It again follows that $\beta^{\Phi} = P^t \beta^{\mathcal{E}} P$. On taking determinants we obtain the following generalization of (55.2)

(55.3)
$$\mathcal{D}\mathrm{isc}_A(B', \Phi) = \det(P)^2 \mathcal{D}\mathrm{isc}_A(B, \mathcal{E}).$$

In this case we clearly have:

(55.4)
$$B' = B \iff \det(P) \in A^*.$$

Now specialize to the case $A = \mathbb{Z}$. As B' has the same $\mathbb{Z}$-rank as B then B' has finite index in B given by $|B/B'| = |\det(P)|$. It follows from (55.3) that

(55.5)
$$|B/B'|^2 = \frac{\mathcal{D}\mathrm{isc}_A(B', \Phi)}{\mathcal{D}\mathrm{isc}_A(B, \mathcal{E})}.$$

To conclude this section, we calculate a few elementary cases. First consider the case $B = A$; then we may take the unit element I_A as an A-basis, from which it follows directly that

(55.6)
$$\mathcal{D}\mathrm{isc}_A(A) = 1 \in A/(A^*)^2.$$

Next suppose $B = B_1 \times B_2$ is the direct product of A-algebras B_1, B_2 both free of finite rank over A. If β (resp. β_i) is the bilinear form of B (resp. B_i) then

$$\beta = \beta_1 \perp \beta_2.$$

Amalgamating A-bases $\mathcal{E}_1$, $\mathcal{E}_2$ for B_1, B_2 to the obvious A-basis $\mathcal{E}$ for B we see that

(55.7) $$\mathcal{D}\mathrm{isc}_A(B, \mathcal{E}) = \mathcal{D}\mathrm{isc}_A(B_1, \mathcal{E}_1)\mathcal{D}\mathrm{isc}_A(B_2, \mathcal{E}_2).$$

In particular, taking the n-fold direct product $A^{(n)} = \underbrace{A \times \cdots \times A}_{n}$ it follows that

(55.8) $$\mathcal{D}\mathrm{isc}_A(A^{(n)}) = 1 \in A/(A^*)^2.$$

Finally, taking $A = \mathbb{Z}$ for convenience, we point out the following degenerate case:

(55.9) If $\mathrm{rad}(B) \neq 0$ and $B/\mathrm{rad}(B)$ is free over $\mathbb{Z}$ then $\mathcal{D}\mathrm{isc}_{\mathbb{Z}}(B) = 0$.

§56: The discriminant of a cyclic algebra:

In what follows S will denote a commutative ring and $\theta : S \to S$ a ring automorphism satisfying $\theta^n = \mathrm{Id}$. We shall denote by A the fixed ring under θ, $A = \{a \in S : \theta(a) = a\}$; evidently $\mathcal{C}_n(S, \theta)$ is an algebra over A. We shall, in addition, assume throughout that S is free of rank n over A and hence:

(56.1) $\mathcal{C}_n(S, \theta)$ is free of rank n^2 over A.

We proceed to compute the discriminant $\mathcal{D}\mathrm{isc}_A(\mathcal{C}_n(S, \theta))$ in terms of S, n and θ. First make the abbreviation $\mathcal{C} = \mathcal{C}_n(S, \theta)$. When $\mathbf{w}, \mathbf{x} \in \mathcal{C}$ write

$$\mathbf{w} = \sum_{k=0}^{n-1} \widetilde{w_k}\mathbf{y}^k, \quad \mathbf{x} = \sum_{l=0}^{n-1} \widetilde{x_l}\mathbf{y}^l$$

where $\widetilde{w_k}, \widetilde{x_l} \in S$. Then multiplication in $\mathcal{C}$ takes the form

$$\mathbf{wx} = \sum_{k,l=0}^{n-1} \widetilde{w_k}\theta^k(\widetilde{x_l})\mathbf{y}^{k+l}$$

where addition of the indices k, l is taken mod n. To compute $\mathcal{D}\mathrm{isc}_A(\mathcal{C})$ we consider the bilinear form $\beta : \mathcal{C} \times \mathcal{C} \to A$

$$\beta(\mathbf{w}, \mathbf{x}) = \mathrm{Tr}_{\mathcal{C}/A}[\mathrm{Ad}_{\mathcal{C}}(\mathbf{wx})].$$

First note that

(56.2) $$\mathrm{Tr}_{\mathcal{C}/A}[\mathrm{Ad}_{\mathcal{C}}(\sigma\mathbf{y}^k)] = \begin{cases} n\mathrm{Tr}_{S/A}[\mathrm{ad}_S(\sigma)] & k = 0 \\ 0 & k \neq 0. \end{cases}$$

If $\Phi = (\varphi)_{1 \leq r \leq n}$ is an A-basis for S we obtain an A-basis $\mathcal{E} = (e_j)_{1 \leq j \leq n^2}$ for $\mathcal{C}$ on putting

$$e_{kn+r} = \varphi_r \mathbf{y}^k (0 \leq k \leq n-1, 1 \leq r \leq n).$$

Let $\beta^{k,l}$ denote the $n \times n$ matrix over R given by $(\beta^{k,l})_{r,s} = \beta(e_{kn+r,lm+s})$. Evidently

(56.3) $$(\beta^{k,l})_{r,s} = \beta(\varphi_r \mathbf{y}^k, \varphi_s \mathbf{y}^l) = \mathrm{Tr}_{\mathcal{C}}[\mathrm{Ad}_{\mathcal{C}}(\varphi_r \theta^k(\varphi_s)\mathbf{y}^{k+l})].$$

Now fix n and denote by ρ the permutation of the set $\{0, 1, \ldots, n-1\}$ given by

$$\rho(k) = -k \pmod n.$$

Then it follows from (56.2) and (56.3) that:

(56.4) $$(\beta^{k,l})_{r,s} = \begin{cases} n\mathrm{Tr}_{S/A}[\mathrm{ad}_S(\varphi_r\theta^k(\varphi_s)] & l = \rho(k) \\ 0 & l \neq \rho(k) \end{cases}.$$

In a subsidiary role we shall also consider the bilinear form $\mu : S \times S \to A$

$$\mu(\varphi, \psi) = \mathrm{Tr}_{S/A}[\mathrm{ad}_S(\varphi\psi)].$$

If we now regard the indices k, l as belonging to the group $\mathbb{Z}/n$ and replace each occurence of the value 0 by n we obtain a decomposition of the matrix

$$\widetilde{\beta} = (\beta(e_i, e_j))(1 \leq i, j \leq n^2)$$

into $n \times n$ blocks $\beta^{k,l}(1 \leq k, l \leq n)$where, from (56.4), it follows that

(56.5) $$(\beta^{k,\rho(k)})_{r,s} = n\mu(\varphi_r, \theta^k(\varphi_s)).$$

and

(56.6) $$(\beta^{k,l}) = 0 \quad \text{if } l \neq \rho(k).$$

In the formalism of §55 this becomes $\beta^{k,\rho(k)} = n\mu^{\Phi,\theta^k(\Phi)}$ so that, by (56.5), on taking determinants we obtain:

(56.7) $$\det(\beta^{k,\rho(k)}) = \det(\theta^k)\mathcal{D}\mathrm{isc}_A(S)n^n.$$

If σ is a permutation of the indices $1 \leq i \leq n^2$ we denote by $P(\sigma)$ the corresponding permutation matrix of size $n^2 \times n^2$. Left multiplication by $P(\sigma)$ then performs the corresponding permutation on the rows of $\widetilde{\beta}$. We denote by $[P(\sigma)\widetilde{\beta}]^{k,l}$ the decomposition of $P(\sigma)\widetilde{\beta}$ into $n \times n$ blocks. It follows from (56.5) that we may choose a specific permutation ω so that

$$[P(\omega)\widetilde{\beta}]^{k,l} = \begin{cases} \beta^{k,\rho(k)} & \text{if } l = k \\ 0 & \text{if } l \neq k. \end{cases}$$

Then $\mathrm{sign}(\omega)\det(\widetilde{\beta}) = \prod_{k=1}^{n} \det(\beta^{k,\rho(k)})$ and from (56.7) we obtain

$$\mathrm{sign}(\omega)\mathcal{D}\mathrm{isc}_A(\mathcal{C}) = \left(\prod_{k=1}^{n} \det(\theta^k)\right) \mathcal{D}\mathrm{isc}_A(S)^n(n^n)^n.$$

Recalling that $\theta^n = \mathrm{Id}$ we calculate that $\prod_{k=1}^{n} \det(\theta^k) = \det(\theta)^{n(n-1)/2}$; multiplying across by $\mathrm{sign}(\omega)$ gives

$$\mathcal{D}\mathrm{isc}_A(\mathcal{C}) = \mathrm{sign}(\omega)\det(\theta)^{n(n-1)/2}\mathcal{D}\mathrm{isc}_A(S)^n n^{n^2}.$$

It remains to calculate $\mathrm{sign}(\omega)$. To do this let f denote the number of indices in $\mathbb{Z}/n$ fixed under ρ. Viewed as a permutation on $\{0, 1, \ldots, n-1\}$, ρ interchanges the elements of precisely $m = (n-f)/2$ pairs, leaving the remaining indices fixed. The permutation ω on $I = \{i | 1 \leq i \leq n^2\}$ may be written

$$\omega = \tau_1 \circ \cdots \circ \tau_m$$

where each τ_r swaps over a pair disjoint subsets of size n taken from I. To swap each such pair requires n transpositions; that is, for each r, $\mathrm{sign}(\tau_r) = (-1)^n$ and so

$$\mathrm{sign}(\omega) = (-1)^{n(n-f)/2}.$$

When n is even the only indices fixed by ρ are $0(= n)$ and $n/2$ so that $f = 2$; in this case $\mathrm{sign}(\omega) = 1$. When n is odd the only index fixed by ρ

is $0(= n)$; then $f = 1$ and $\mathrm{sign}(\omega) = (-1)^{n(n-1)/2}$. To summarize we have calculated that

(56.8) $$\mathcal{D}\mathrm{isc}_A(\mathcal{C}) = \omega_n \det(\theta)^{n(n-1)/2} \mathcal{D}\mathrm{isc}_A(S)^n n^{n^2}$$

where

(56.9) $$\omega_n = \mathrm{sign}(\omega) = \begin{cases} 1 & \text{if } n \text{ is even} \\ (-1)^{n(n-1)/2} & \text{if } n \text{ is odd.} \end{cases}$$

§57: The discriminant of a quasi-triangular algebra:

In this section we shall compute the discriminant of $\mathcal{T}_n(A, (\pi))$ when A is a commutative ring and $\pi \in A$. We begin by computing the discriminant of the full matrix ring $M_n(A)$. To this end we first describe $M_n(A)$ as a cyclic algebra. Let c denote the cyclic permutation of $\{1, \dots, n\}$

$$c(r) = \begin{cases} r+1 & r < n \\ 1 & r = n \end{cases}$$

and let $c_* : A^{(n)} \to A^{(n)}$ denote the ring automorphism of the n-fold direct product which permutes the co-ordinates via c;

$$c_*(x_1, \dots, x_n) = (x_{c(1)}, \dots, x_{c(n)}).$$

Then A is isomorphic to the subring of $A^{(n)}$ fixed under c_* via the diagonal imbedding $A \to A^{(n)}; x \mapsto (x, \dots x, x)$. In particular, $\mathcal{C}_n(A^{(n)}, c_*)$ is an A-algebra.

Proposition 57.1: *There is a ring isomorphism*

$$\nu : \mathcal{C}_n(A^{(n)}, c_*) \xrightarrow{\simeq} M_n(A).$$

Proof. There is a ring homomorphism $\nu : A^{(n)} \to M_n(A)$ given by

$$\nu(x_1, x_2, \dots, x_n) = \begin{pmatrix} x_1 & & & \\ & x_2 & & \\ & & \ddots & \\ & & & x_n \end{pmatrix}.$$

Consider the permutation matrix $\eta \in M_n(A)$ given by

$$\eta_{r,s} = \delta_{c(r),s}$$

where δ is the Kronecker delta. On assigning $\nu(\mathbf{y}) = \eta$ we see that ν extends to a ring homomorphism $\nu : \mathcal{C}_n(A^{(n)}, c_*) \to M_n(A)$. We claim that ν is bijective.

As both $\mathcal{C}_n(A^{(n)}, c_*)$ and $M_n(A)$ are free of rank n^2 over the commutative ring A, any A-linear surjection $\mathcal{C}_n(A^{(n)}, c_*) \to M_n(A)$ is necessarily injective. It now suffices to show that ν is surjective. To see this let $\mathcal{E} = \{\epsilon(i,j)\}(1 \leq i,j \leq n)$ denote the canonical A-basis for $M_n(A)$ given by

$$\epsilon(i,j)_{r,s} = \delta_{i,r}\delta_{j,s}.$$

One checks that $\epsilon(i,j) = \eta^{(1-i)}\epsilon(1,1)\eta^{(j-1)}$. Taking $E = (1,0,\ldots,0) \in A^{(n)}$ one has $\nu(E) = \epsilon(1,1)$ and so $\epsilon(i,j) = \nu[\mathbf{y}^{(1-i)}E\mathbf{y}^{(j-1)}]$. Hence ν is surjective. □

By (55.8),$\mathcal{D}\mathrm{isc}_A(A^{(n)}) = 1$. Moreover, $\det(c_*) = \mathrm{sign}(c)$; thus it follows from (57.1) and (56.8) that:

(57.2) $$\mathcal{D}\mathrm{isc}_A(M_n(A)) = \omega_n \mathrm{sign}(c)^{n(n-1)/2} n^{n^2}.$$

However $\mathrm{sign}(c) = (-1)$ if n is even whilst $\mathrm{sign}(c) = 1$ if n is odd. Thus

(57.3) $$\mathrm{sign}(c)^{n(n-1)/2} = \begin{cases} (-1)^{n(n-1)/2} & \text{if } n \text{ is even} \\ 1 & \text{if } n \text{ is odd} \end{cases}.$$

Comparing (56.9) and (57.3) we see that for all n,

(57.4) $$\omega_n \mathrm{sign}(c)^{n(n-1)/2} = (-1)^{n(n-1)/2}.$$

Hence we see that from (57.2) and (57.4) that:

(57.5) $$\mathcal{D}\mathrm{isc}_A(M_n(A)) = (-1)^{n(n-1)/2} n^{n^2}.$$

We proceed to compute $\mathcal{D}\mathrm{isc}_A(\mathcal{T}_n(A,\pi))$ for arbitrary $\pi \in A$. First abbreviate $\mathcal{T}_n(A,\pi)$ to $\mathcal{T}_n$. If $\mathcal{F}$ is an A-basis for $\mathcal{T}_n$ then by (55.2)

$$\mathcal{D}\mathrm{isc}_A(\mathcal{T}_n, \mathcal{F}) = \det(P)^2 \mathcal{D}\mathrm{isc}_A(M_n(A), \mathcal{E})$$

where P is the $n^2 \times n^2$ matrix which expresses $\mathcal{F}$ in terms of the canonical basis $\mathcal{E}$ of $M_n(A)$. It follows from (57.5) that

(57.6) $$\mathcal{D}\mathrm{isc}_A(\mathcal{T}_n, \mathcal{F}) = (-1)^{n(n-1)/2} \det(P)^2 n^{n^2}.$$

However, for the specific basis $\mathcal{F} = \{\varphi(i,j)\}$ $(1 \le i,j \le n)$ for $\mathcal{T}_n(A,\pi)$ given by

$$\varphi(i,j) = \begin{cases} \epsilon(i,j) & \text{if } i \le j \\ \pi\epsilon(i,j) & \text{if } j < i \end{cases}$$

we see that $\det(P) = \pi^{n(n-1)/2}$. Substituting in (57.6) we obtain:

(57.7) $$\mathcal{D}\mathrm{isc}_A(\mathcal{T}_n(A,\pi)) = (-1)^{n(n-1)/2}\pi^{n(n-1)}n^{n^2}.$$

§58: A quasi-triangular representation of $G(p,q)$:

We maintain the notation of §57. In particular, A will denote a commutative ring and $I \lhd A$ an ideal in A. Moreover $\natural : M_q(A) \to M_q(A/I)$ will denote the surjective ring homomorphism induced from the canonical mapping $A \to A/I$.

If $\psi : G \to GL_n(A)$ is a group representation we shall say that ψ is *quasi-triangular with respect to* I when for each $g \in G$, $\psi(g) \in \mathcal{T}_n(A,I)$. When I is clear from context, we say simply that ψ is *quasi-triangular.* If $\psi : G \to GL_n(A)$ is a group representation we shall denote by $\overline{\psi} : G \to GL_q(A/I)$ the representation $\overline{\psi}(g) = \natural(\psi(g))$. In consequence of (48.3) we have:

(58.1) $$\psi(g) \in \mathcal{T}_q(A,I) \iff \overline{\psi}(g) \in \mathcal{T}_q(A/I).$$

We recall that a matrix $X \in M_n(A)$ is said to be *uni-triangular* when X is upper triangular and, in addition, $X_{i,i} = 1$ for all i. We denote by J_n the following uni-triangular $n \times n$ matrix;

$$J_n = \begin{pmatrix} 1 & 1 & 0 & 0 & 0 & \dots & 0 & 0 & 0 & 0 \\ 0 & 1 & 1 & 0 & 0 & \dots & 0 & 0 & 0 & 0 \\ 0 & 0 & 1 & 1 & 0 & \dots & 0 & 0 & 0 & 0 \\ & & & \ddots & & & & & & \\ & & & & \ddots & & & & & \\ & & & & & \ddots & & & & \\ 0 & 0 & 0 & 0 & 0 & \dots & 0 & 1 & 1 & 0 \\ 0 & 0 & 0 & 0 & 0 & \dots & 0 & 0 & 1 & 1 \\ 0 & 0 & 0 & 0 & 0 & \dots & 0 & 0 & 0 & 1 \end{pmatrix}.$$

The following is easily proved by induction on n.

Proposition 58.2: *Let A be a commutative integral domain and let $Y, Z \in M_n(A)$ satisfy $YJ_n = ZY$; if Z is uni-triangular then Y is upper triangular.*

Let p be an odd prime and $q \geq 2$ a divisor of $p-1$ and, as before, put $\zeta = \exp(\frac{2\pi i}{p})$. Then the natural action of $\mathrm{Gal}(\mathbb{Q}(\zeta)/\mathbb{Q})$ on $\mathbb{Z}(\zeta)$ restricts to one of the cyclic group C_q via the imbedding $C_q \subset C_{p-1} \cong \mathrm{Gal}(\mathbb{Q}(\zeta)/\mathbb{Q})$. Let $\mathbb{E} = \mathbb{Q}[\zeta]^{C_q}$ denote the subfield fixed by C_q and put $A = \mathbb{Z}(\zeta)^{C_q}$. Then A is the ring of integers in $\mathbb{E}$. It is known that ([6], p. 87) p ramifies completely in $\mathbb{Z}(\zeta)$. Consequently, we also have:

(58.3) $\qquad p$ ramifies completely in $A = \mathbb{Z}(\zeta)^{C_q}$.

Denote by $\widetilde{\pi}$ the unique prime in $\mathbb{Z}(\zeta)$ over p so that $\widetilde{\pi}^{p-1} = p$. Likewise, let π be the unique prime in A over p so that $\pi^{(p-1)/q} = p$. Then, as ideals in $\mathbb{Z}(\zeta)$, we have :

(58.4) $\qquad (\widetilde{\pi}^q) = (\pi).$

It is well known ([6] p. 87) that $(\zeta - 1)^{p-1} = pu$ for some unit $u \in \mathbb{Z}(\zeta)^*$. Hence

$$\left(\frac{(\zeta - 1)}{\widetilde{\pi}}\right)^{p-1} = u$$

so that $\dfrac{(\zeta-1)}{\widetilde{\pi}}$ is a unit in $\mathbb{Z}(\zeta)$ and hence $(\zeta - 1) = (\widetilde{\pi})$. Consequently

(58.5) $\qquad ((\zeta - 1)^q) = (\pi).$

Let y be a generator of C_q. On choosing a specific integer $a \in \{2, \ldots, p-1\}$ with the property that the residue class $[a]$ has order q in the unit group $\mathbb{F}_p^*$ we make the identification $y \longleftrightarrow [a]$. Then the Galois action of the generator $y \in C_q$ on the roots of unity $\{\zeta, \zeta^2, \ldots, \zeta^{p-1}\}$ is given by $y(\zeta^r) = \zeta^{ar}$. In particular, this C_q action partitions the powers $\{\zeta, \zeta^2, \ldots, \zeta^{p-1}\}$ into $d = (p-1)/q$ orbits each with cardinal q, the orbit of ζ being $\{\zeta, \zeta^a, \ldots, \zeta^{a^{q-1}}\}$. Consider the polynomial:

$$\gamma(\lambda) = \prod_{r=0}^{q-1} (\lambda - \zeta^{a^r}).$$

Let $\sigma_r(x_1, \ldots, x_q)$ denote the rth symmetric function of the indeterminates $x_1, \ldots, x_q$. Then the Lagrange expansion gives

$$\gamma(\lambda) = \lambda^q + \sum_{r=0}^{q-1} (-1)^r \sigma_{q-r} \lambda^r$$

where $\sigma_k = \sigma_k(\zeta, \zeta^a, \ldots, \zeta^{a^{q-1}})$. Evidently σ_k depends only on the orbit $\{\zeta, \zeta^a, \ldots, \zeta^{a^{q-1}}\}$ of ζ and is invariant under the action of y. Thus each $\sigma_k \in A = \mathbb{Z}(\zeta)^{C_q}$. Hence

(58.6) $$\gamma(\lambda) \in A[\lambda].$$

Furthermore, standard Galois theory shows that $\gamma(\lambda)$ is irreducible over $\mathbb{E}$. As $\gamma(\zeta) = 0$ and $\mathbb{Q}(\zeta)$ has degree q over $\mathbb{E}$ we see also that:

(58.7) $\gamma(\lambda)$ is the minimal polynomial of ζ over $\mathbb{E}$.

On taking $a_k = (-1)^{k+1}\sigma_{q-k}(\zeta, \zeta^a, \ldots, \zeta^{a^{q-1}}) \in A$ we note that:

(58.8) $$\zeta^q = a_{q-1}\zeta^{q-1} + a_{q-2}\zeta^{q-2} + \cdots + a_1\zeta + a_0.$$

As $\mathbb{Q}(\zeta)$ has dimension q over $\mathbb{E}$ it follows from (58.8) that $\mathbb{Z}(\zeta)$ is a free module of rank q over A. Moreover

(58.9) $\{1, \zeta, \zeta^2, \ldots \zeta^{q-1}\}$ is an A-basis for $\mathbb{Z}(\zeta)$.

By means of the binomial formulae

$$\begin{aligned}
\zeta &= (\zeta - 1) + \quad 1 \\
\zeta^2 &= (\zeta - 1)^2 + \quad 2(\zeta - 1) \quad + 1 \\
\zeta^r &= (\zeta - 1)^r - \textstyle\sum_{k=0}^{r-1}(-1)^{r-k}\binom{r}{k}\zeta^k
\end{aligned}$$

by successive elementary basis transformations we conclude that:

(58.10) $\{(\zeta - 1)^{q-1}, (\zeta - 1)^{q-2}, \ldots, (\zeta - 1), 1\}$ is an A-basis for $\mathbb{Z}(\zeta)$.

Now let $\psi : G(p, q) \to GL_{p-1}(\mathbb{Z})$ be the representation obtained from the Galois action of C_q on $I^*_{C_p} = \mathbb{Z}(\zeta)$ and let $i : \mathbb{Z} \hookrightarrow A = \mathbb{Z}(\zeta)^{C_q}$ denote the canonical inclusion of rings. With this notation we have:

Theorem 58.11: *There exists a representation* $\rho : G(p, q) \to T_q(A, \pi)^*$ *such that*

$$i^*(\rho) \cong \psi.$$

Proof. $G(p, q)$ acts on the right of $\mathbb{Z}(\zeta)$ by $\mathbf{z} \cdot (x^r y^s) = \theta^{-s}(\mathbf{z} \cdot \zeta^r)$. As in (58.10), $\mathbb{Z}(\zeta)$ is a free module of rank q over A with basis

$$\{(\zeta - 1)^{q-1}, (\zeta - 1)^{q-2}, \ldots, (\zeta - 1), 1\}.$$

With respect to this basis let $\rho : G(p,q) \to M_q(A)$ be the representation associated to the above right action. We claim that $\rho(g) \in \mathcal{T}_q(A,\pi)^*$ for all $g \in G(p,q)$.

To see this, observe that the right action of x is described by $\rho(x^{-1})$ where

$$\rho(x^{-1})[(\zeta-1)^r] = \begin{cases} (\zeta-1)^{r+1} + (\zeta-1)^r & 1 \le r \le q-2 \\ (\zeta-1)^q + (\zeta-1)^{q-1} & r = q-1. \end{cases}$$

However, from (58.5) and (58.10) for some $u_0, u_1, \dots u_{q-1} \in A$ we have

(58.12) $$(\zeta-1)^q = \pi\{\textstyle\sum_{r=0}^{q-1} u_r(\zeta-1)^{q-r}\}.$$

Hence with respect to the basis of (58.10), $\rho(x^{-1})$ has the quasi-triangular form

$$\rho(x^{-1}) = \begin{pmatrix} 1+\pi u_1 & 1 & 0 & 0 & \dots & 0 & 0 \\ \pi u_2 & 1 & 1 & 0 & \dots & 0 & 0 \\ & & & \ddots & & & \\ & & & & \ddots & & \\ & & & & & \ddots & \\ \pi u_{q-1} & 0 & 0 & 0 & \dots & 1 & 1 \\ \pi u_q & 0 & 0 & 0 & \dots & 0 & 1 \end{pmatrix}.$$

Put $\widehat{X} = \rho(x^{-1})$ and put $X = \natural(\widehat{X})$ where $\natural : GL_q(A) \to GL_q(A/\pi)$is the homomorphism induced from the canonical ring homomorphism $A \to A/\pi$. The above expression for $\rho(x^{-1})$ shows that X is the Jordan block $X = J_q$. Now put $\widehat{Y} = \rho(y) \in GL_q(A)$ and $Y = \natural(\widehat{Y}) \in GL_q(A/\pi)$. From the group relation $yx^{-1} = x^{-a}y$ and the identity $X = J_q$ it follows that

$$Y \cdot J_q = X^a \cdot Y.$$

However, $X^a = J_q^a$ is uni-triangular. It thereby follows from (58.2) that

(58.13) Y is upper triangular.

Now let $\overline{\psi} : G(p,q) \to GL_q(A/\pi)$ be the representation $\overline{\psi}(g) = \natural(\rho(g))$. As x^{-1} and y generate $G(p,q)$ and both $\overline{\psi}(x^{-1}) = X$ and $\overline{\psi}(y) = Y$ are upper triangular we see that $\overline{\psi}(g)$ is upper triangular for each $g \in G(p,q)$; that is:

(58.14) $\natural(\rho(g))$ is upper triangular for each $g \in G(p,q)$.

It follows from (58.1) that:

(58.15) $\rho(g)$ is quasi-triangular for each $g \in G(p,q)$.

§59: The isomorphism $\mathcal{C}_q(I^*,\theta) \cong \mathcal{T}_q(A,\pi)$:

In this section we show that the cyclic algebra $\mathcal{C}_q(I^*,\theta)$ of §56 can be described as a ring of quasi-triangular matrices. In what follows we adhere to the notation of §56 and §58. In particular, π will denote the unique prime in A over p. However, for ease of reading, we write $S = \mathbb{Z}(\zeta)$. As S is free of rank q over A it follows by a theorem of Dedekind, (cf [25] p. 444) that for some unit $\eta \in A^*$

(59.1) $$\mathcal{D}\mathrm{isc}_A(S) = \eta\pi^{q-1}.$$

Put $u = \det(\theta)^{q(q-1)/2} \cdot \eta$. As $\det(\theta)$ is a unit in A then so also is u. Comparing (56.8) with (57.2) we obtain the following as an equation in $A/(A^*)^2$.

(59.2) $$\frac{\mathcal{D}\mathrm{isc}_A[\mathcal{C}_q(S,\theta)]}{\mathcal{D}\mathrm{isc}_A[M_q(A)]} = \pm u\pi^{q(q-1)}.$$

Likewise, comparing (57.2) and (57.7):

(59.3) $$\frac{\mathcal{D}\mathrm{isc}_A[\mathcal{T}_q(A,\pi)]}{\mathcal{D}\mathrm{isc}_A[M_q(A)]} = \pm\pi^{q(q-1)}.$$

Hence, as u is a unit, we have the following equation in $A^*/(A^*)^2$:

(59.4) $$\frac{\mathcal{D}\mathrm{isc}_A[\mathcal{C}_q(S,\theta)]}{\mathcal{D}\mathrm{isc}_A[\mathcal{T}_q(A,\pi)]} = \pm u.$$

Proposition 59.5: *Any injective ring homomorphism $\iota : \mathcal{C}_q(S,\theta) \to \mathcal{T}_q(A,\pi)$ is automatically an isomorphism.*

Proof. Put $B = \mathrm{Im}(\iota)$; we claim that $B = \mathcal{T}_q(A,\pi)$. As ι is injective and $\mathcal{C}_q(S,\theta)$ is free of rank q^2 over A then B is free of rank q^2 over A and $\mathcal{D}\mathrm{isc}_A(B) = \mathcal{D}\mathrm{isc}_A[\mathcal{C}_q(S,\theta)]$. As $\mathcal{T}_q(A,\pi)$ is also free of rank q^2 over A there is an $q^2 \times q^2$ matrix P over A which expresses some basis for B in terms of the standard basis for $\mathcal{T}_q(A,\pi)$. By (55.3) and (59.4) we have the following equation in $A^*/(A^*)^2$:

$$\det(P)^2 = \frac{\mathcal{D}\mathrm{isc}_A(B)}{\mathcal{D}\mathrm{isc}_A[\mathcal{T}_q(A,\pi)]} = \frac{\mathcal{D}\mathrm{isc}_A[\mathcal{C}_q(S,\theta)]}{\mathcal{D}\mathrm{isc}_A[\mathcal{T}_q(A,\pi)]} = \pm u.$$

Hence P is invertible and $B = \mathcal{T}_q(A,\pi)$ as claimed. □

Let $\rho_* : \mathbb{Z}[G(p,q)] \to \mathcal{T}_q(A,\pi)$ be the ring homomorphism induced by the representation ρ of §58. We note

(59.6) $$\rho_*(\Sigma_x) = 0.$$

Hence ρ induces a ring homomorphism $\rho_* : \mathcal{C}_q(I^*,\theta) \to \mathcal{T}_q(A,\pi)$. As $\mathbb{Q}(\zeta)^\theta \cong A \otimes \mathbb{Q}$ then ρ also induces a ring homomorphism $\rho_* \otimes \mathrm{Id} : \mathcal{C}_q(\mathbb{Q}(\zeta),\theta) \to M_q(A \otimes \mathbb{Q})$. However, $\mathcal{C}_q(\mathbb{Q}(\zeta),\theta)$ is a simple $\mathbb{Q}$-algebra so $\rho_* \otimes \mathrm{Id} : \mathcal{C}_q(\mathbb{Q}(\zeta),\theta) \to M_q(A \otimes \mathbb{Q})$ is injective. Hence $\rho_* : \mathcal{C}_q(I^*,\theta) \to \mathcal{T}_q(A,\pi)$ is also injective. It follows from (59.5) that:

Theorem 59.7: *$\rho_* : \mathcal{C}_q(I^*,\theta) \to \mathcal{T}_q(A,\pi)$ is a ring isomorphism.*

We may thus re-interpret (54.3) as a fibre square of the form

(59.8)
$$\begin{array}{ccc} \mathbb{Z}[G(p,q)] & \to & \mathcal{T}_q(A,\pi) \\ \downarrow & & \downarrow \\ \mathbb{Z}[C_q] & \to & \mathbb{F}_p[C_q]. \end{array}$$

Tensoring the above fibre square with $\mathbb{R}$ we see that

$$\begin{aligned} \mathbb{R}[G(p,q)] &\cong \mathbb{R}[C_q] \times \mathcal{T}_q(A,\pi) \otimes \mathbb{R} \\ &\cong \mathbb{R}[C_q] \times M_q(A \otimes \mathbb{R}) \end{aligned}$$

and as $\mathbb{R}[C_q]$ and $A \otimes \mathbb{R}$ are products of fields, either $\mathbb{R}$ or $\mathbb{C}$, we see that:

(59.9) $\mathbb{Z}[G(p,q)]$ satisfies the Eichler condition.

Writing $\Lambda = \mathbb{Z}[G(p,q)]$, we note that the above surjective ring homomorphism

$$\rho : \Lambda \to \mathcal{T}_q(A, \pi)$$

induces a functor $\rho^* : \mathcal{M}od_{\mathcal{T}_q} \to \mathcal{M}od_\Lambda$ by which any $\mathcal{T}_q$ module acquires the structure of a Λ-module. If M is a Λ-lattice we shall say that M is *coinduced* when $M \cdot \Sigma_x = 0$. In view of the isomorphism $\mathcal{T}_q(A,\pi) \cong \mathcal{C}_q(I^*, \theta)$, it is straightforward to see that, up to Λ-isomorphism, a Λ-lattice M lies in $\mathrm{Im}(\rho^*)$ precisely when M is coinduced. In this case we shall make no distinction between M considered as $\mathcal{T}_q(A,\pi)$ module and M considered as Λ module. This notational blurring is justified by the following general principle:

(59.10) If M, N are coinduced Λ-lattices then

$$M \cong_\Lambda N \Longleftrightarrow M \cong_{\mathcal{T}_q(A,\pi)} N.$$

§60: A worked example:

To illustrate the foregoing, we construct an explicit quasi-triangular representation (cf [54]) for the group $G(7,3)$. Recall that this is the semidirect product

$$G(7,3) = C_7 \rtimes C_3 = \langle x \rangle \rtimes \langle y \rangle$$

where y acts on $\langle x \rangle \cong C_7$ by $yxy^{-1} = x^2$. We begin by taking $S = \mathbb{Z}[x]/c_7(x)$ where $c_7(x)$ is the cyclotomic polynomial

$$c_7(x) = x^6 + x^5 + x^4 + x^3 + x^2 + x + 1.$$

Under the correspondence $x \longleftrightarrow \zeta$, S can be identified with the ring of algebraic integers $\mathbb{Z}(\zeta) \subset \mathbb{Q}(\zeta)$ where $\zeta = \exp(\frac{2\pi i}{7})$. Then $\mathrm{Gal}(\mathbb{Q}(\zeta)/\mathbb{Q}) \cong C_6$, generated by the automorphism $\varphi(\zeta) = \zeta^3$. Putting $\theta = \varphi^2$, then θ generates the subgroup of order 3 in $\mathrm{Gal}(\mathbb{Q}(\zeta))/\mathbb{Q}$. Now take the fixed point ring under $\langle\theta\rangle = C_3$;

$$A = S^{C_3} = \{\mathbf{z} \in S | \theta(\mathbf{z}) = \mathbf{z}\}.$$

The action of θ on the powers ζ^k is as follows:

$$\theta(\zeta) = \zeta^2; \quad \theta(\zeta^2) = \zeta^4; \quad \theta(\zeta^3) = \zeta^6;$$
$$\theta(\zeta^4) = \zeta; \quad \theta(\zeta^5) = \zeta^3; \quad \theta(\zeta^6) = \zeta^5.$$

Observe that $\mathrm{Gal}(A \otimes \mathbb{Q}/\mathbb{Q}) \cong C_2$, wherein the generator corresponds to complex conjugation '$\mathbf{z} \mapsto \overline{\mathbf{z}}$'. We now put

$$\alpha = \zeta + \zeta^2 + \zeta^4 \quad \text{so that } \overline{\alpha} = \zeta^6 + \zeta^5 + \zeta^3.$$

Then α satisfies the minimal equation $\alpha^2 + \alpha + 2 = 0$ so that we may make the identifications

(60.1)
$$A = \mathbb{Z}[\alpha]/(\alpha^2 + \alpha + 2)$$

(60.2)
$$\alpha = \frac{-1 + i\sqrt{7}}{2}; \quad \overline{\alpha} = \frac{-1 - i\sqrt{7}}{2}.$$

Putting $\pi = 2\alpha + 1$ we see that

(60.3)
$$\pi^2 = -7.$$

It follows that π is the unique prime in A over 7 and that 7 ramifies completely in A. For future reference we note that

(60.4) there is a ring isomorphism $A/\pi \xrightarrow{\simeq} \mathbb{F}_7$ determined by $\alpha \mapsto 3$.

Put
$$G = G(7,3) = \langle x, y \mid x^7 = 1, y^3 = 1, yx = x^2 y \rangle.$$

Take $\{1, \zeta, \zeta^2\}$ as a basis for $S = \mathbb{Z}[\zeta]$ over the fixed point ring $A = \mathbb{Z}[\zeta]^{C_3}$ and note that $\zeta^3 = 1 + (\alpha+1)\cdot\zeta + \alpha\cdot\zeta^2$. Consequently x acts on the right of S by

$$\begin{cases} 1 \cdot x = \zeta \\ \zeta \cdot x = \zeta^2 \\ \zeta^2 \cdot x = 1 + (\alpha+1)\cdot\zeta + \alpha\cdot\zeta^2 \end{cases}$$

giving a representation $\psi : C_7 \to \mathrm{GL}_3(A)$ as follows:

$$\psi(x^{-1}) = \begin{pmatrix} 0 & 0 & 1 \\ 1 & 0 & \alpha+1 \\ 0 & 1 & \alpha \end{pmatrix}.$$

We calculate that
$$\psi(x) = \psi(x^{-6}) = \begin{pmatrix} -(\alpha+1) & 1 & 0 \\ -\alpha & 0 & 1 \\ 1 & 0 & 0 \end{pmatrix}$$

so that
$$\psi(x^2) = \begin{pmatrix} -1 & -(\alpha+1) & 1 \\ -1 & -\alpha & 0 \\ -(\alpha+1) & 1 & 0 \end{pmatrix}.$$

Moreover, y acts on the left of S as a Galois operator thus:

$$y(1) = 1; \quad y(\zeta) = \zeta^2; \quad y(\zeta^2) = \zeta^4.$$

As $\zeta^4 = \alpha - \zeta - \zeta^2$ we obtain a representation $\psi : C_3 \to \mathrm{GL}_3(A)$ thus:

$$\psi(y) = \begin{pmatrix} 1 & 0 & \alpha \\ 0 & 0 & -1 \\ 0 & 1 & -1 \end{pmatrix}.$$

As
$$\psi(y)\psi(x) = \psi(x^2)\psi(y) = \begin{pmatrix} -1 & 1 & 0 \\ -1 & 0 & 0 \\ -(\alpha+1) & 0 & 1 \end{pmatrix}$$

we obtain a representation $\psi : G(7,3) \to GL_3(A)$ by assigning

$$\psi(x^a y^b) = \psi(x)^a \psi(y)^b.$$

We proceed to put the restriction of ψ to C_7 into quasi-triangular form. For this, we note that, with respect to the isomorphism $A/\pi \cong \mathbb{F}_7$, the canonical surjection $A \to A/\pi$ is equivalent to the mapping $A \to \mathbb{F}_7$; $\alpha \to 3$; thus

$$\overline{\psi}(x^{-1}) = \begin{pmatrix} 0 & 0 & 1 \\ 1 & 0 & 4 \\ 0 & 1 & 3 \end{pmatrix} \pmod{\pi}.$$

Now consider the representation $\rho : C_7 \to GL_3(\mathbb{F}_7)$

$$\rho(g) = \overline{Q}\,\overline{\psi}(g)\overline{Q}^{-1}$$

where $\overline{Q}, \overline{Q}^{-1} \in SL_3(\mathbb{F}_7)$ are given by

$$\overline{Q} = \begin{pmatrix} 6 & 0 & 0 \\ 1 & 0 & 6 \\ 6 & 6 & 6 \end{pmatrix}; \quad \overline{Q}^{-1} = \begin{pmatrix} 6 & 0 & 0 \\ 2 & 1 & 6 \\ 6 & 6 & 0 \end{pmatrix}$$

so that $\rho(x^{-1}) = \begin{pmatrix} 1 & 1 & 0 \\ 0 & 1 & 1 \\ 0 & 0 & 1 \end{pmatrix} \pmod 7$

giving $\rho(x^{-1}) = J_3(1)$ as required. For any field $\mathbb{F}$, $E_3(\mathbb{F}) = SL_3(\mathbb{F})$. Hence $\overline{Q} \in E_3(\mathbb{F}_7)$. As the canonical map $\mathbb{Z} \to \mathbb{F}_7$ is surjective, there exists $Q \in E_3(\mathbb{Z})$ such that $Q \equiv \overline{Q} \pmod 7$. In fact, we may take

$$Q = \begin{pmatrix} -1 & 0 & 0 \\ 1 & 0 & -1 \\ -1 & -1 & -1 \end{pmatrix}; \quad Q^{-1} = \begin{pmatrix} -1 & 0 & 0 \\ 2 & 1 & -1 \\ -1 & -1 & 0 \end{pmatrix}.$$

Then $Q \in SL_3(\mathbb{Z}) = E_3(\mathbb{Z})$ and $Q \equiv \overline{Q} \pmod 7$. Take $\rho : G(p,q) \to GL_3(A)$ to be the conjugation of ψ by Q thus: $\rho(g) = Q\psi(g)Q^{-1}$. Calculation then gives

$$\rho(x) = \begin{pmatrix} -(\alpha+3) & -1 & 1 \\ (\alpha+4) & 1 & -1 \\ -(2\alpha+1) & 0 & 1 \end{pmatrix}; \quad \rho(y) = \begin{pmatrix} \alpha+1 & \alpha & 0 \\ -(\alpha+4) & -(\alpha+2) & 1 \\ \alpha-3 & \alpha-3 & 1 \end{pmatrix}.$$

Noting that $\pi = 2\alpha + 1$, $\alpha - 3 = (\alpha+1)\pi$, $\alpha + 4 = -\alpha\pi$ then

$$\rho(x) = \begin{pmatrix} -(\alpha+3) & -1 & 1 \\ -\alpha\pi & 1 & -1 \\ -\pi & 0 & 1 \end{pmatrix}; \quad \rho(y) = \begin{pmatrix} \alpha+1 & \alpha & 0 \\ \alpha\pi & -(\alpha+2) & 1 \\ (\alpha+1)\pi & (\alpha+1)\pi & 1 \end{pmatrix}$$

so that $\mathrm{Im}(\rho) \subset \mathcal{T}_3(A,\pi)^*$ as predicted. We note also that

$$\rho(y^{-1}) = \begin{pmatrix} -(2\alpha-1) & -\alpha & \alpha \\ 2\alpha+1 & \alpha+1 & -(\alpha+1) \\ -(\alpha+2) & -(\alpha-3) & 1 \end{pmatrix}.$$

On reducing mod π via the ring homomorphism $A \to \mathbb{F}_7$; $\alpha \mapsto 3$, we see that

$$\rho(y^{-1}) = \begin{pmatrix} 2 & 4 & 3 \\ 0 & 4 & 3 \\ 0 & 0 & 1 \end{pmatrix} \pmod \pi.$$

The chosen residue mod 7 which satisfies $yxy^{-1} = x^a$ is $a = 2$. We note that:

$$\rho(y^{-1}) = \begin{pmatrix} a & * & * \\ 0 & a^2 & * \\ 0 & 0 & a^3 \end{pmatrix} \pmod{\pi};$$

that is, when computed mod π, y acts on the right of the row module $R(k)$ as $\mathbf{z} \mapsto \mathbf{z}a^k$. This point will be pursued further in the next chapter.

§61: Lifting units to cyclotomic rings:

As before we put

$$\begin{aligned} p &: \text{ an odd prime} \\ q &: \text{ a divisor of } p-1 \\ d &= \left(\frac{p-1}{q}\right) \\ \zeta &= \exp(2\pi i/p) \\ K &= \mathbb{Q}(\zeta) \\ R &= \mathbb{Z}[\zeta]. \end{aligned}$$

Then $R \subset K$ is the subring of algebraic integers of K. As $\mathrm{Gal}(K/\mathbb{Q}) \cong C_{p-1}$ then $\mathrm{Gal}(K/\mathbb{Q})$ contains a unique subgroup C_2 of order 2 and a unique subgroup C_q of order q. We put

$$\begin{aligned} K_+ &= K^{C_2} \\ R_+ &= R^{C_2} \\ L &= K^{C_q} \\ A &= R^{C_q}. \end{aligned}$$

K_+ is the maximal real subfield of K and $R_+ = R \cap K_+$ is the subring of algebraic integers in K_+. Likewise A is the subring of algebraic integers in L. The prime p ramifies completely in R and we denote by

$$\widehat{\pi} : \text{ the unique prime in } R \text{ over } p.$$

Likewise p ramifies completely in both R_+ and A; putting $\pi_+ = \widehat{\pi}^2$ and $\pi = \widehat{\pi}^q$ then π_+ represents the unique prime in R_+ over p and π represents the unique prime in A over p. Moreover,

(61.1) if q is even then $A \subset R_+$ and $\pi = (\pi_+)^{q/2}$.

We next put

(61.2) $$\mathcal{K} = \{x \in \mathbb{F}_p^* | x^{2d} = 1\}.$$

As $\mathcal{K}$ is a multiplicative subgroup of $\mathbb{F}_p^* \cong C_{p-1}$ and so $\mathcal{K}$ is cyclic; moreover:

(61.3) $$|\mathcal{K}| = \begin{cases} d & q \text{ odd} \\ 2d & q \text{ even.} \end{cases}$$

Choose $h \in \{1, \ldots, p-1\} \subset \mathbb{Z}$ such that the image $\overline{h} \in \mathbb{F}_p^*$ is a multiplicative generator. As $p - 1 = dq$ it is a straightforward deduction from (61.3) that

(61.4) $$\mathcal{K} \text{ is generated by } \begin{cases} \overline{h}^q & q \text{ odd;} \\ \overline{h}^{q/2} & q \text{ even.} \end{cases}$$

Noting that $A/\pi \cong \mathbb{F}_p$ we denote by $\natural : U(A) \to \mathbb{F}_p^*$ the restriction to unit groups of the canonical ring homomorphism $A \to \mathbb{F}_p$. We claim:

Proposition 61.5: $\mathrm{Im}(\natural) = \mathcal{K}$.

Proof. We first prove that $\mathrm{Im}(\natural) \subset \mathcal{K}$. From the Galois exact sequence

$$1 \to \mathrm{Gal}(K/L) \to \mathrm{Gal}(K/\mathbb{Q}) \to \mathrm{Gal}(L/\mathbb{Q}) \to 1$$

together with the isomorphisms $\mathrm{Gal}(K/\mathbb{Q}) \cong C_{p-1}$, $\mathrm{Gal}(K/L) \cong C_q$ we see that $\mathrm{Gal}(L/\mathbb{Q}) \cong C_d$. Let $\tau \in \mathrm{Gal}(L/\mathbb{Q})$ be a generator. If $a \in A$ then

$$\tau(a) = a \pmod{\pi}$$

so that $$N_{L/\mathbb{Q}}(a) = \textstyle\prod_{r=0}^{d-1} \tau^r(a) = a^d \pmod{\pi}.$$

If $u \in U(A)$ then $N_{L/\mathbb{Q}}(u) = \pm 1 \in \mathbb{Z}^*$ so that $u^d = \pm 1 \bmod (\pi)$ and hence $u^{2d} = 1 \bmod (\pi)$. It follows that $\natural(u) \in \mathcal{K}$ and so $\mathrm{Im}(\natural) \subset \mathcal{K}$.

To prove the reverse inclusion, it suffices to show that $\mathrm{Im}(\natural)$ contains a generator of the cyclic group $\mathcal{K}$. Put

$$\xi_h = \left(\frac{\zeta^h - \zeta^{-h}}{\zeta - \zeta^{-1}}\right) \in U(R_+).$$

As is well known (cf [25], p. 525–526) $\xi_h \in U(R_+) \subset U(R)$. Moreover

(61.6) $$\xi_h \equiv h \bmod \pi_+.$$

Now define
$$u_h = \begin{cases} N_{K/L}(\xi_b) & \text{if } q \text{ is odd} \\ N_{K_+/L}(\xi_h) & \text{if } q \text{ is even} \end{cases}$$

so that $u_h \in U(A)$. Moreover, from (61.6) we see that

(61.7)
$$\natural(u_h) = \begin{cases} \overline{h}^q & \text{if } q \text{ is odd} \\ \overline{h}^{q/2} & \text{if } q \text{ is even} \end{cases}.$$

Either way $\natural(u_h)$ generates $\mathcal{K}$ and $\mathcal{K} \subset \mathrm{Im}(\natural)$ so completing the proof. □

Clearly the quotient $\mathbb{F}_p^*/\mathcal{K}$ is cyclic. As $p-1 = dq$ it follows from (61.3) that

(61.8)
$$\mathbb{F}_p^*/\mathcal{K} \cong \begin{cases} C_q & q \text{ odd} \\ C_{q/2} & q \text{ even.} \end{cases}$$

§62: Liftable subgroups of $\underbrace{\mathbb{F}_p^* \times \cdots \times \mathbb{F}_p^*}_{q}$:

Let $\varphi : R \to S$ be a ring homomorphism and let $H \subset S^*$ be a subgroup of the unit group S^*; we say that H is *liftable relative to* φ if there is a subgroup $\widetilde{H} \subset R^*$ such that $\varphi(\widetilde{H}) = H$. We say that H is a *maximal liftable subgroup relative to* φ when H is liftable relative to φ and when, given a proper inclusion of subgroups $H \subsetneq H'$ of S^*, H' is not liftable relative to φ.

Proposition 62.1: *Let $\varphi : R \to S$ be a ring homomorphism; if S^* is finitely generated abelian then S^* has a unique maximal liftable subgroup relative to φ.*

Proof. If H is a subgroup of S^* which is liftable to R^* we claim that there is a normal subgroup $\widehat{H} \lhd G$ such that $\varphi(\widehat{H}) = H$. For, as S^* is abelian then $\varphi(\gamma\widetilde{H}\gamma^{-1}) = H$ for each $\gamma \in R^*$. Taking $\widehat{H}$ to be the subgroup of R^* generated by the set $\{\gamma h \gamma^{-1} | \gamma \in R^*, h \in \widetilde{H}\}$ we see that $\widehat{H}$ is normal in R^* and $\varphi(\widehat{H}) = H$.

Clearly S^* has at least one subgroup, namely the trivial subgroup, which is liftable to R^*. Moreover, if H_1, H_2 are liftable subgroups of S^* then, choosing normal subgroups $\widehat{H}_1, \widehat{H}_2$ of R^* such that $\varphi(\widehat{H}_i) = H_i$, it follows that $\widehat{H}_1 \cdot \widehat{H}_2 \lhd R^*$ and $\varphi(\widehat{H}_1 \cdot \widehat{H}_2) = H_1 \cdot H_2$. Hence $H_1 \cdot H_2$ is also liftable. The conclusion now follows as S^* satisfies the ascending chain condition on subgroups. □

For the remainder of this section we focus on the surjective ring homomorphism $\nu : \mathcal{T}_q(A, \pi) \longrightarrow \mathbb{F}_p[C_q]$ which occurs in the fibre product diagram in (59.8). By the Wedderburn-Maschke Theorem we have $\mathbb{F}_p[C_q] \cong \mathbb{F}_p^{(q)}$ and, with that identification, we may factorize ν as a composition $\nu = \natural_2 \circ \natural_1$ where, after making the identification $A/\pi \cong \mathbb{F}_p$, $\natural_1$ and $\natural_2$ are the natural surjections indicated in the following diagram:

62.2

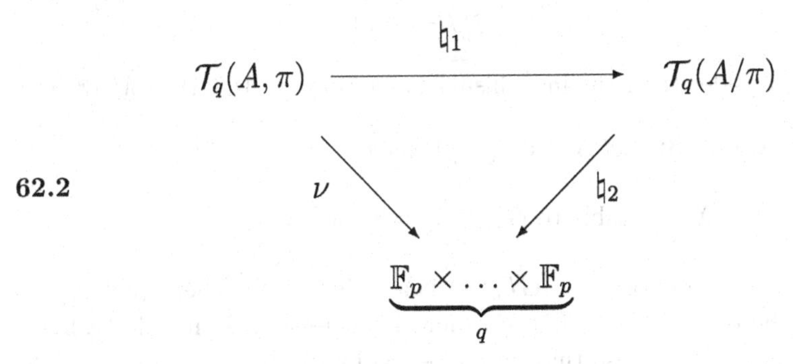

As $\underbrace{\mathbb{F}_p^* \times \cdots \times \mathbb{F}_p^*}_{q}$ is finite and abelian, the hypothesis of (62.1) is satisfied. Hence relative to ν, $\underbrace{\mathbb{F}_p^* \times \cdots \times \mathbb{F}_p^*}_{q}$ has a unique maximal liftable subgroup which we proceed to describe. We first treat a simpler case. Recall that $A = \mathbb{Z}[\zeta_p]^{C_q}$ and that $A/\pi \cong \mathbb{F}_p$ so that there is a surjective ring homomorphism

$$A \longrightarrow \mathbb{F}_p.$$

Noting that $\mathbb{F}_p^* \cong C_{p-1}$, choose a specific integer h in the range $1 \le h \le p-1$ such that the residue class $[h]$ generates $\mathbb{F}_p^*$ as a multiplicative group and define

(62.3)
$$\gamma = \begin{cases} [h^q] & q \text{ odd} \\ [h^{q/2}] & q \text{ even.} \end{cases}$$

Clearly γ generates a subgroup $\langle \gamma \rangle \subset \mathbb{F}_p^*$ and for $u \in \mathbb{F}_p^*$ we have:

(62.4)
$$u \text{ is liftable to } A^* \Longleftrightarrow u \in \langle \gamma \rangle.$$

At this point, for future reference, it is convenient to introduce the following notation:

(62.5)
$$\breve{q} = \begin{cases} q & \text{if } q \text{ is odd} \\ q/2 & \text{if } q \text{ is even} \end{cases}$$

and put

$$\hat{d} = \begin{cases} (p-1)/q & \text{if } q \text{ is odd} \\ 2(p-1)/q & \text{if } q \text{ is even} \end{cases} \tag{62.6}$$

so that $(p-1) = \hat{d} \cdot \check{q}$. Consequently $\langle \gamma \rangle \cong C_{\hat{d}}$ and hence

$$\mathbb{F}_p^* / \langle \gamma \rangle \cong C_{\check{q}}. \tag{62.7}$$

Next we consider the homomorphism of matrix rings $M_q(A) \to M_q(\mathbb{F}_p)$.

Proposition 62.8: *Let $X \in GL_q(\mathbb{F}_p)$; then*

$$X \text{ is liftable to } GL_q(A) \iff \det(X) \in \langle \gamma \rangle.$$

Proof. ($\Longrightarrow$) Suppose $\widehat{X} \in GL_q(A)$ is a lifting of X. Then $\natural(\det(\widehat{X})) = \det(\widehat{X})$ where $\natural : A \to \mathbb{F}_p$ is the canonical surjection. Hence $\det(\widehat{X})$ is a lifting of $\det(X)$ to A^* so that $\det(X) \in \langle \gamma \rangle$ by (62.4).

($\Longleftarrow$) Write$X = D \cdot E$ where E is a product of elementary transvections and D is the diagonal matrix

$$D = \begin{pmatrix} \det(X) & & & & \\ & 1 & & & \\ & & \ddots & & \\ & & & 1 & \\ & & & & 1 \end{pmatrix}.$$

By hypothesis $\det(X) \in \langle \gamma \rangle$ so we may choose $\widehat{u} \in A^*$ such that $\natural(\widehat{u}) = \det(X)$. As $\natural : A \to \mathbb{F}_p$ is surjective then there exists $\widehat{E} \in E_q(A)$ such that $\natural_*(\widehat{E}) = E$. Putting $\widehat{X} = \widehat{D} \cdot \widehat{E}$ where

$$\widehat{D} = \begin{pmatrix} \widehat{u} & & & & \\ & 1 & & & \\ & & \ddots & & \\ & & & 1 & \\ & & & & 1 \end{pmatrix}$$

we see that $\widehat{X}$ is a lifting of X to $GL_q(A)$. □

Let $\upsilon = (\upsilon_1, \ldots, \upsilon_q) \in \underbrace{\mathbb{F}_p^* \times \cdots \times \mathbb{F}_p^*}_{q}$ be the element defined by taking

$$\upsilon_r = \begin{cases} [h] & r = 1 \\ 1 & r \neq 1. \end{cases}$$

Likewise define elements $\psi_2, \ldots, \psi_q$ in $\underbrace{\mathbb{F}_p^* \times \cdots \times \mathbb{F}_p^*}_{q}$ as follows:

$$(\psi_k)_r = \begin{cases} [h]^{-1} & r = 1 \\ [h] & r = k \\ 1 & r \notin \{1, k\}. \end{cases}$$

It is straightforward to see that:

(62.9) $\upsilon, \psi_2, \ldots, \psi_q$ is a minimal generating set for $\underbrace{\mathbb{F}_p^* \times \cdots \times \mathbb{F}_p^*}_{q}$.

For internal accounting purposes, which will perhaps become clearer, we find it convenient also to consider the identity element ψ_1; that is, $(\psi_1)_r = 1$ for all r. Now define subgroups Δ, Ψ of $\underbrace{\mathbb{F}_p^* \times \cdots \times \mathbb{F}_p^*}_{q}$ by

$$\Delta = \langle \upsilon \rangle; \quad \Psi = \langle \psi_1, \ldots, \psi_q \rangle.$$

It follows directly from (62.9) that $\underbrace{\mathbb{F}_p^* \times \cdots \times \mathbb{F}_p^*}_{q}$ is an internal direct product.

(62.10) $$\underbrace{\mathbb{F}_p^* \times \cdots \times \mathbb{F}_p^*}_{q} = \Delta \cdot \Psi.$$

To proceed, observe that we have a commutative diagram of ring homomorphisms

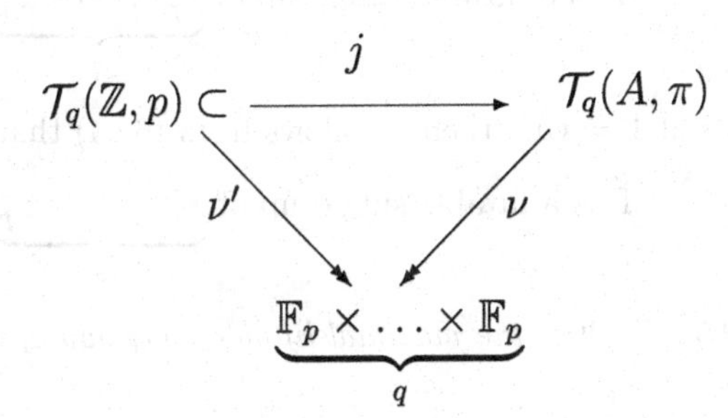

Maintaining our previous choice of h in the range $1 \le h \le p-1$ then as h and p are coprime, appealing to Bezout's Theorem we can find integers b, d such that

(62.11) $$hd + pb = 1$$

so that, in particular, the residue class of d represents $[h]^{-1}$ in $\mathbb{F}_p^*$. For j in the range $2 \le j \le q$ we define elements $\widehat{\psi}_2, \dots, \widehat{\psi}_q, \widehat{\eta}_2, \dots, \widehat{\eta}_q$ in $\mathcal{T}_q(\mathbb{Z}, p)$ as follows:

$$(\widehat{\psi}_j)_{rs} = \begin{cases} d & r=1, s=1 \\ h & r=j, s=j \\ b & r=1, s=j \\ -p & r=j, s=1 \\ \delta_{rs} & \{r,s\} \neq \{1,j\} \end{cases} \quad ; \quad (\widehat{\eta}_j)_{rs} = \begin{cases} h & r=1, s=1 \\ d & r=j, s=j \\ -b & r=1, s=j \\ p & r=j, s=1 \\ \delta_{rs} & \{r,s\} \neq \{1,j\} \end{cases}$$

where δ_{rs} denotes the Kronecker delta. It is straightforward to calculate that

$$\widehat{\psi}_j \cdot \widehat{\eta}_j = \widehat{\eta}_j \cdot \widehat{\psi}_j = I_q.$$

Consequently $\psi_j \in \mathcal{T}_q(\mathbb{Z}, p)^*$. Straightfoward calculation now shows that:

(62.12) $$\nu'(\widehat{\psi}_j) = \psi_j;$$

Again, taking $\widehat{\psi}_1$ to be the identity element of $\mathcal{T}_q(\mathbb{Z}, p)^*$ we define

$$\widehat{\Psi} = \langle \widehat{\psi}_1, \dots, \widehat{\psi}_q \rangle \subset \mathcal{T}_q(\mathbb{Z}, p)^*.$$

In the present case, the inclusion $\mathbb{Z} \to A$ has the property that $\mathbb{Z} \cap (\pi) = (p)$ and $A/\pi = \mathbb{Z}/p = \mathbb{F}_p$. We can equally consider $\widehat{\Psi}$ as a subgroup of $\mathcal{T}_q(A, \pi)^*$ in which regard it follows from (62.12) that $\nu(\widehat{\psi}_j) = \psi_j$. Thus $\nu(\widehat{\Psi}) = \Psi$; that is:

(62.13) $$\Psi \text{ is a liftable subgroup of } \underbrace{\mathbb{F}_p^* \times \cdots \times \mathbb{F}_p^*}_{q}.$$

Moreover, if we put $\Gamma = \langle \upsilon^{\check{q}} \rangle$ then it follows from (62.4) that

(62.14) $$\Gamma \text{ is a liftable subgroup of } \underbrace{\mathbb{F}_p^* \times \cdots \times \mathbb{F}_p^*}_{q}.$$

Theorem 62.15: *$\Gamma \cdot \Psi$ is the maximal liftable subgroup of* $\underbrace{\mathbb{F}_p^* \times \cdots \times \mathbb{F}_p^*}_{q}$.

Proof. From (62.13) and (62.14), it follows that $\Gamma \cdot \Psi$ is a liftable subgroup of $\underbrace{\mathbb{F}_p^* \times \cdots \times \mathbb{F}_p^*}_{q}$. In the following the commutative diagram of unit groups

$$\begin{array}{ccc} \mathcal{T}_q(A,\pi)^* & \xrightarrow{\natural_1} & \mathcal{T}_q(A/\pi)^* \\ & \searrow^{\nu} \quad \swarrow_{\natural_2} & \\ & \underbrace{\mathbb{F}_p^* \times \ldots \times \mathbb{F}_p^*}_{q} & \end{array}$$

suppose that $\widehat{\lambda} \in \mathcal{T}_q(A,\pi)^*$ is a lifting of $\lambda \in \underbrace{\mathbb{F}_p^* \times \cdots \times \mathbb{F}_p^*}_{q} = \Delta \cdot \Psi$.

We must show that $\lambda \in \Gamma \cdot \Psi$. We first consider the special case where $\lambda = (u, 1, \ldots, 1) \in \Delta$. We claim that $u \in \langle \gamma \rangle$ so that $\lambda \in \Gamma$. As $\lambda \in \Delta$ then $\natural_2 \circ \natural_1(\widehat{\lambda}) = \lambda$ so that $\natural_1(\widehat{\lambda})$ is a lifting of λ to $\mathcal{T}_q(A/\pi)^*$. Hence $\natural_1(\widehat{\lambda})$ is an upper triangular matrix of the form

$$\natural_1(\widehat{\lambda}) = \begin{pmatrix} u & & & \\ & 1 & & * \\ & & \ddots & \\ & & & \ddots & \\ & & & & 1 \end{pmatrix}.$$

As $\widehat{\lambda}$ is a lifting of $\natural_1(\widehat{\lambda}) \in GL_q(A/\pi)$ to $GL_q(A)$ then $\det(\natural_1(\widehat{\lambda})) \in \langle \gamma \rangle$, by (62.8). However $\det(\natural_1(\widehat{\lambda})) = u$ so that $u \in \langle \gamma \rangle$ as claimed and hence $\lambda \in \Gamma$.

In general, suppose that $\widehat{\lambda}$ is a lifting of $\lambda = \delta \cdot \psi$ where $\delta \in \Delta$ and $\psi \in \Psi$. Then ψ has a lifting $\widehat{\psi} \in \Psi \subset \mathcal{T}_q(A,\pi)^*$ so that $\widehat{\lambda} \cdot \widehat{\psi}^{-1}$ is a lifting of $\delta \in \Delta$. Thus $\delta \in \Gamma$ by the above special case and hence $\lambda = \delta \cdot \psi \in \Gamma \cdot \Psi$. □

The subgroup $\Gamma \cdot \Psi$ has alternative descriptions which it is necessary to consider. For $1 \le k \le q$ we define $v(k) \in \underbrace{\mathbb{F}_p^* \times \cdots \times \mathbb{F}_p^*}_{q}$ by

$$(v(k))_r = \begin{cases} [h] & r = k \\ 1 & r \neq k. \end{cases}$$

For each integer k in the range $1 \le k \le q$ we define elements $\psi_1^{(k)}, \ldots, \psi_q^{(k)}$ in $\underbrace{\mathbb{F}_p^* \times \cdots \times \mathbb{F}_p^*}_{q}$ as follows: if $j \neq k$ we put

$$\psi_j^{(k)})_r = \begin{cases} [h] & r = j \\ [h]^{-1} & r = k \\ 1 & r \notin \{j, k\}. \end{cases}$$

Moreover, we define $(\psi_k^{(k)})_r = 1$ for all r; that is, $(\psi_k^{(k)})$ is simply the identity of the ring $\underbrace{\mathbb{F}_p \times \cdots \times \mathbb{F}_p}_{q}$. Now define subgroups $\Delta(k)$, $\Psi(k)$ of $\underbrace{\mathbb{F}_p^* \times \cdots \times \mathbb{F}_p^*}_{q}$ by

$$\Delta(k) = \langle \upsilon(k) \rangle; \quad \Psi(k) = \langle \psi_1^{(k)}, \ldots, \psi_q^{(k)} \rangle.$$

Clearly $\Delta = \Delta(1)$ and $\Psi = \Psi(1)$. Moreover it is straightforward to see we again have an internal direct product decomposition:

(62.16)
$$\Delta(k) \cdot \Psi(k) = \underbrace{\mathbb{F}_p^* \times \cdots \times \mathbb{F}_p^*}_{q}.$$

Define $\Gamma(k) = \langle \upsilon(k)^{\check{q}} \rangle$. Noting that $\upsilon(k) = \upsilon \cdot \psi_k$ we see that

$$\upsilon(k)^{\check{q}} = \upsilon^{\check{q}} \cdot \psi_k^{\check{q}} \in \Gamma \cdot \Psi.$$

Hence $\Gamma(k) \subset \Gamma \cdot \Psi$ and so

(62.17) $\Gamma(k)$ is liftable to $\mathcal{T}_q(A, \pi)^*$.

Analogous to the liftings $\widehat{\psi}_j$ of ψ_j we may construct liftings of $\widehat{\psi}_j^{(k)}$ of $\psi_j^{(k)}$: for $j < k$ define

$$(\widehat{\psi}_j^{(k)})_{rs} = \begin{cases} d & r = k, s = k \\ h & r = j, s = j \\ -p & r = k, s = j \\ b & r = j, s = k \\ \delta_{rs} & \{r, s\} \neq \{k, j\} \end{cases} ; \quad (\widehat{\eta}_j^{(k)})_{rs} = \begin{cases} h & r = k, s = k \\ d & r = j, s = j \\ p & r = k, s = j \\ -b & r = j, s = k \\ \delta_{rs} & \{r, s\} \neq \{k, j\} \end{cases}$$

whilst for $k < j$ we put;

$$(\widehat{\psi}_j^{(k)})_{rs} = \begin{cases} d & r = k, s = k \\ h & r = j, s = j \\ b & r = k, s = j \\ -p & r = j, s = k \\ \delta_{rs} & \{r,s\} \neq \{k,j\} \end{cases} \quad ; \quad (\widehat{\eta}_j^{(k)})_{rs} = \begin{cases} h & r = k, s = k \\ d & r = j, s = j \\ -b & r = k, s = j \\ p & r = j, s = k \\ \delta_{rs} & \{r,s\} \neq \{k,j\} \end{cases}$$

It is again straightforward to calculate that $\widehat{\psi}_j^{(k)} \cdot \widehat{\eta}_j^{(k)} = \widehat{\eta}_j^{(k)} \cdot \widehat{\psi}_j^{(k)} = I_q$. Consequently $\widehat{\psi}_j^{(k)} \in \mathcal{T}_q(A,\pi)^*$. Straightfoward calculation now shows that:

$$\nu(\widehat{\psi}_j^{(k)}) = \psi_j^{(k)}.$$

Again, taking $\widehat{\psi}_k^{(k)}$ to be the identity element of $\mathcal{T}_q(A,\pi)$ we define

$$\widehat{\Psi}(k) = \langle \widehat{\psi}_1^{(k)}, \ldots, \widehat{\psi}_q^{(k)} \rangle \subset \mathcal{T}_q(A,\pi)^*.$$

Hence we also have:

(62.18) $\Psi(k)$ is liftable to $\mathcal{T}_q(A,\pi)^*$.

It follows, as in the proof of (62.15), that $\Gamma(k) \cdot \Psi(k)$ is liftable to $\mathcal{T}_q(A,\pi)^*$. As $\Gamma \cdot \Psi$ is the unique maximal liftable subgroup we see that $\Gamma(k) \cdot \Psi(k) \subset \Gamma \cdot \Psi$. However, it is straightforward to calculate that

$$|\Gamma(k) \cdot \Psi(k)| = \widehat{d} \cdot (p-1)^{q-1} = |\Gamma \cdot \Psi|.$$

Hence $\Gamma(k) \cdot \Psi(k) = \Gamma \cdot \Psi$ and so, for each k, we also have:

Theorem 62.19: *$\Gamma(k) \cdot \Psi(k)$ is the maximal liftable subgroup of $\underbrace{\mathbb{F}_p^* \times \cdots \times \mathbb{F}_p^*}_{q}$.*

For future reference we conclude this section by giving a simple numerical criterion for liftability; thus recall that $\breve{q} = \begin{cases} q & q \text{ odd} \\ q/2 & q \text{ even.} \end{cases}$

As in §60 put $\mathcal{K} = \{\mathbf{x} \in \mathbb{F}_p^* | \mathbf{x}^{2d} = 1\}$. Let $N : \underbrace{\mathbb{F}_p^* \times \cdots \times \mathbb{F}_p^*}_{q} \to \mathbb{F}_p^*$ be the mapping $N(a_1, \ldots, a_q) = \prod_{i=1}^q a_i$; then:

Proposition 62.20: *$N^{-1}(\mathcal{K})$ is the maximal liftable subgroup of* $\underbrace{\mathbb{F}_p^* \times \cdots \times \mathbb{F}_p^*}_{q}$.

Proof. Observe that, $\Gamma(k) \cdot \Psi(k) \subset N^{-1}(\mathcal{K})$ for any k. The conclusion follows as $\Gamma(k) \cdot \Psi(k)$ and $N^{-1}(\mathcal{K})$ both have index $\check{q}$ in $\underbrace{\mathbb{F}_p^* \times \cdots \times \mathbb{F}_p^*}_{q}$. □

Let $\mathbf{a} = (a_1, \ldots, a_q) \in \underbrace{\mathbb{F}_p^* \times \cdots \times \mathbb{F}_p^*}_{q}$. As a consequence of (62.20) we have:

(62.21) If $N(\mathbf{a}) = \pm 1$ then $\mathbf{a}$ is liftable to $\mathcal{T}_q(A, \pi)^*$.

§63: p-adic analogues:

Let $\widehat{\mathbb{Z}}$ denote the ring of p-adic integers, let $\widehat{\mathbb{Q}}$ denote its field of fractions and take $S = \mathbb{Z}(\zeta_p)$. Put $c_p(x) = x^{p-1} + \cdots + x + 1$ so that $S \cong \mathbb{Z}[x]/c_p(x)$. The standard proof that $c_p(x)$ is irreducible over $\mathbb{Q}$ likewise shows that $c_p(x)$ remains irreducible over $\widehat{\mathbb{Q}}$. As we have previously observed, S is a Dedekind domain in which p ramifies completely. Let $\widetilde{\pi} \in S$ be the unique prime over p and let $\widehat{S}$ denote the complete local ring obtained from S by localizing $\widetilde{\pi}$. As $S \cong \mathbb{Z}[x]/c_p(x)$ is the ring of integers in $\mathbb{Q}[x]/c_p(x)$ then $\widehat{S}$ is the ring of integers in $\widehat{\mathbb{Q}}[x]/c_p(x)$. Hence $\widehat{S} \cong \widehat{\mathbb{Z}}[x]/c_p(x) \cong (\mathbb{Z}[x]/c_p(x)) \otimes_{\mathbb{Z}} \widehat{\mathbb{Z}} \cong S \otimes_{\mathbb{Z}} \widehat{\mathbb{Z}}$ and so

(63.1) $$\widehat{S} \cong S \otimes_{\mathbb{Z}} \widehat{\mathbb{Z}}.$$

As before, put $A = S^{C_q}$; then again p ramifies completely in A and $\pi = (\widetilde{\pi})^q$ is the unique prime in A over p. Now take $\widehat{A}$ to be the complete local ring obtained from A by localizing π; then $\widehat{A} = (\widehat{S})^{C_q} = (S \otimes_{\mathbb{Z}} \widehat{\mathbb{Z}})^{C_q} = S^{C_q} \otimes_{\mathbb{Z}} \widehat{\mathbb{Z}}$ and so

(63.2) $$\widehat{A} \cong A \otimes_{\mathbb{Z}} \widehat{\mathbb{Z}}.$$

The theorems of §57 and §58 now have straightforward p-adic analogues; specifically if $\widehat{\psi} : G(p,q) \to GL_{p-1}(\widehat{\mathbb{Z}})$ is the representation obtained from the Galois action of C_q on $\widehat{S} = \widehat{\mathbb{Z}}(\zeta)$ and $i : \widehat{\mathbb{Z}} \hookrightarrow \widehat{A} = \widehat{\mathbb{Z}}(\zeta)^{C_q}$ is the canonical inclusion then:

(63.3) There exists a representation $\widehat{\rho} : G(p,q) \to \mathcal{T}_q(\widehat{A}, \pi)^*$ such that

$$i^*(\widehat{\rho}) \cong \widehat{\psi}.$$

As a consequence, the group ring $\widehat{\mathbb{Z}}[G(p,q)]$ of $G(p,q)$ with p-adic coefficients can be described as the fibre product:

(63.4)
$$\begin{cases} \widehat{\mathbb{Z}}[G(p,q)] & \xrightarrow{\widehat{\rho}_*} & \mathcal{T}_q(\widehat{A},\pi) \\ \downarrow & & \downarrow \\ \widehat{\mathbb{Z}}[C_q] & \longrightarrow & \mathbb{F}_p[C_q]. \end{cases}$$

We note that the ring homomorphism $\widehat{\rho}_* : \widehat{\mathbb{Z}}[G(p,q)] \to \mathcal{T}_q(\widehat{A},\pi)$ in (63.4) can also be regarded as the continous extension, in the p-adic topology, of the ring homomorphism $\rho_* : \mathbb{Z}[G(p,q)] \to \mathcal{T}_q(A,\pi)$ induced by the representation ρ.

Chapter Eight

Galois modules

In the last chapter we showed the existence of a ring isomorphism

$$\rho_* : \mathcal{C}_q(I^*, \theta) \to \mathcal{T}_q(A, \pi).$$

It follows that the module categories of these rings are equivalent. As we shall see, both rings are hereditary so that, in particular, submodules of either ring are projective. These two descriptions are complementary. For most purposes we prefer to work with the row module description of submodules of $\mathcal{T}_q(A, \pi)$. However, for some purposes the corresponding submodules of $\mathcal{C}_q(I^*, \theta)$ are more convenient. These are best described by means of the Galois action of C_q on $\mathbb{Z}[C_p]$ and various of its submodules and quotient modules.

§64: Galois modules:

If M is a module over $\mathbb{Z}[C_p]$ then by a *Galois structure* on M we mean an additive automorphism $\Theta : M \to M$ such that $\Theta^q = \mathrm{Id}_M$ and $\Theta(m \cdot x) = \Theta(m) \cdot \theta(x)$ for all $m \in M$ where θ is our chosen automorphism of C_p. By a *Galois lattice* we shall mean a pair (M, Θ) where M is a lattice over $\mathbb{Z}[C_p]$ and Θ is a Galois structure on M. The Galois lattice (M, Θ) becomes a (right) lattice over Λ via the action

$$m \cdot x^r y^s = \Theta^{-s}(m \cdot x^r).$$

For any Galois lattice (M, Θ) there is an isomorphism of abelian groups

$$\Psi : \mathbb{Z}[C_q] \otimes (M, \Theta) \xrightarrow{\simeq} i_*(M)(= M \otimes_{\mathbb{Z}[C_p]} \Lambda)$$

defined by taking $\Psi(y^b \otimes m) = \Theta^{-b}(m) \otimes y^b$. It is straightforward to check that Ψ is also a homomorphism of (right) Λ-modules. We obtain:

Proposition 64.1: *$\mathbb{Z}[C_q] \otimes (M, \Theta) \cong i_*(M)$ for any Galois lattice (M, Θ).*

$\mathbb{Z}[C_p]$ acquires the structure of a Λ-module on extending the action of C_p to one of $G(p,q)$ by requiring y to act on the right by conjugation thus;

$$x \cdot y = y^{-1}xy = x^{-a}.$$

Extending this action linearly, we get

$$\left(\sum_{r=0}^{p-1} c_r x^r\right) \cdot y = \sum_{r=0}^{p-1} c_r x^{-ar}.$$

We denote the Λ-module so obtained by $\overline{\mathbb{Z}[C_p]}$. More generally, if $J \subset \mathbb{Z}[C_p]$ is a $\mathbb{Z}[C_p]$-submodule invariant under conjugation by y, that is,

$$y^{-1}Jy = J$$

then J becomes a Λ-module, denoted by $\overline{J}$, with y acting by the same formula. Finally, if $J \subset \mathbb{Z}[C_p]$ is an invariant submodule, the quotient $\mathbb{Z}[C_p]/J$ is also a Λ-module. As an example, put $I = \mathrm{Ker}(\epsilon)$ where $\epsilon : \mathbb{Z}[C_p] \to \mathbb{Z}$ is the augmentation homomorphism; then

(64.2) I is invariant under conjugation by y giving the Λ-module $\overline{I}$.

Evidently the set $\{(x^b - 1)\}_{1 \le b \le p-1}$ is a $\mathbb{Z}$-basis for $\overline{I}$. For future reference we note that the right action of y on this basis is given by:

(64.3) $$(x^b - 1) \cdot y = (x^{-ab} - 1).$$

(64.4) $(x-1)^k I$ is invariant under conjugation by y.

In consequence of (64.4) we obtain the Λ-module $\overline{(x-1)^k I}$. The ideal (Σ) is also invariant where $\Sigma = \sum_{r=0}^{p-1} x^r$; hence

(64.5) $I^* = \mathbb{Z}[C_p]/(\Sigma)$ acquires the structure of a Λ module, denoted by $\overline{I^*}$.

As we shall see in the next section, in contrast to the $\mathbb{Z}[C_p]$-isomorphism $I_C^* \cong I_C$, $\overline{I_C^*}$ is *not isomorphic to* $\overline{I_C}$ *over* Λ. Likewise, $\overline{(x-1)^k I_C}$ is *not, in general, isomorphic* to either $\overline{I_C^*}$ or $\overline{I_C}$. We note that $\overline{I_C^*}$ can be recognized up to isomorphism by the following property:

Proposition 64.6: *Let M be a Λ-lattice satisfying the following three conditions:*

(i) *there exists $\mu \in M$ such that $\mu \cdot y = \mu$ and $M = \mathrm{span}_{\mathbb{Z}}\{\mu \cdot x^r | 0 \le r \le p-1\}$;*
(ii) $\mathrm{rk}_{\mathbb{Z}}(M) = p-1$;
(iii) $m \cdot \Sigma_x = 0$ *for each $m \in M$.*

Then $M \cong_\Lambda \overline{I^*_C}$ and $\{\mu \cdot x^r | 0 \le r \le p-2\}$ is a $\mathbb{Z}$-basis for M.

Proof. Conditions (ii) and (iii) above are satisfied for $\overline{I^*_C}$. Let $\natural : \overline{\mathbb{Z}}[C_p] \to \overline{I^*_C}$ be the natural mapping and put $\eta = \natural(1)$. Then $\eta \cdot y = \eta$ and $\{\eta \cdot x^r | 0 \le r \le p-2\}$ is a $\mathbb{Z}$-basis for $\overline{I^*_C}$. Now suppose that M is a Λ-lattice satisfying conditions (i), (ii) and (iii) and consider the homomorphism of abelian groups $\Psi : \overline{I^*_C} \to M$ defined on the basis $\{\eta \cdot x^r | 0 \le r \le p-1\}$ by $\Psi(\eta \cdot x^r) = \mu \cdot x^r$. As $M = \mathrm{span}_{\mathbb{Z}}\{\mu \cdot x^r | 0 \le r \le p-1\}$ then Ψ is necessarily surjective and as $\mathrm{rk}_{\mathbb{Z}}(\overline{I^*_C}) = \mathrm{rk}_{\mathbb{Z}}(M) = p-1$ then Ψ is bijective and $\{\mu \cdot x^r | 0 \le r \le p-2\}$ is a $\mathbb{Z}$-basis for M. Evidently Ψ is now an isomorphism of $\mathbb{Z}[C_p]$-modules. Moreover from the identities $\eta \cdot y = \eta$ and $\mu \cdot y = \mu$ it follows easily that Ψ is also an isomorphism over Λ. □

Noting that $i_*(\mathbb{Z}[C_p]) = \Lambda$ we now see from (64.1) that:

(64.7) $$\mathbb{Z}[C_q] \otimes \overline{\mathbb{Z}[C_p]} \cong \Lambda.$$

Let Z be a set with $|Z| = q$ on which $\widehat{C_q} = \{1, \theta, \ldots, \theta^{q-1}\}$ acts transitively on the left via $z \mapsto \theta_*(z)$; for each $z \in Z$ let $F(z)$ be the free $\mathbb{Z}[C_p]$-module of rank 1 with basis element $[z]$ and put $F(Z) = \bigoplus_{z \in Z} F(z)$. Then $F(Z)$ is a Galois module with Galois structure Θ where

$$\Theta([z] \cdot x^r) = [\theta_*(z)] \cdot \theta(x^r)$$

and it is straightforward to see that, as Λ-modules, $F(Z) \cong \Lambda$. More generally, suppose that Z is a finite set on which $\widehat{C_q}$ acts freely on the left and denote by $Z = Z_1 \coprod \ldots \cdots \coprod Z_m$ the partition of Z into disjoint orbits where each $|Z_i| = q$. By the above, $F(Z_i) \cong \Lambda$ for each i so that $F(Z) = \bigoplus_{i=1}^m F(Z_i) \cong \Lambda^m$; that is:

(64.8) If Z is a finite set on which $\widehat{C_q}$ acts freely with m orbits then $F(Z) \cong \Lambda^m$.

As an example of (64.8) we have:

Proposition 64.9: $\overline{I_C} \otimes [\Sigma_y) \cong \Lambda^d$.

Proof. Note that $i^*(\overline{I_C} \otimes [\Sigma_y)) \cong I_C \otimes \mathbb{Z}[C_p] \cong \bigoplus_{e=1}^{p-1} F(e)$ where $F(e)$ is the free module of rank 1 over $\mathbb{Z}[C_p]$ on the basis element $(x^e - 1) \otimes \Sigma_y$. Now $\widehat{C_q} = \{\mathrm{Id}, \theta, \theta^2, \ldots, \theta^{q-1}\}$ acts freely on $Z = \{(x^e - 1) \otimes \Sigma_y | 1 \le e \le p-1\}$ via the action

$$\theta_*((x^e - 1) \otimes \Sigma_y) = (\theta(x^e) - 1) \otimes \Sigma_y$$

under which Z decomposes as a disjoint union $Z_1 \coprod \ldots\ldots\ldots \coprod Z_d$ of $d = \frac{(p-1)}{q}$ cyclic orbits. In the above notation, $\overline{I_C} \otimes [\Sigma_y) \cong \bigoplus_{r=1}^{d} F(Z_r) \cong \Lambda^d$. □

Corollary 64.10: $\overline{I_C} \otimes [y - 1) \cong \Lambda^{d(q-1)}$.

Proof. Applying $\overline{I_C} \otimes -$ to the sequence $0 \to [y - 1) \to \Lambda \to [\Sigma_y) \to 0$ gives an exact sequence

$$0 \to \overline{I_C} \otimes [y-1) \to \overline{I_C} \otimes \Lambda \to \overline{I_C} \otimes [\Sigma_y) \to 0.$$

As $\overline{I_C} \otimes [\Sigma_y) \cong \Lambda^d$ this latter sequence splits. Hence $\overline{I_C} \otimes [y-1) \oplus \Lambda^d \cong \Lambda^{p-1}$ so that $\overline{I_C} \otimes [y-1)$ is stably free of rank $p-d-1$. As Λ satisfies the Eichler condition then, by the Swan-Jacobinski Theorem $\overline{I_C} \otimes [y - 1) \cong \Lambda^{p-d-1}$. However $p - d - 1 = d(q - 1)$ and so $\overline{I_C} \otimes [y - 1) \cong \Lambda^{d(q-1)}$ as claimed. □

For any Λ-lattices A, B, $(A \otimes B)^* \cong A^* \otimes B^*$. As Λ and $[y-1)$ are self-dual then:

Corollary 64.11: $\overline{I_C^*} \otimes [y - 1) \cong \Lambda^{d(q-1)}$.

It is a consequence of Frobenius reciprocity, (35.4), that $M \otimes \Lambda \cong \Lambda^m$ whenever M is a Λ-lattice with $\mathrm{rk}_{\mathbb{Z}}(M) = m$. In particular:

(64.12) $$\overline{I_C^*} \otimes \Lambda \cong \Lambda^{(p-1)}.$$

There is a variation on the argument of (64.8) which we record for future reference. Again we note that the action by conjugation of C_q on the non-trivial elements $\{x^r | 1 \le r \le p-1\}$ of C_p is free with $d = (p-1)/q$ orbits whilst $1 = x^0$ is the unique fixed point. Consequently the action by conjugation of C_q on the basis elements $\{(x^r - 1)_{1 \le r \le p-1}\}$ of $\overline{I_C}$ is also free with $d = (p-1)/q$ orbits. As a consequence if $j : \mathbb{Z}[C_q] \hookrightarrow \Lambda$ denotes the inclusion we have the following:

(64.13) $$j^*(\overline{I_C}) \cong \mathbb{Z}[C_q]^{(p-1)/q}.$$

§65: The Galois action of y:

In the representation ρ of (58.11), whilst the quasi-triangularity of $\rho(x^{-1})$ is evident by construction, that of $\rho(y^{-1})$ is known only implicitly from (58.15). We elicit some explicit information on the form of $\rho(y^{-1})$. For $0 \leq k \leq q-2$ define

$$U(k) = \text{span}_A\{(\zeta-1)^r | k+1 \leq r \leq q-1\}$$

and put $U(k) = 0$ for $q-1 \leq k$. Recalling that $(\zeta-1)^q \in (\pi)$, it is straightforward to check that:

(65.1) $$U(k)U(l) \subset U(k+l+1) + (\pi).$$

Now consider the Galois action given by $\Theta(\zeta) = \zeta^a$.

Proposition 65.2: *For each k, $1 \leq k \leq q-1$ there are elements $\upsilon(k) \in U(k)$ and $\pi(k) \in (\pi)$ such that $\Theta[(\zeta-1)^k] = a^k(\zeta-1)^k + \upsilon(k) + \pi(k)$.*

Proof. Observe that $\Theta(\zeta-1) = \Theta(\zeta) - 1 = \zeta^a - 1$ and that

$$\begin{aligned}\zeta^a - 1 &= ((\zeta-1)+1)^a - 1 \\ &= a(\zeta-1) + \sum_{s=2}^{a} \binom{a}{s} (\zeta-1)^s.\end{aligned}$$

Let $\mathcal{P}(k)$ be the statement for $\Theta[(\zeta-1)^k]$. Then $\mathcal{P}(1)$ is verified on putting

$$\upsilon(1) = \sum_{s=2}^{a} \binom{a}{s} (\zeta-1)^s \quad \text{and} \quad \pi(1) = 0.$$

Suppose $\mathcal{P}(r)$ is true for $1 \leq r \leq k < q-1$. As Θ is a ring homomorphism then

$$\begin{aligned}\Theta[(\zeta-1)^{k+1}] &= \Theta(\zeta-1)\cdot\Theta[(\zeta-1)^k] \\ &= [a(\zeta-1)+\upsilon(1)]\cdot[a^k(\zeta-1)^k + \upsilon(k) + \pi(k)] \\ &= a^{k+1}(\zeta-1)^{k+1} + \Upsilon + \Psi\end{aligned}$$

where $$\begin{cases}\Upsilon = a^k\upsilon(1)(\zeta-1)^k + a(\zeta-1)\upsilon(k) + \upsilon(1)\upsilon(k) \\ \Psi = [a(\zeta-1)+\upsilon(1)]\pi(k).\end{cases}$$

Clearly $\Psi \in (\pi)$ whilst $\Upsilon \in U(k+1) + (\pi)$ by (65.1). Thus we have

$$\Upsilon + \Psi = v(k+1) + \pi(k+1)$$

for some $v(k+1) \in U(k+1)$ and $\pi(k+1) \in (\pi)$. Hence

$$\Theta[(\zeta-1)^{k+1}] = a^{k+1}(\zeta-1)^{k+1} + v(k+1) + \pi(k+1)$$

verifying $\mathcal{P}(k+1)$. □

Any $Y \in M_q(A, \pi)$ can be written uniquely as a sum

(65.3) $$Y = \Delta(Y) + U(Y) + L(Y)$$

where $\Delta(Y)$ is diagonal, $U(Y)$ is strictly upper triangular and $L(Y)$ is strictly lower triangular. Moreover, as $Y \in \mathcal{T}_q(A, \pi)$ then $L(Y) = \pi L'(Y)$ for some strictly lower triangular matrix $L'(Y)$. If $\mu_0, \mu_1, \ldots, \mu_{q-1} \in A$ we denote by $\Delta(\mu_{q-1}, \ldots, \mu_0)$ the diagonal $q \times q$ matrix

$$\Delta(\mu_{q-1}, \ldots, \mu_0) = \begin{pmatrix} \mu_{q-1} & & & & \\ & \mu_{q-2} & & & \\ & & \ddots & & \\ & & & \mu_1 & \\ & & & & \mu_0 \end{pmatrix}.$$

It follows from (65.2) that, with respect to the basis $\{(\zeta-1)^{q-k}\}_{1\le k\le q}$for I_C^*, the matrix $M(\Theta)$ of Θ takes the form $M(\Theta) = \Delta(a^{q-1}, a^{q-2}, \ldots, a, 1) + U + \Pi$ where U is strictly upper triangular and $\Pi = \pi \cdot X$ for some $X \in M_q(A)$. Let $X = \Delta' + U' + L'$ be the decomposition of X given in (65.3) and write $\Delta' = \Delta(\xi_{q-1}, \xi_{q-2}, \ldots, \xi_1, \xi_0)$ for some $\xi_i \in A$. Writing $U(\Theta) = U + \pi U'$ and $L(\Theta) = \pi L'$ we see that with respect to the basis $\{(\zeta-1)^{q-k}\}_{1\le k\le q}$for I_C^*, the matrix $M(\Theta)$ takes the form

(65.4) $$M(\Theta) = \Delta(a^{q-1} + \pi\xi_{q-1}, \ldots\ldots, a + \pi\xi_1, 1 + \pi\xi_0) + U(\Theta) + L(\Theta)$$

where $U(\Theta)$ is strictly upper triangular and $L(\Theta)$ is strictly lower triangular. Denoting by $\overline{M}(\Theta)$ the reduction of $M(\Theta)$ mod π we see that:

$$\overline{M}(\Theta) = \begin{pmatrix} a^{q-1} & * & * & * & * & * \\ & a^{q-2} & * & * & * & * \\ & & & \ddots & & \\ & & & & a^1 & * \\ & & & & & 1 \end{pmatrix}.$$

As $a^{-r} = a^{q-r} \bmod p$ and in particular, $a^q = 1 \bmod p$ then:

$$\text{(65.5)} \qquad \overline{M}(\Theta^{-1}) = \begin{pmatrix} a & * & * & * & * & * \\ & a^2 & * & * & * & * \\ & & & \ddots & & \\ & & & & a^{q-1} & * \\ & & & & & a^q \end{pmatrix}.$$

§66: $R(1)$ and $R(q)$ as Galois modules:

We adhere to the notational distinction between $\mathbb{Z}[C_p]$ duals, decorated with '$*$', and Λ-duals, decorated with '$\bullet$'. We first note that $i^*(\overline{\mathbb{Z}[C_p]}) \cong \mathbb{Z}[C_p]$. Hence

$$\text{(66.1)} \qquad \mathbb{Z}[C_p] \cong \mathrm{Hom}_{\mathbb{Z}[C_p]}(\mathbb{Z}[C_p], \mathbb{Z}[C_p]) \cong \mathrm{Hom}_{\mathbb{Z}[C_p]}(i^*(\overline{\mathbb{Z}[C_p]}, \mathbb{Z}[C_p]))$$

whilst the Eckmann-Shapiro Lemma gives an isomorphism of $\mathbb{Z}[C_p]$-modules

$$\text{(66.2)} \qquad \mathrm{Hom}_{\mathbb{Z}[C_p]}(i^*(\overline{\mathbb{Z}[C_p]}), \mathbb{Z}[C_p]) \cong \mathrm{Hom}_{\Lambda}(\overline{\mathbb{Z}[C_p]}, i_*(\mathbb{Z}[C_p])).$$

Furthermore, $i_*(\mathbb{Z}[C_p]) \cong \Lambda$ so that

$$\text{(66.3)} \qquad \mathrm{Hom}_{\Lambda}(\overline{\mathbb{Z}[C_p]}, i_*(\mathbb{Z}[C_p])) \cong \mathrm{Hom}_{\Lambda}(\overline{\mathbb{Z}[C_p]}, \Lambda) \cong (\overline{\mathbb{Z}[C_p]})^{\bullet}$$

giving an isomorphism of $\mathbb{Z}[C_p]$-modules

$$\text{(66.4)} \qquad \mathbb{Z}[C_p] \cong \mathrm{Hom}_{\Lambda}(\overline{\mathbb{Z}[C_p]}, \Lambda).$$

It is straightforward now to check that under this isomorphism, the y-action on $\mathrm{Hom}_{\Lambda}(\overline{\mathbb{Z}[C_p]}, \Lambda)$ corresponds to that in $\overline{\mathbb{Z}[C_p]}$. Thus we get:

$$\text{(66.5)} \qquad (\overline{\mathbb{Z}[C_p]})^{\bullet} \cong_{\Lambda} \overline{\mathbb{Z}[C_p]}.$$

As $(\overline{\mathbb{Z}[C_p]})^{\bullet} \cong \overline{\mathbb{Z}[C_p]}$ and $(\overline{\mathbb{Z}})^* \cong \overline{\mathbb{Z}}$, the Λ-dual of the exact sequence

$$0 \to \overline{I} \to \overline{\mathbb{Z}[C_p]} \to \overline{\mathbb{Z}} \to 0$$

takes the form $(0 \to \overline{\mathbb{Z}} \to \overline{\mathbb{Z}[C_p]} \to (\overline{I})^{\bullet} \to 0)$. Comparison with the exact sequence $(0 \to \overline{\mathbb{Z}} \to \overline{\mathbb{Z}[C_p]} \to \overline{I^*} \to 0)$ gives $\overline{I^*} \cong \overline{\mathbb{Z}[C_p]}/\overline{\mathbb{Z}} \cong (\overline{I})^{\bullet}$. Hence

$$\text{(66.6)} \qquad (\overline{I^*})^{\bullet} \cong (\overline{I})^{\bullet\bullet}.$$

As $(\overline{I})^{\bullet\bullet} \cong \overline{I}$ then:

(66.7) $$(\overline{I^*})^{\bullet} \cong \overline{I}.$$

Proposition 66.8: $R(q) \cong \overline{I^*}$.

Proof. Define $\mu = (\mu_{ij}) \in \mathcal{T}_q(A, \pi)$ where $\mu_{ij} = \delta_{iq}\delta_{jq}$. Clearly $\mu \in R(q)$ and, appealing to (65.5) regarding the action of y, it straightforward to check that $(R(q), \mu)$ satisfies the conditions of (64.6) above. □

It follows from (66.8) that $R(q)^{\bullet} \cong (\overline{I^*})^{\bullet}$. However, by (53.3), $R(1) \cong R(q)^{\bullet}$ and $(\overline{I^*})^{\bullet} \cong \overline{I}$. Consequently we have:

(66.9) $$R(1) \cong \overline{I}.$$

If (M, Θ) is a Galois module then $\mathrm{End}_{\mathbf{Z}[C_p]}(M)$ itself becomes a Galois module under the action

$$^{\Theta}T = \Theta \circ T \circ \Theta^{-1}.$$

Moreover we then see that:

(66.10) $$\mathrm{End}_{\Lambda}(M, \Theta) = \mathrm{End}_{\mathbf{Z}[C_p]}(M)^{\Theta} = \{T \in \mathrm{End}_{\mathbf{Z}[C_p]} : {}^{\Theta}T = T\}.$$

We know that $\mathrm{End}_{\mathbf{Z}[C_p]}(I_C^*) \cong I_C^* \cong \mathbb{Z}(\zeta_p)$ and that $A = \mathbb{Z}(\zeta_p)^{C_q}$. Hence

(66.11) $$\mathrm{End}_{\Lambda}(\overline{I_C^*}) \cong A.$$

§67: The theorem of Auslander-Rim:

A ring $\mathcal{T}$ is said to be *hereditary* when every right ideal of $\mathcal{T}$ is a projective module. An equivalent formulation (cf [39], [10] p. 14) is:

(67.1) $\mathcal{T}$ is hereditary$\Longleftrightarrow$ every right submodule of a free $\mathcal{T}$-module is projective.

We first show:

Proposition 67.2: *There exists $c \in R$ such that $\sum_{r=0}^{q-1} \theta^r(c) = 1$.*

Proof. There is a natural free action of θ on the set $\{\zeta, \zeta^2, \ldots, \zeta^{p-1}\}$. Choose a set of orbit representatives $\{c_1, \ldots, c_d\} \subset \{\zeta, \zeta^2, \ldots, \zeta^{p-1}\}$ for this action; then

$$\sum_{r=0}^{q-1} \theta^r \left(\sum_{s=1}^{d} c_s \right) = \sum_{t=1}^{p-1} \zeta^t = -1.$$

Thus putting $c = -\sum_{s=1}^{d} c_s$ we see that $\sum_{r=0}^{q-1} \theta^r(c) = 1$. □

We shall write $\mathcal{C} = \mathcal{C}_q(R, \theta)$. In what follows M will denote a $\mathcal{C}$-module such that:

(*) M admits an imbedding $M \hookrightarrow F$ as a $\mathcal{C}$-submodule of a free $\mathcal{C}$-module F.

We denote by $i : R \hookrightarrow \mathcal{C}$ the obvious inclusion of R as a subring of $\mathcal{C}$. Then $i^*(M)$ is an R-submodule of the free R-module $i^*(\mathcal{F})$. As R is an hereditary ring then $i^*(M)$ is projective over R. Let

$$(\mathcal{E}) \qquad 0 \to X \to E \overset{p}{\to} M \to 0$$

be an exact sequence of $\mathcal{C}$-modules in which E is free over $\mathcal{C}$. In particular, the sequence $i^*(\mathcal{E})$ obtained by restriction of scalars is also exact. As $i^*(M)$ is projective over R then $i^*(\mathcal{E})$ splits and there exists an R-homomorphism $s : i^*(M) \to i^*(E)$ such that $p \circ s = \mathrm{Id}_M$. For $0 \le r \le q-1$ define $s^{[r]} : i^*(M) \to i^*(E)$ by

$$s^{[r]}(m) = s[m\mathbf{y}^r c]\mathbf{y}^{-r}.$$

We claim that:

Proposition 67.3: *$s^{[r]} : i^*(M) \to i^*(E)$ is R-linear.*

Proof. Clearly $s^{[r]}$ preserves addition. Now let $m \in M$ and $\lambda \in R$; then

$$\begin{aligned} s^{[r]}(m\lambda) &= s[m\lambda\mathbf{y}^r c]\mathbf{y}^{-r} \\ &= s[m\mathbf{y}^r\mathbf{y}^{-r}\lambda\mathbf{y}^r c]\mathbf{y}^{-r} \\ &= s[m\mathbf{y}^r\theta^{-r}(\lambda)c]\mathbf{y}^{-r}. \end{aligned}$$

As R is commutative then $\theta^{-r}(\lambda)c = c\theta^{-r}(\lambda)$ so that

$$s^{[r]}(m\lambda) = s[m\mathbf{y}^r c\theta^{-r}(\lambda)]\mathbf{y}^{-r}.$$

However, s is R-linear so that

$$\begin{aligned} s^{[r]}(m\lambda) &= s[m\mathbf{y}^r c]\theta^{-r}(\lambda)\mathbf{y}^{-r} \\ &= s[m\mathbf{y}^r c]\mathbf{y}^{-r}\lambda\mathbf{y}^r\mathbf{y}^{-r} \\ &= s[m\mathbf{y}^r c]\mathbf{y}^{-r}\lambda \\ &= s^{[r]}(m)\lambda. \end{aligned}$$

Thus $s^{[r]}$ is R-linear as claimed. □

Observe further that:

(67.4) $$s^{[r]}(m\mathbf{y}) = \begin{cases} s^{[r+1]}(m)\mathbf{y} & 0 \le r \le q-2 \\ s^{[0]}(m)\mathbf{y} & r = q-1. \end{cases}$$

Consequently, putting$\sigma(m) = \sum_{r=0}^{q-1} s^{[r]}(m)$ we see that:

(67.5) $\sigma : M \to E$ is $\mathcal{C}$-linear.

Computing we see that

$$\begin{aligned} p \circ \sigma(m) &= p\left(\sum_{r=0}^{q-1} s(m\mathbf{y}^r c)\mathbf{y}^{-r}\right) \\ &= \sum_{r=0}^{q-1} p \circ s(m)\mathbf{y}^r c\mathbf{y}^{-r} \\ &= m \cdot \left(\sum_{r=0}^{q-1} \mathbf{y}^r c\mathbf{y}^{-r}\right) \\ &= m \cdot \left(\sum_{r=0}^{q-1} \theta^r(c)\right). \end{aligned}$$

From (67.2) we see that $p \circ \sigma(m) = m$ and hence:

(67.6) $p \circ \sigma = \mathrm{Id}_M$.

Thus the sequence $(\mathcal{E})$ splits over $\mathcal{C}$ and so we have:

(67.7) $E \cong_{\mathcal{C}} M \oplus X$.

As M is thereby a direct summand of the free $\mathcal{C}$-module E we have shown

(67.8) If M is a submodule of a free $\mathcal{C}$-module then M is projective over $\mathcal{C}$.

It follows from Kaplansky's Theorem (67.1) that:

(67.9) The cyclic ring $\mathcal{C}_q(R,\theta)$ is hereditary.

Consequently, from the isomorphism $\mathcal{C}_q(R,\theta) \cong \mathcal{T}_q(A,\pi)$ it follows that

(67.10) $\mathcal{T}_q(A,\pi)$is an hereditary ring.

In this particular case we may regard (67.10) as an explicit form of the theorem of Auslander and Rim ([3], [14]).

§68: Galois module description of the row modules:

In this section we interpret the row modules as Galois modules by proving that

$$(\clubsuit) \qquad R(k) \cong \overline{(x-1)^{k-1}I} \text{ for } 1 \leq k \leq q-1.$$

We have seen that $\mathcal{C}_q(I^*)$ is hereditary. Hence

(68.1) Each $\overline{(x-1)^k I}$ is projective over $\mathcal{C}_q(I^*)$.

$\mathcal{T}_q(A,\pi)$ is also hereditary. Consequently:

(68.2) Each $R(i)$ is projective over $\mathcal{T}_q(A,\pi)$.

We choose to regard each $\overline{(x-1)^k I}$ as a projective module over $\mathcal{T}_q(A,\pi)$ via the isomorphism $\mathcal{C}_q(I^*) \cong \mathcal{T}_q(A,\pi)$. Before proceeding we note that:

Proposition 68.3: *Let M be a Λ-module whose underlying abelian group is the field $\mathbb{F}_p$. Then x acts trivially on M.*

Let $\mathbb{F}_p[a^k]$ denote the Λ-module whose underlying abelian group is $\mathbb{F}_p$ and on which x acts trivially and y acts as multiplication by a^k. Now let $\Gamma \in \mathcal{T}_q(A,\pi)$ be the matrix

$$\Gamma = \begin{pmatrix} 0 & 1 & 0 & 0 & \dots & 0 & 0 \\ 0 & 0 & 1 & 0 & \dots & 0 & 0 \\ \vdots & \vdots & \vdots & \vdots & & \vdots & \vdots \\ 0 & 0 & 0 & 0 & \dots & 1 & 0 \\ 0 & 0 & 0 & 0 & \dots & 0 & 1 \\ \pi & 0 & 0 & 0 & \dots & 0 & 0 \end{pmatrix}.$$

Formally $\Gamma = (\gamma_{ij})$ where $\gamma_{ij} = \begin{cases} \delta_{i+1,j} & 1 \leq i \leq q-1 \\ \pi\delta_{1,j} & i = q \end{cases}$.

Let $\Gamma_* : \mathcal{T}_q(A,\pi) \to \mathcal{T}_q(A,\pi)$ be the $\mathcal{T}_q(A,\pi)$-homomorphism $\Gamma_*(\beta) = \Gamma \cdot \beta$. Clearly $\Gamma^q = \pi \cdot \mathrm{I}_q$. Define $\Gamma_* : \mathcal{T}_q(A,\pi) \to \mathcal{T}_q(A,\pi)$ by $\Gamma_*(\beta) = \Gamma \cdot \beta$. Then Γ_* is evidently injective as π is a nonzero element of the integral domain I_C^*.

Proposition 68.4: *There is an exact sequence of Λ-modules*

$$0 \to R(1) \xrightarrow{\Gamma_*} R(q) \longrightarrow \mathbb{F}_p[1] \to 0.$$

Proof. Write a typical element $\beta \in R(1)$ as

$$\beta = \begin{pmatrix} b_1 & b_2 & \dots & b_{q-1} & b_q \\ 0 & 0 & \dots & 0 & 0 \\ \vdots & \vdots & & \vdots & \vdots \\ 0 & 0 & \dots & 0 & 0 \\ 0 & 0 & \dots & 0 & 0 \\ 0 & 0 & \dots & 0 & 0 \end{pmatrix}$$

so that

$$\Gamma_*(\beta) = \begin{pmatrix} 0 & 0 & \dots & 0 & 0 \\ 0 & 0 & \dots & 0 & 0 \\ \vdots & \vdots & & \vdots & \vdots \\ 0 & 0 & \dots & 0 & 0 \\ 0 & 0 & \dots & 0 & 0 \\ \pi b_1 & \pi b_2 & \dots & \pi b_{q-1} & \pi b_q \end{pmatrix}.$$

Thus $R(1) \cong \Gamma_*(R(1)) \subset R(q)$. However, a typical element $\gamma \in R(q)$ has the form

$$\gamma = \begin{pmatrix} 0 & 0 & \dots & 0 & 0 \\ 0 & 0 & \dots & 0 & 0 \\ \vdots & \vdots & & \vdots & \vdots \\ 0 & 0 & \dots & 0 & 0 \\ \pi c_1 & \pi c_2 & \dots & \pi c_{q-1} & c_q \end{pmatrix} \in R(q)$$

which differs from an element of $\Gamma_*(R(1))$ only in the $(q,q)^{th}$ entry. As abelian groups, $R(q)/\Gamma_*(R(1)) \cong A/\pi \cong \mathbb{F}_p$. Finally, $\rho(y^{-1})$ takes the following form in $\mathcal{T}_q(A/\pi)$,

$$\rho(y^{-1}) = \begin{pmatrix} \overline{a} & * & * & * & * & * \\ & \overline{a}^2 & * & * & * & * \\ & & \overline{a}^3 & * & * & * \\ & & & \ddots & & \\ & & & & \overline{a}^{q-1} & * \\ & & & & & 1 \end{pmatrix}.$$

Hence y acts trivially on the right of the $(q,q)^{th}$ entry and $R(q)/\Gamma_*(R(1)) \cong_\Lambda \mathbb{F}_p(1)$ giving an exact sequence $0 \to R(1) \overset{\Gamma_*}{\hookrightarrow} R(q) \to \mathbb{F}_p(1) \to 0$ as claimed. □

Rather more obviously we have:

Proposition 68.5: *There is an exact sequence of Λ-modules*

$$0 \to \overline{(\zeta-1)I^*} \xrightarrow{\zeta-1} \overline{I^*} \longrightarrow \mathbb{F}_p[1] \to 0$$

Proposition 68.6: $R(1) \cong_\Lambda \overline{(\zeta-1)I^*}$.

Proof. Regarding (68.4) and (68.5) as sequences of $\mathcal{T}_q(A,\pi)$-modules then $R(q)$ and $\overline{I^*}$ are projective. From Schanuel's Lemma we conclude that

$$R(1) \oplus \overline{I^*} \cong_{\mathcal{T}_q(A,\pi)} \overline{(\zeta-1)I^*} \oplus R(q).$$

By (66.8) above, $R(q) \cong \overline{I^*}$. Hence $R(1) \cong_{\mathcal{T}_q(A,\pi)} \overline{(\zeta-1)I^*}$ by the cancellation theorem of (51.10). It follows from (59.10) that $R(1) \cong_\Lambda \overline{(\zeta-1)I^*}$. □

In similar fashion to (68.4) we have:

Proposition 68.7: *For $1 \le k \le q-1$ there is an exact sequence of Λ-modules*

$$0 \to R(k+1) \xrightarrow{\Gamma_*} R(k) \longrightarrow \mathbb{F}_p[a^k] \to 0.$$

Proof. First note that $\Gamma_*(R(k+1)) \subset R(k)$ for $1 \le k \le q-1$. To see this in detail, consider a typical element

$$\beta = \begin{pmatrix} 0 & \dots & 0 & 0 & 0 & 0 & \dots & 0 \\ \vdots & & \vdots & \vdots & \vdots & \vdots & & \vdots \\ 0 & \dots & 0 & 0 & 0 & 0 & \dots & 0 \\ 0 & \dots & 0 & 0 & 0 & 0 & \dots & 0 \\ \pi b_1 & \dots & \pi b_{k-1} & \pi b_k & b_{k+1} & b_{k+2} & \dots & b_q \\ \vdots & & \vdots & \vdots & \vdots & \vdots & & \vdots \\ 0 & \dots & 0 & 0 & 0 & 0 & \dots & 0 \end{pmatrix} \in R(k+1).$$

Then

$$\Gamma_*(\beta) = \begin{pmatrix} 0 & \dots & 0 & 0 & 0 & 0 & \dots & 0 \\ \vdots & & \vdots & \vdots & \vdots & \vdots & & \vdots \\ 0 & \dots & 0 & 0 & 0 & 0 & \dots & 0 \\ \pi b_1 & \dots & \pi b_{k-1} & \pi b_k & b_{k+1} & b_{k+2} & \dots & b_q \\ 0 & \dots & 0 & 0 & 0 & 0 & \dots & 0 \\ \vdots & & \vdots & \vdots & \vdots & \vdots & & \vdots \\ 0 & \dots & 0 & 0 & 0 & 0 & \dots & 0 \end{pmatrix} \in R(k).$$

Thus $R(k+1) \cong \Gamma_*(R(k+1)) \subset R(k)$. A typical element $\gamma \in R(k)$ has the form

$$\gamma = \begin{pmatrix} 0 & \dots & 0 & 0 & 0 & 0 & \dots & 0 \\ \vdots & & \vdots & \vdots & \vdots & \vdots & & \vdots \\ 0 & \dots & 0 & 0 & 0 & 0 & \dots & 0 \\ \pi c_1 & \dots & \pi c_{k-1} & c_k & c_{k+1} & c_{k+2} & \dots & c_q \\ 0 & \dots & 0 & 0 & 0 & 0 & \dots & 0 \\ 0 & \dots & 0 & 0 & 0 & 0 & \dots & 0 \\ \vdots & & \vdots & \vdots & \vdots & \vdots & & \vdots \\ 0 & \dots & 0 & 0 & 0 & 0 & \dots & 0 \end{pmatrix} \in R(k)$$

which differs from a typical element of $\Gamma_*(R(k+1))$ only in the $(k,k)^{th}$ entry, showing that, as abelian groups, $R(k)/\Gamma_*(R(k+1)) \cong A/\pi \cong \mathbb{F}_p$.

Finally, from (65.5) the reduction $\rho(y^{-1}) \in \mathcal{T}_q(A/\pi)$ takes the form

$$\rho(y^{-1}) = \begin{pmatrix} \overline{a} & * & * & * & * & * & * & * & * \\ & \overline{a}^2 & * & * & * & * & * & * & * \\ & & \ddots & & & & & & \\ & & & & \overline{a}^k & * & * & * & * \\ & & & & & & \ddots & & \\ & & & & & & & \overline{a}^{q-1} & * \\ & & & & & & & & 1 \end{pmatrix}.$$

Hence in the right action in the quotient, y acts on the $(k,k)^{th}$ entry as multiplication by $\overline{a}^k$. Thus, as Λ-modules, $R(k)/\Gamma_*(R(k+1)) \cong \mathbb{F}_p(\overline{a}^k)$ so, as claimed, we get an exact sequence $0 \to R(k+1) \overset{\Gamma_*}{\hookrightarrow} R(k) \to \mathbb{F}_p(\overline{a}^k) \to 0$. □

Theorem 68.8: *$R(k) \cong_\Lambda \overline{(\zeta-1)^k I^*}$ for $1 \le k \le q-1$.*

Proof. By induction on k. The statement is true for $k = 1$ by (68.6). As in §65, for $0 \le k \le q-2$ define

$$U(k) = \mathrm{span}_A\{(\zeta-1)^r | k+1 \le r \le q-1\}$$

and put $U(k) = 0$ for $q-1 \le k$. Now consider the Galois action given by $\Theta(\zeta) = \zeta^a$ and recall that, in (65.2) we showed that for each k, $1 \le k \le q-1$ there are elements $v(k) \in U(k)$ and $\pi(k) \in (\pi)$ such that

$$\Theta[(\zeta-1)^k] = a^k(\zeta-1)^k + v(k) + \pi(k).$$

It follows that for $1 \le k \le q-1$ there is an exact sequence of Λ-modules

(68.9) $$0 \to \overline{(\zeta-1)^{k+1}I^*} \xrightarrow{\zeta-1} \overline{(\zeta-1)^k I^*} \longrightarrow \mathbb{F}_p[a^k] \to 0.$$

We compare this with the exact sequence

(68.10) $$0 \to R(k+1) \xrightarrow{\Gamma_*} R(k) \longrightarrow \mathbb{F}_p[a^k] \to 0.$$

In the first instance, we regard both of these sequences as defined over $\mathcal{T}_q(A,\pi)$ in which context both $R(k)$ and $\overline{(\zeta-1)^k I^*}$ are projective. By Schanuel's Lemma

$$R(k+1) \oplus \overline{(\zeta-1)^k I^*} \cong \overline{(\zeta-1)^{k+1} I^*} \oplus R(k).$$

However, by induction, $R(k) \cong \overline{(\zeta - 1)^k I^*}$. It now follows from (51.10) that $R(k+1) \cong_{\mathcal{T}_q(A,\pi)} \overline{(\zeta - 1)^{k+1} I^*}$. Hence by (59.10) $R(k+1) \cong_\Lambda \overline{(\zeta - 1)^{k+1} I^*}$.

□

By (68.6) and (68.8) we now see that

(68.11) $$\overline{I} \cong \overline{(\zeta - 1) I^*}$$

and hence

(68.12) $$\overline{(x-1)^k I} \cong \overline{(\zeta - 1)^{k+1} I^*}.$$

From (68.8) and (68.10) we now obtain the following, which is the statement (♣) at the beginning of this section:

(68.13) $$R(k) \cong \overline{(x-1)^{k-1} I} \quad \text{for } 1 \le k \le q-1.$$

Finally note that there is an exact sequence of Λ-modules $0 \to \mathcal{T}_q \to \Lambda \to \mathbb{Z}[C_q] \to 0$. Applying the restriction of scalars functor j^* from Λ-modules to $\mathbb{Z}[C_q]$-module we obtain an exact sequence over $\mathbb{Z}[C_q]$

$$0 \to j^*(\mathcal{T}_q) \to j^*(\Lambda) \to j^*(\mathbb{Z}[C_q]) \to 0$$

in which $j^*(\mathbb{Z}[C_q]) = \mathbb{Z}[C_q]$. In particular this sequence splits so that $j^*(\mathcal{T}_q)$ is a direct summand of the free $\mathbb{Z}[C_q]$-module $j^*(\Lambda)$. As $R(i)$ is a direct summand of $\mathcal{T}_q$ we see that:

(68.14) Each $j^*(R(i))$ is projective over $\mathbb{Z}[C_q]$.

Chapter Nine

The sequencing theorem

For any group G the *augmentation ideal* I_G is defined by $I_G = \mathrm{Ker}(\epsilon)$ where $\epsilon : \mathbb{Z}[G] \to \mathbb{Z}$ is the *augmentation homomorphism* defined by the correspondence $g \mapsto 1$. For $G = G(p,q)$ we will show that there is a direct sum decomposition $I_G \cong \overline{I_C} \oplus [y-1)$. In consequence, will deduce the existence of an exact sequence

$$0 \to R(1) \to \Lambda \longrightarrow \Lambda \to R(q) \to 0.$$

We term this the *basic sequence*; from it we infer the existence, for $1 \leq k \leq q-1$, of derived sequences $0 \to R(k+1) \to P(k) \longrightarrow \Lambda \to R(k) \to 0$ where $P(k)$ is a projective module of rank 1 and for which $\bigoplus_{r=1}^{q-1} P(r) \cong \Lambda^{(q-1)}$. As a consequence it follows that $R(k)$ is a generalized syzygy of $R(1)$; in fact:

$$R(k) \in D_{2k-2}(R(1)).$$

In Chapter Fourteen, we shall give sufficient conditions under which $R(k)$ is a genuine syzygy of $R(1)$. However, as we shall see in Chapter Thirteen, this is not the case in general. Finally, as an advance on the established fact that $G(p,q)$ has cohomological period $2q$ it also follows that each $G(p,q)$ actually has free period $2q$.

§69: Basic identities:

Let $\Lambda = \mathbb{Z}[G(p,q)]$ and let $j : \mathbb{Z}[C_q] \to \Lambda$ be the ring homomorphism induced from the inclusion $C_q \hookrightarrow G(p,q)$; then we have extension and restriction of scalars functors $j_* : \mathcal{M}od_{\mathbb{Z}[C_q]} \to \mathcal{M}od_\Lambda; j^* : \mathcal{M}od_\Lambda \to \mathcal{M}od_{\mathbb{Z}[C_q]}$. Moreover, in the fibre product diagram

$$\mathfrak{S} = \begin{cases} \Lambda & \stackrel{\pi_+}{\to} & \mathcal{T}_q(A,\pi) \\ \downarrow \pi_- & & \downarrow \varphi_+ \\ \mathbb{Z}[C_q] & \stackrel{\varphi_-}{\to} & \mathbb{F}_p[C_q] \end{cases}$$

the ring $\mathcal{T}_q(A, \pi)$ acquires the structure of a Λ-module by *'coinduction'* from the ring homomorphism π_+, coinduction being the same process as 'restriction of scalars' except for the fact that π_+ is not injective. We also have a coinduction functor $\pi_-^* : \mathcal{M}od_{\mathbb{Z}[C_q]} \to \mathcal{M}od_\Lambda$ obtained from the ring homomorphism $\pi_- : \Lambda \to \mathbb{Z}[C_q]$. In most circumstances it should cause no confusion if simply write $M = \pi_+^*(M)$. However, for the purposes of clarity, when it is necessary to emphasise their Λ-module structures we shall write

$$\breve{M} = \pi_-^*(M).$$

As $\pi_- \circ j = \mathrm{Id}$ then $j^* \circ \pi_-^* = \mathrm{Id}$ so that

(69.1) $$j^*(\breve{M}) = M.$$

In particular, if $M, N \in \mathcal{M}od_{\mathbb{Z}[C_q]}$ then

(69.2) $$\breve{M} \cong_\Lambda \breve{N} \iff M \cong_{\mathbb{Z}[C_q]} N.$$

The $\mathbb{Z}[C_q]$ modules that we shall need to consider in this way are $\mathbb{Z}[C_q]$ itself, the augmentation ideal $I_q = I(C_q)$ and its $\mathbb{Z}[C_q]$ dual which we write as I_q^*. Thus we have $\breve{\mathbb{Z}}[C_q] = \pi_-^*(\mathbb{Z}[C_q])$, $\breve{I}_q = \pi_-^*(I_q)$ and $\breve{I}_q^* = \pi_-^*(I_q^*)$. Whilst I_q and I_q^* are not actually identical they are, nevertheless, isomorphic; thus we see that:

(69.3) $$\breve{I}_q \cong_\Lambda \breve{I}_q^*.$$

Whilst there is a conceptual difference between the two, for the purposes of computation we shall, when convenient, ignore the distinction between them.

Next consider the ring homomorphism $i : \mathbb{Z}[C_p] \to \Lambda$ induced from the inclusion $C_p \hookrightarrow G(p, q)$; we again have extension and restriction of scalars functors

$$i_* : \mathcal{M}od_{\mathbb{Z}[C_p]} \to \mathcal{M}od_\Lambda; \quad i^* : \mathcal{M}od_\Lambda \to \mathcal{M}od_{\mathbb{Z}[C_p]}.$$

We note the following identities where '$\cong$' means '$\cong_\Lambda$' unless otherwise qualified:

(69.4) $$i_*(\mathbb{Z}) = \breve{\mathbb{Z}}[C_q]$$

(69.5) $$i_*(\mathbb{Z}[C_p]) = \Lambda$$

(69.6) $$i_*(I^*(C_p)) \cong \mathcal{C}_q(\mathbb{Z}(\zeta_p))$$

However, $I(C_p) \cong_{\mathbb{Z}[C_p]} I^*(C_p)$ and $\mathcal{T}_q(A,\pi) \cong \mathcal{C}_q(\mathbb{Z}(\zeta_p))$ so that

(69.7) $$i_*(I(C_p)) \cong \mathcal{T}_q(A,\pi).$$

Application of i_* to the exact sequence $0 \to I(C_p) \to \mathbb{Z}[C_p] \to \mathbb{Z} \to 0$ gives the exact sequence of Λ-modules

(69.8) $$0 \to i_*(I(C_p)) \to \Lambda \to \breve{\mathbb{Z}}[C_q] \to 0$$

and hence also the exact sequence

(69.9) $$0 \to \mathcal{T}_q(A,\pi) \to \Lambda \to \breve{\mathbb{Z}}[C_q] \to 0.$$

Making the abbreviation $\mathcal{T}_q(A,\pi) = \mathcal{T}_q$, we first establish

Proposition 69.10: $\mathrm{Hom}_\Lambda(\mathcal{T}_q, \breve{\mathbb{Z}}[C_q]) = 0.$

Proof. We have $i_*(I[C_p]) \cong \mathcal{T}_q, i_*(\mathbb{Z}) \cong \mathbb{Z}[C_q]$ and $i^*(\breve{\mathbb{Z}}[C_q]) \cong \underbrace{\mathbb{Z} \oplus \cdots \oplus \mathbb{Z}}_{q}$. In particular, $i^*i_*(\mathbb{Z}) \cong \underbrace{\mathbb{Z} \oplus \cdots \oplus \mathbb{Z}}_{q}$. The Eckmann-Shapiro Theorem now gives

$$\begin{aligned}\mathrm{Hom}_\Lambda(\mathcal{T}_q, \breve{\mathbb{Z}}[C_q]) &= \mathrm{Hom}_\Lambda(i_*(I[C_p]), i_*(\mathbb{Z}))\\ &= \mathrm{Hom}_{\mathbb{Z}[C_p]}(I[C_p], i^*i_*(\mathbb{Z}))\\ &= \mathrm{Hom}_{\mathbb{Z}[C_p]}(I[C_p], \mathbb{Z})^{(q)}.\end{aligned}$$

The conclusion follows as $\mathrm{Hom}_{\mathbb{Z}[C_p]}(I[C_p], \mathbb{Z}) = 0$. □

Returning again to j_* and j^* we have:

(69.11) $$j_*(\mathbb{Z}[C_q]) = \Lambda$$

(69.12) $$j_*(\mathbb{Z}) = \overline{\mathbb{Z}[C_p]}$$

(69.13) $$j_*(I(C_q)) \cong [y-1)$$

(69.14) $$j^*(\breve{\mathbb{Z}}[C_q] = \mathbb{Z}[C_q]$$

(69.15) $$j^*(\Lambda) = \mathbb{Z}[C_q]^{(p)}$$

Applying j^* to (69.9) and appealing to (69.14), (69.15) gives an exact sequence

$$0 \to j^*(\mathcal{T}_q(A,\pi)) \to \mathbb{Z}[C_q]^{(p)} \to \mathbb{Z}[C_q] \to 0$$

which necessarily splits to give an isomorphism $j^*(\mathcal{T}_q(A,\pi)) \oplus \mathbb{Z}[C_q] \cong_{\mathbb{Z}[C_q]} \mathbb{Z}[C_q]^{(p)}$. Thus $j^*(\mathcal{T}_q(A,\pi))$ is stably free of rank $(p-1)$, so that, by (4.12) we see that:

(69.16) $$j^*(\mathcal{T}_q(A,\pi)) \cong \mathbb{Z}[C_q]^{(p-1)}.$$

As $\mathcal{T}_q(A,\pi) = \oplus_{k=1}^q R(k)$ then $j^*(\mathcal{T}_q(A,\pi)) \cong \oplus_{k=1}^q j^*(R(k))$ and hence

(69.17) $j^*(R(k))$ is projective of rank $(p-1)/q$ over $\mathbb{Z}[C_q]$.

We saw in (64.13) that $j^*(\overline{I(C_p)}) \cong \mathbb{Z}[C_q]^{(p-1)/q}$. By (66.9), $R(1) \cong \overline{I(C_p)}$; thus:

(69.18) $$j^*(R(1)) \cong \mathbb{Z}[C_q]^{(p-1)/q}.$$

We saw in (53.3) that $R(q)$ is the $\mathcal{T}_q(A,\pi)$-dual to $R(1)$; by (59.10) $R(q)$ is also Λ-dual to $R(1)$ and $j^*(R(q))$ is $\mathbb{Z}[C_q]$-dual to $j^*(R(1)) \cong \mathbb{Z}[C_q]^{(p-1)/q}$. As this latter is self-dual we have:

(69.19) $$j^*(R(q)) \cong \mathbb{Z}[C_q]^{(p-1)/q}.$$

By contrast to (69.18) and (69.19), we shall see in Chapter Thirteen that, in general, $j^*(R(k))$ is not free when $2 \le k \le q-1$.

Application of j^* to the exact sequence $0 \to \overline{I(C_p)} \to \overline{\mathbb{Z}[C_p]} \to \mathbb{Z} \to 0$ gives the exact sequence $0 \to j^*(\overline{I(C_p)}) \to j^*(\overline{\mathbb{Z}[C_p]}) \to \mathbb{Z} \to 0$. As $j^*(\overline{I(C_p)}) \cong \mathbb{Z}[C_q]^{(p-1)/q}$ and $\mathbb{Z}$ is 1-coprojective this sequence splits to give an isomorphism

(69.20) $$j^*(\overline{\mathbb{Z}[C_p]}) \cong \mathbb{Z} \oplus \mathbb{Z}[C_q]^{(p-1)/q}.$$

However, as we have seen in (69.12), $\overline{\mathbb{Z}[C_p]} = j_*(\mathbb{Z})$ so that

(69.21) $$j^*(j_*(\mathbb{Z})) \cong \mathbb{Z} \oplus \mathbb{Z}[C_q]^{(p-1)/q}.$$

§70: Cohomological calculations:

We proceed to calculate the cohomology of the row modules $R(i)$. To this end, we shall use boldface symbols **Hom**, **End** and **Ext** when describing homomorphisms, endomorphisms and extensions of Λ-modules and standard Roman font, Hom, End and Ext, when referring to the corresponding notions over $\mathbb{Z}[C_p]$. In addition, to simplify notation we write $\mathcal{T}_q = \mathcal{T}_q(A,\pi)$.

It is an elementary consequence of Wedderburn theory that $\mathrm{Hom}(\mathbb{Z}, I_C) = 0$. From the Eckmann-Shapiro isomorphism $\mathbf{Hom}(\mathbb{Z}, i_*(I_C)) \cong \mathrm{Hom}(\mathbb{Z}, I_C)$ it follows that $\mathbf{Hom}(\mathbb{Z}, i_*(I_C)) = 0$. However $\mathcal{T}_q \cong i_*(I_C)$ so that:

$$\textbf{(70.1)} \qquad \mathbf{Hom}(\mathbb{Z}, \mathcal{T}_q) = 0.$$

We note that $i^*(\breve{I}_Q) \cong \mathbb{Z}^{(q-1)}$ so that $\mathrm{Hom}(i^*(\breve{I}_Q), I_C) \cong \mathrm{Hom}(\mathbb{Z}, I_C)^{(q-1)} = 0$. From the Eckmann-Shapiro isomorphism $\mathbf{Hom}(\breve{I}_Q, i_*(I_C)) \cong \mathrm{Hom}(i^*(\breve{I}_Q), I_C)$ it follows that $\mathbf{Hom}(\breve{I}_Q, i_*(I_C)) = 0$. Again $i_*(I_C) \cong \mathcal{T}_q$ so that:

$$\textbf{(70.2)} \qquad \mathbf{Hom}(\breve{I}_Q, \mathcal{T}_q) = 0.$$

As $\mathcal{T}_q = \bigoplus_{k=1}^{q} R(k)$ it follows that for all k.

$$\textbf{(70.3)} \qquad \mathbf{Hom}(\breve{I}_Q, R(k)) = 0.$$

It is a standard calculation, as in §17, to show that

$$\textbf{(70.4)} \qquad \mathrm{Ext}^k(\mathbb{Z}, I_C) \cong \begin{cases} \mathbb{F}_p & k = 1 \\ 0 & k = 2. \end{cases}$$

We note that $\mathbf{Ext}^k(\mathbb{Z}, \mathcal{T}_q) = \mathbf{Ext}^k(\mathbb{Z}, i_*(I_C))$ whilst from the Eckmann-Shapiro Lemma we have $\mathbf{Ext}^k(\mathbb{Z}, i_*(I_C)) = \mathrm{Ext}^k(i^*(\mathbb{Z}), I_C) = \mathrm{Ext}^k((\mathbb{Z}), I_C)$. Hence

$$\textbf{(70.5)} \qquad \mathbf{Ext}^k(\mathbb{Z}, \mathcal{T}_q) = \mathrm{Ext}^k(\mathbb{Z}, I_C).$$

It now follows from (70.4) that

$$\textbf{(70.6)} \qquad \mathbf{Ext}^k(\mathbb{Z}, \mathcal{T}_q) \cong \begin{cases} \mathbb{F}_p & k = 1 \\ 0 & k = 2. \end{cases}$$

Noting that $\mathbf{Ext}^2(\mathbb{Z}, \mathcal{T}_q) \cong \bigoplus_{k=1}^{q} \mathbf{Ext}^2(\mathbb{Z}, R(k))$ it follows from (70.6) that:

$$\textbf{(70.7)} \qquad \mathbf{Ext}^2(\mathbb{Z}, R(k)) = 0 \text{ for all } k \ (1 \le k \le q).$$

As $\mathbb{Z}[C_p]$ is the integral group ring of a finite group then by (4.7), it is indecomposable as a module over itself. Hence $\overline{\mathbb{Z}[C_p]}$ is indecomposable over Λ. The existence of the exact sequence $0 \to \overline{I_C} \to \overline{\mathbb{Z}[C_p]} \to \mathbb{Z} \to 0$

then implies that $\mathbf{Ext}^1(\mathbb{Z}, \overline{I_C}) \neq 0$. As $\overline{I_C} \cong R(1)$ then

(70.8) $$\mathbf{Ext}^1(\mathbb{Z}, R(1)) \neq 0.$$

From (70.6) we see that $\bigoplus_{k=1}^{q} \mathbf{Ext}^1(\mathbb{Z}, R(k)) \cong \mathbb{F}_p$. As $\mathbf{Ext}^1(\mathbb{Z}, R(1)) \neq 0$ then:

(70.9) $$\mathbf{Ext}^1(\mathbb{Z}, R(k)) \cong \begin{cases} \mathbb{F}_p & k = 1 \\ 0 & k \neq 1. \end{cases}$$

As $i^*(\breve{\mathbb{Z}}[C_q]) = \mathbb{Z}^{(q)}$ it follows from the Eckmann-Shapiro Theorem that

$$\mathbf{Ext}^k(\breve{\mathbb{Z}}[C_q], i_*(I_C)) \cong \mathrm{Ext}^k(i^*(\mathbb{Z}[C_q]), I_C) \cong \bigoplus_{i=1}^{q} \mathrm{Ext}^k(\mathbb{Z}, I_C)$$

so that

(70.10) $$\mathbf{Ext}^k(\breve{\mathbb{Z}}[C_q], \mathcal{T}_q) \cong \begin{cases} \underbrace{\mathbb{F}_p \oplus \cdots \oplus \mathbb{F}_p}_{q} & k = 1 \\ 0 & k = 2. \end{cases}$$

From the exact sequence (69.9) it follows by the Corepresentation Theorem (21.12) that $\mathbf{End}_{\mathcal{D}\mathrm{er}}(\mathcal{T}_q) \cong \mathbf{Ext}^1(\breve{\mathbb{Z}}[C_q], \mathcal{T}_q)$ and so:

(70.11) $$\mathbf{End}_{\mathcal{D}\mathrm{er}}(\mathcal{T}_q) \cong \underbrace{\mathbb{F}_p \times \cdots \times \mathbb{F}_p}_{q}.$$

From the decomposition $\mathcal{T}_q \cong \bigoplus_{i=1}^{q} R(i)$ it follows from (70.11) that:

$$\bigoplus_{i,j=1}^{q} \mathbf{Hom}_{\mathcal{D}\mathrm{er}}(R(i), R(j)) \cong \underbrace{\mathbb{F}_p \times \cdots \times \mathbb{F}_p}_{q}.$$

As $R(i)$ is not projective over Λ then, by (18.11), $\mathbf{Hom}_{\mathcal{D}\mathrm{er}}(R(i), R(i)) \neq 0$. Hence:

(70.12) $$\mathbf{Hom}_{\mathcal{D}\mathrm{er}}(R(i), R(j)) \cong \begin{cases} \mathbb{F}_p & i = j \\ 0 & i \neq j. \end{cases}$$

Note that

Proposition 70.13: $\mathbf{Ext}^1(\breve{\mathbb{Z}}[C_q], R(k)) \cong \mathbb{F}_p$ *for all* k $(1 \leq k \leq q)$.

Proof. As $\mathcal{T}_q \cong \bigoplus_{k=1}^{q} R(k)$ we can re-write (69.9) as an extension

$$0 \to \bigoplus_{k=1}^{q} R(k) \to \Lambda \to \breve{\mathbb{Z}}[C_q] \to 0$$

which is classified by cohomology classes $c = (c_k)_{1 \le k \le q}$ where $c_k \in \mathbf{Ext}^1(\breve{\mathbb{Z}}[C_q], R(k))$. If $\mathbf{Ext}^1(\breve{\mathbb{Z}}[C_q], R(k)) = 0$ then Λ decomposes as a direct sum $\Lambda \cong R(k) \oplus X$ where the module X occurs in the extension

$$0 \to \bigoplus_{t \neq k} R(t) \to X \to \mathbb{Z}[C_q] \to 0$$

classified by the sequence $(c_t)_{t \neq k}$. However Λ, being the integral group ring of a finite group, is indecomposable ([14], vol. 1, p. 678). Consequently each $c_k \neq 0$ and hence $\mathbf{Ext}^1(\mathbb{Z}[C_q], R(k)) \neq 0$. As $\mathcal{T}_q \cong \bigoplus_{k=1}^{q} R(k)$ it follows from (70.10) that

$$\bigoplus_{k=1}^{q} \mathbf{Ext}^1(\breve{\mathbb{Z}}[C_q], R(k)) \cong \underbrace{\mathbb{F}_p \oplus \cdots \oplus \mathbb{F}_p}_{q}.$$

As $\mathbf{Ext}^1(\breve{\mathbb{Z}}[C_q], R(k)) \neq 0$ then $\mathbf{Ext}^1(\breve{\mathbb{Z}}[C_q], R(k)) \cong \mathbb{F}_p$ as claimed. □

Applying $\mathbf{Hom}(-, R(k))$ to the exact sequence $0 \to \breve{I}_Q \to \breve{\mathbb{Z}}[C_q] \to \mathbb{Z} \to 0$ we obtain a long exact sequence in cohomology, from which, in conjunction with (70.3), (70.13) and (70.7), we construct the following commutative diagram:

$$\begin{array}{ccccccccc} \mathbf{Hom}(\breve{I}_Q, R(k)) & \to & \mathbf{Ext}^1(\mathbb{Z}, R(k)) & \to & \mathbf{Ext}^1(\breve{\mathbb{Z}}[C_q], R(k)) & \to & \mathbf{Ext}^1(\breve{I}_Q, R(k)) & \to & \mathbf{Ext}^2(\mathbb{Z}, R(k)) \\ \| & & \| & & \| & & \| & & \| \\ 0 & \to & \mathbf{Ext}^1(\mathbb{Z}, R(k)) & \to & \mathbb{F}_p & \to & \mathbf{Ext}^1(\breve{I}_Q, R(k)) & \to & 0. \end{array}$$

When $k = 1$ then $\mathbf{Ext}^1(\mathbb{Z}, R(1)) \cong \mathbb{F}_p$ so that $\mathbf{Ext}^1(\breve{I}_Q, R(1)) = 0$ whilst if $k \neq 1$ then $\mathbf{Ext}^1(\mathbb{Z}, R(k)) = 0$ so that $\mathbf{Ext}^1(\breve{I}_Q, R(k)) \cong \mathbb{F}_p$; that is:

(70.14) $\mathbf{Ext}^1(\breve{I}_Q, R(k)) \cong \begin{cases} 0 & k = 1 \\ \mathbb{F}_p & k \neq 1. \end{cases}$

Again as $\mathcal{T}_q = \bigoplus_{k=1}^{q} R(k)$ it now follows that:

(70.15) $$\mathbf{Ext}^1(\breve{I}_Q, \mathcal{T}_q) \cong \mathbb{F}_p^{(q-1)}.$$

Observe that $i_*(I_C^*) \cong \mathcal{T}_q \cong \bigoplus_{r=1}^q R(r)$ and $i^*(R(r)) \cong I_C^*$. From the first Eckmann-Shapiro relation we obtain:

$$\begin{aligned}\mathbf{Ext}^2(\mathcal{T}_q, \mathcal{T}_q) &\cong \bigoplus_{r=1}^{q} \mathbf{Ext}^2(i_*(I_C^*), R(r)) \\ &\cong \bigoplus_{r=1}^{q} \mathrm{Ext}^2(I_C^*, i^*(R(r))) \\ &\cong \bigoplus_{r=1}^{q} \mathrm{Ext}^2(I_C^*, I_C^*).\end{aligned}$$

Noting that $\mathrm{Ext}^2(I_C^*, I_C^*) \cong \mathbb{Z}/p$ then $\mathbf{Ext}^2(\mathcal{T}_q, \mathcal{T}_q) \cong \underbrace{\mathbb{F}_p \oplus \cdots \oplus \mathbb{F}_p}_{q}$. Likewise from the second Eckmann-Shapiro relation we deduce that

$$\begin{aligned}\mathbf{Ext}^2(R(r), \mathcal{T}_q) &\cong \mathbf{Ext}^2(R(r), i_*(I_C^*)) \\ &\cong \mathrm{Ext}^2(i^*(R(r), I_C^*) \\ &\cong \mathrm{Ext}^2(I_C^*, I_C^*).\end{aligned}$$

Hence we see that $\mathbf{Ext}^2(R(r), \mathcal{T}_q) \cong \mathbb{F}_p$. Writing $\mathcal{T}_q \cong \bigoplus_{s=1}^q R(s)$ we have $\bigoplus_{s=1}^q \mathbf{Ext}^2(R(r), R(s)) \cong \mathbb{F}_p$. As $\mathbb{F}_p$ is indecomposable then for each $r \in \{1, \ldots, q\}$ there exists $\sigma(r) \in \{1, \ldots, q\}$ such that:

(70.16)
$$\mathbf{Ext}^2(R(r), R(s)) \cong \begin{cases} \mathbb{F}_p & s = \sigma(r) \\ 0 & s \neq \sigma(r). \end{cases}$$

The correspondence $i \mapsto \sigma(i)$ evidently defines a mapping $\sigma : \{1, \ldots, q\} \to \{1, \ldots, q\}$. We claim that σ is bijective. Otherwise, it fails to be surjective and there exists $k \in \{1, \ldots, q\}$ such that for all $i \in \{1, \ldots, q\}$ $\mathbf{Ext}^2(R(i), R(k)) = 0$. Thus $\mathbf{Ext}^2(\mathcal{T}_q, R(k)) = \bigoplus_{i=1}^q \mathbf{Ext}^2(R(i), R(k)) = 0$. By duality

$$\mathbf{Ext}^2(R(k)^\bullet, \mathcal{T}_q^\bullet) = 0.$$

However, $R(k)^\bullet \cong R(q+1-k)$ and $\mathcal{T}_q^\bullet \cong \mathcal{T}_q \cong \bigoplus_{s=1}^q R(s)$ so that, for all $s \in \{1, \ldots, q\}$

$$\mathbf{Ext}^2(R(q+1-k), R(s)) = 0.$$

This contradicts (70.16) and so σ is bijective; to summarize:

Proposition 70.17: *There exists a (necessarily unique) permutation σ of $\{1, \ldots, q\}$ with the property that, for each $i \in \{1, \ldots, q\}$,*

$$\mathbf{Ext}^2(R(i), R(j)) \cong \begin{cases} \mathbb{F}_p & j = \sigma(i) \\ 0 & j \neq \sigma(i). \end{cases}$$

The permutation σ is called the *sequencing permutation*; the rest of this chapter is concerned with its exact description.

§71: Decomposing the augmentation ideal of Λ:

Put
$$\begin{cases} \mathbf{E} = \{x^b - 1 : 1 \leq b \leq p-1\} \\ \Phi = \{y^a - 1 : 1 \leq a \leq q-1\} \\ \Psi = \{y^a x^b - 1 : 1 \leq a \leq q-1, 1 \leq b \leq p-1\} \\ \Psi' = \{(y^a - 1)x^b : 1 \leq a \leq q-1, 1 \leq b \leq p-1\} \end{cases}$$

Then $\mathbf{E} \cup \Phi \cup \Psi$ is an integral basis for I_G. Observing that

$$(y^a - 1)x^b = (y^a x^b - 1) - (x^b - 1)$$

then by an elementary basis change we see also that $\mathbf{E} \cup \Phi \cup \Psi'$ is also an integral basis for I_G. However $\Phi \cup \Psi'$ is clearly an integral basis for the right ideal $[y-1)$. As this extends to a basis for I_G it follows that $I_G/[y-1)$ is free over $\mathbb{Z}$. Moreover if $\natural : I_G \to I_G/[y-1)$ is the identification map then

(71.1) $\{\natural(x^b - 1) : 1 \leq b \leq p-1\}$ is an integral basis for $I_G/[y-1)$.

It follows immediately that $I_G/[y-1)$ is isomorphic to I_C as a module over $\mathbb{Z}[C_p]$. Computing the right action of y on I_G we find

$$\begin{aligned} (x^b - 1) \cdot y &= y[y^{-1}x^b y - 1] \\ &= y[x^{-ab} - 1] \\ &= (y-1)(x^{-ab} - 1) + (x^{-ab} - 1). \end{aligned}$$

As $(y-1)(x^{-ab} - 1) \in [y-1)$ the above calculation thereby shows that

$$\natural(x^b - 1) \cdot y = \natural(x^{-ab} - 1)$$

which, by (64.3), coincides with the Galois action on $\overline{I_C}$. Thus $I_G/[y-1) \cong \overline{I_C}$, from which we see that:

(71.2) There exists an exact sequence $0 \to [y-1) \to I_G \to \overline{I_C} \to 0$.

We proceed to show that the exact sequence of (71.2) splits. We continue to use boldface symbols **Hom**, $\mathbf{Ext}^k$ when describing homomorphisms and extensions of Λ-modules but we now use italics *Hom*, Ext^a when referring to homomorphisms and extensions of modules over $\mathbb{Z}[C_q]$. Let $j : \mathbb{Z}[C_q] \hookrightarrow \Lambda$ denote the inclusion; note that $[y-1) = j_*(I_Q)$ and that $j^*(\overline{I_C})$ is free of rank $(p-1)/q$ over $Z[C_q]$; that is;

(71.3) $$j^*(\overline{I_C}) \cong \mathbb{Z}[C_q]^{(p-1)/q}.$$

It follows that

(71.4) $Ext^1(j^*(\overline{I_C}), L) = 0$ for any lattice L over $Z[C_q]$.

We obtain the following decomposition theorem (cf [24], [38]).

Theorem 71.5: *I_G decomposes as a direct sum $I_G \cong \overline{I_C} \oplus [y-1)$.*

Proof. Writing $[y-1) = j_*(I_Q)$ we have

$$\mathbf{Ext}^1(\overline{I_C}, [y-1)) = \mathbf{Ext}^1(\overline{I_C}, j_*(I_Q)) = Ext^1(j^*(\overline{I_C}), I_Q).$$

Then $\mathbf{Ext}^1(\overline{I_C}, [y-1)) = 0$ follows from the Eckmann-Shapiro Theorem applied to (71.4). Hence (71.2) splits and $I_G \cong \overline{I_C} \oplus [y-1)$ as claimed. □

As $R(1) \cong \overline{I_C}$ then for future convenience we also write:

(71.6) $$I_G \cong R(1) \oplus [y-1).$$

§72: The basic sequence:

From (71.5) there is an exact sequence $0 \to \overline{I_C} \oplus [y-1) \to \Lambda \to \mathbb{Z} \to 0$. Application of $\overline{I_C^*} \otimes -$ gives an exact sequence

$$0 \to (\overline{I_C^*} \otimes \overline{I_C}) \oplus (\overline{I_C^*} \otimes [y-1)) \to \overline{I_C^*} \otimes \Lambda \to \overline{I_C^*} \otimes \mathbb{Z} \to 0$$

which, by (35.4), (64.11) we may write more conveniently as

(72.1) $$0 \to (\overline{I_C^*} \otimes \overline{I_C}) \oplus \Lambda^{d(q-1)} \to \Lambda^{p-1} \to \overline{I_C^*} \to 0.$$

As $\Lambda^{d(q-1)}$ and $\overline{I_C^*} \otimes \overline{I_C}$ are self-dual, then dualization of (72.1) gives

(72.2) $$0 \to \overline{I_C} \to \Lambda^{p-1} \to (\overline{I_C^*} \otimes \overline{I_C}) \oplus \Lambda^{d(q-1)} \to 0.$$

Splicing (72.1) and (72.2) together gives an exact sequence

(72.3) $$0 \to \overline{I_C} \longrightarrow \Lambda^{(p-1)} \longrightarrow \Lambda^{(p-1)} \longrightarrow \overline{I_C^*} \to 0.$$

However, $\overline{I_C^*}$ is monogenic and finitely presented so there is an exact sequence

(72.4) $$0 \to K \longrightarrow \Lambda^b \longrightarrow \Lambda \longrightarrow \overline{I_C^*} \to 0.$$

Comparing (72.3) and (72.4) by iterating Schanuel's Lemma we see that:

(72.5) $$\overline{I_C} \oplus \Lambda^{p+b-1} \cong K \oplus \Lambda^p.$$

We may modify (72.4) successively, first to an exact sequence

$$0 \to K \oplus \Lambda^p \longrightarrow \Lambda^{p+b} \longrightarrow \Lambda \longrightarrow \overline{I_C^*} \to 0.$$

Then, using (72.5), to an exact sequence

$$0 \to \overline{I_C} \oplus \Lambda^{p+b-1} \xrightarrow{j} \Lambda^{p+b} \longrightarrow \Lambda \longrightarrow \overline{I_C^*} \to 0.$$

Finally to an exact sequence

(72.6) $$0 \to \overline{I_C} \longrightarrow S \longrightarrow \Lambda \longrightarrow \overline{I_C^*} \to 0$$

where $S = \Lambda^{p+b}/j(\Lambda^{p+b-1})$. It follows from the 'de-stabilization theorem' (22.1) that S is projective. Moreover, from the exact sequence

$$0 \to \Lambda^{p+b-1} \xrightarrow{j} \Lambda^{p+b} \to S \to 0$$

we see that $S \oplus \Lambda^{p+b-1} \cong \Lambda^{p+b}$. However, Λ satisfies the Eichler condition so that, by (4.12) $S \cong \Lambda$. Substitution into (72.6) gives an exact sequence

(72.7) $$\begin{array}{ccccccccc} & & & & K(q) & & & & \\ & & & \nearrow & & \searrow & & & \\ 0 & \longrightarrow & \overline{I_C} & \longrightarrow \Lambda & \longrightarrow & \Lambda & \twoheadrightarrow & \overline{I_C^*} & \longrightarrow 0 \end{array}$$

where $K(q)$ is the kernel of the surjection $\Lambda \twoheadrightarrow \overline{I_C^*}$. The sequence (72.7) is called the *basic sequence for* Λ.

The existence of the basic sequence is fundamental in what follows. Clearly any such sequence takes the form $0 \to \overline{I_C} \to \Lambda \xrightarrow{\alpha} \Lambda \to \overline{I_C^*} \to 0$ where $\alpha \in \Lambda$ generates the right ideal $K(q)$. The computationally minded reader will find that, in any particular case, the task of finding a convenient α is nontrivial.

§73: The derived sequences:

Each $R(i)$ is monogenic; hence for each $i \in \{1, \ldots, q\}$ there is an exact sequence

(73.1)
$$\mathcal{S}(i) = (0 \to K(i) \to \Lambda \to R(i) \to 0).$$

By dimension shifting from (70.17), $\mathrm{Ext}^1(K(i), R(j)) \cong \begin{cases} \mathbb{F}_p & j = \sigma(i) \\ 0 & j \neq \sigma(i). \end{cases}$

Recall from (64.1) that $\mathbb{Z}[C_q] \otimes \overline{I_C} \cong i_*(I_C) \cong i_*(I_C^*) \cong \mathbb{Z}[C_q] \otimes \overline{I_C^*}$ and that $\mathbb{Z}[C_q] \otimes \Lambda \cong \Lambda^q$. Applying the functor $\mathbb{Z}[C_q] \otimes -$ to (72.7) gives an exact sequence

$$0 \longrightarrow i_*(I_C) \longrightarrow \Lambda^q \xrightarrow{\ \nearrow K \searrow\ } \Lambda^q \longrightarrow i_*(I_C) \longrightarrow 0$$

where $K = \mathbb{Z}[C_q] \otimes K(q)$. By (69.7), $i_*(I_C) \cong \mathcal{T}_q(A, \pi) \cong \bigoplus_{i=1}^q R(i)$. Moreover $\bigoplus_{i=1}^q R(i) \cong \bigoplus_{i=1}^q R(\sigma(i))$ so that we have an exact sequence

(73.2)
$$0 \longrightarrow \bigoplus_{i=1}^q R(\sigma(i)) \longrightarrow \Lambda^q \xrightarrow{\ \nearrow K \searrow\ } \Lambda^q \longrightarrow \bigoplus_{i=1}^q R(i) \longrightarrow 0.$$

On comparing the portion $0 \to K \to \Lambda^q \to \bigoplus_{i=1}^q R(i) \to 0$ of (73.2) with

$$\bigoplus_{i=1}^q \mathcal{S}(i) = \left(0 \to \bigoplus_{i=1}^q K(i) \to \Lambda^q \to \bigoplus_{i=1}^q R(i) \to 0\right)$$

it follows from Schanuel's Lemma that

(73.3)
$$K \oplus \Lambda^q \cong \left(\bigoplus_{i=1}^q K(i)\right) \oplus \Lambda^q.$$

Taking σ to be the sequencing permutation we claim

Proposition 73.4: *There exists an exact sequence of the form*

$$\mathcal{W} = \left(0 \to \bigoplus_{i=1}^q R(\sigma(i)) \to \Lambda^q \to \bigoplus_{i=1}^q K(i) \to 0\right).$$

Proof. Modify $0 \to \bigoplus_{i=1}^{q} R(\sigma(i)) \to \Lambda^q \to K \to 0$ of (73.2) first to $0 \to \bigoplus_{i=1}^{q} R(\sigma(i)) \to \Lambda^q \oplus \Lambda^q \to K \oplus \Lambda^q \to 0$, then, using (73.3), to

$$0 \to \bigoplus_{i=1}^{q} R(\sigma(i)) \longrightarrow \Lambda^{2q} \longrightarrow \left(\bigoplus_{i=1}^{q} K(i)\right) \oplus \Lambda^q \to 0.$$

Dualization gives $0 \to (\bigoplus_{i=1}^{q} K(i)^*) \oplus \Lambda^q \xrightarrow{\iota} \Lambda^{2q} \longrightarrow \bigoplus_{i=1}^{q} R(\sigma(i))^* \to 0$ which we modify again to

$$0 \to \bigoplus_{i=1}^{q} K(i)^* \to \Lambda^{2q}/(\iota(\Lambda^q)) \to \bigoplus_{i=1}^{q} R(\sigma(i))^* \to 0.$$

Again by the 'de-stabilization theorem' (22.1) we see that $\Lambda^{2q}/\iota(\Lambda^q)$ is stably free of rank q over Λ. By the Swan-Jacobinski Theorem, $\Lambda^{2q}/\iota(\Lambda^q) \cong \Lambda^q$ so there is an exact sequence

$$0 \to \bigoplus_{i=1}^{q} K(i)^* \longrightarrow \Lambda^q \longrightarrow \bigoplus_{i=1}^{q} R(\sigma(i))^* \to 0.$$

Re-dualization gives the sequence

$$0 \to \bigoplus_{i=1}^{q} R(\sigma(i)) \to \Lambda^q \to \bigoplus_{i=1}^{q} K(i) \to 0.$$ □

Theorem 73.5: *For each i there exists an exact sequence*

$$\mathcal{W}(i) = (0 \to R(\sigma(i)) \longrightarrow P(i) \longrightarrow K(i) \to 0)$$

in which $P(i)$ is projective of rank 1 over Λ. Moreover, $\bigoplus_{i=1}^{q} P(i) \cong \Lambda^q$.

Proof. Let $[\mathcal{W}]$ denote the congruence class of the extension constructed in (73.4). Then $[\mathcal{W}] \in \mathrm{Ext}^1(\bigoplus_{i=1}^{q} K(i), \bigoplus_{j=1}^{q} R(\sigma(j))) \cong \bigoplus_{i,j=1}^{q} \mathrm{Ext}^1(K(i), R(\sigma(j)))$. Dimension shifting applied to (70.17) shows that $\mathrm{Ext}^1(K(i), R(j)) = 0$ when $j \neq \sigma(i)$ so that

$$\mathrm{Ext}^1\left(\bigoplus_{i=1}^{q} K(i), \bigoplus_{j=1}^{q} R(\sigma(j))\right) \cong \bigoplus_{i=1}^{q} \mathrm{Ext}^1(K(i), R(\sigma(i)))$$

and $\mathcal{W}$ is congruent to a direct sum $\mathcal{W} \approx \mathcal{W}(1) \oplus \cdots \oplus \mathcal{W}(q)$ where $\mathcal{W}(i)$ has the form $\mathcal{W}(i) = (0 \to R(\sigma(i)) \to P(i) \to K(i) \to 0)$. In particular,

$$\Lambda^q \cong P(1) \oplus \cdots \oplus P(q)$$

so that each $P(i)$ is projective. By Swan's 'local freeness' theorem ([14], vol. 1, p. 676) each $P(i) \otimes \mathbb{Q}$ is free over $\Lambda \otimes \mathbb{Q}$. As each $P(i)$ is nonzero, a straightforward calculation of $\mathbb{Z}$-ranks shows that $\mathrm{rk}_\Lambda(P(i)) = 1$. □

Splicing the exact sequence $\mathcal{S}(i)$ of (73.1) with $\mathcal{W}(i)$ of (73.5) gives an extension

(73.6) $$\mathcal{Z}(i) = (0 \to R(\sigma(i)) \to P(i) \to \Lambda \to R(i) \to 0).$$

Observe that for $\mathcal{Z}(q)$ we may take the basic sequence of (72.7) so that:

(73.7) $$\sigma(q) = 1 \text{ and } P(q) \cong \Lambda.$$

§74: Computing the sequencing permutation:

Put $c_p(x) = x^{p-1} + \cdots + x + 1$ and put $S = \mathbb{Z}[x]/c_p(x)$ where $\zeta_p = \exp(2\pi i/p)$. Then S is a Dedekind domain in which p ramifies completely (cf [6], p. 87). Let $\widehat{\mathbb{Z}}$ denote the ring of p-adic integers, let $\widehat{\mathbb{Q}}$ denote its field of fractions. The standard proof that $c_p(x)$ is irreducible over $\mathbb{Q}$ likewise shows that $c_p(x)$ remains irreducible over $\widehat{\mathbb{Q}}$. Let $\widetilde{\pi} \in S$ be the unique prime over p and let $\widehat{S}$ denote the complete local ring obtained from S by localizing $\widetilde{\pi}$. As $S \cong \mathbb{Z}[x]/c_p(x)$ is the ring of integers in $\mathbb{Q}[x]/c_p(x)$ then $\widehat{S}$ is the ring of integers in $\widehat{\mathbb{Q}}[x]/c_p(x)$. Hence $\widehat{S} \cong \widehat{\mathbb{Z}}[x]/c_p(x) \cong (\mathbb{Z}[x]/c_p(x)) \otimes_{\mathbb{Z}} \widehat{\mathbb{Z}} \cong S \otimes_{\mathbb{Z}} \widehat{\mathbb{Z}}$. As before, put $A = S^{C_q}$; then again p ramifies completely in A and $\pi = (\widetilde{\pi})^q$ is the unique prime in A over p. Now take $\widehat{A}$ to be the complete local ring obtained from A by localizing at π; then $\widehat{A} = (\widehat{S})^{C_q} = (S \otimes_{\mathbb{Z}} \widehat{\mathbb{Z}})^{C_q} = S^{C_q} \otimes_{\mathbb{Z}} \widehat{\mathbb{Z}}$ and so $\widehat{A} \cong A \otimes_{\mathbb{Z}} \widehat{\mathbb{Z}}$. For any Λ-lattice M, we denote by $\widehat{M} = M \otimes_\Lambda \widehat{\Lambda}$. the corresponding $\widehat{\Lambda}$-lattice.

As we observed in (73.7), $\sigma(q) = 1$ and $P(q) = \Lambda$. We proceed to show that $\sigma(i) = i + 1$ for $1 \le i \le q - 1$. On applying the exact functor $- \otimes_{\mathbb{Z}} \widehat{\mathbb{Z}}$ to (73.6) we obtain an exact sequence of $\widehat{\Lambda}$ modules

(74.1) $$\widehat{\mathcal{Z}}(i) = \mathcal{Z}(i) \otimes_{\mathbb{Z}} \widehat{\mathbb{Z}} = (0 \to \widehat{R}(\sigma(i)) \to \widehat{P}(i) \to \widehat{\Lambda} \to \widehat{R}(i) \to 0).$$

We have a p-adic analogue of (70.17) namely:

(74.2) $$\mathbf{Ext}^2(\widehat{R}(i), \widehat{R}(j)) \cong \begin{cases} \mathbb{F}_p & j = \sigma(i) \\ 0 & j \ne \sigma(i). \end{cases}$$

We also have p-adic analogues of (68.4) and (68.7):

(74.3) There is an exact sequence of $\widehat{\Lambda}$-modules

$$0 \to \widehat{R}(1) \hookrightarrow \widehat{R}(q) \to \mathbb{F}_p(1) \to 0.$$

(74.4) For $1 \le k \le q - 1$ there are exact sequences of $\widehat{\Lambda}$-modules

$$0 \to \widehat{R}(k+1) \hookrightarrow \widehat{R}(k) \to \mathbb{F}_p(\overline{a}^k) \to 0.$$

Let $\natural : \widehat{\mathbb{Z}} \to \mathbb{F}_p$ be the canonical mapping. There exists a qth root of unity $\widehat{a} \in \widehat{\mathbb{Z}}$ such that $\natural(\widehat{a}) = \overline{a}$. so that $\widehat{\rho}(y^{-1})$ takes the form

$$\widehat{\rho}(y^{-1}) = \begin{pmatrix} \widehat{a} & * & * & * & * & * \\ & \widehat{a}^2 & * & * & * & * \\ & & \widehat{a}^3 & * & * & * \\ & & & \ddots & & \\ & & & & \widehat{a}^{q-1} & * \\ & & & & & 1 \end{pmatrix}.$$

Let $\widehat{\mathbb{Z}}(\widehat{a}^k)$ denote the $\widehat{\mathbb{Z}}[C_q]$ module whose underlying $\widehat{\mathbb{Z}}$ module is $\widehat{\mathbb{Z}}$ on which y acts, on the right, as multiplication by $\widehat{a}^k$. As $\widehat{R}(k) \twoheadrightarrow \mathbb{F}_p[\overline{a}^k]$ it follows that $\widehat{R}(k) \otimes \widehat{\mathbb{Z}}[\widehat{a}] \twoheadrightarrow \mathbb{F}_p[\overline{a}^k] \otimes \widehat{\mathbb{Z}}[\widehat{a}] \cong \mathbb{F}_p[\overline{a}^{k+1}]$. As $\widehat{R}(k+1) \twoheadrightarrow \mathbb{F}_p[\overline{a}^{k+1}]$ we see that:

(74.5) $$\widehat{R}(k+1) \cong_{\widehat{\Lambda}} \widehat{R}(k) \otimes_{\widehat{\mathbb{Z}}} \widehat{\mathbb{Z}}(\widehat{a}) \quad \text{for } 1 \leq k \leq q-1.$$

(74.6) $$\widehat{R}(1) \cong_{\widehat{\Lambda}} \widehat{R}(q) \otimes_{\widehat{\mathbb{Z}}} \widehat{\mathbb{Z}}(\widehat{a}).$$

Start with the basic sequence $0 \to \overline{I_C} \to \Lambda \longrightarrow \Lambda \to \overline{I_C^*} \to 0$ and, using (66.8), (66.9) rewrite in 'row notation' thus

(74.7) $$0 \to R(1) \to \Lambda \longrightarrow \Lambda \to R(q) \to 0.$$

Applying $- \otimes_{\mathbb{Z}} \widehat{\mathbb{Z}}$ to (74.7) gives an exact sequence

(74.8) $$0 \to \widehat{R}(1) \to \widehat{\Lambda} \longrightarrow \widehat{\Lambda} \to \widehat{R}(q) \to 0.$$

On applying $- \otimes_{\widehat{\mathbb{Z}}} \widehat{\mathbb{Z}}(\widehat{a})$ to (74.8) iteratively and appealing to (74.5) and (74.6) we generate exact sequences $\widehat{\mathbf{S(i)}}$ with $1 \leq k \leq q-1$ thus,

$\widehat{\mathbf{S(i)}}$ $$0 \to \widehat{R}(i+1) \to \widehat{\Lambda} \longrightarrow \widehat{\Lambda} \to \widehat{R}(i) \to 0.$$

Comparing the sequences $\widehat{\mathcal{Z}}(i)$ and $\widehat{\mathbf{S(i)}}$ by means of the generalized form of Schanuel's Lemma we see that

$$\widehat{R(\sigma(i))} \oplus \widehat{\Lambda}^{(2)} \cong \widehat{R(i+1)} \oplus \widehat{\Lambda} \oplus \widehat{P}.$$

Thus $\mathbf{Ext}^2(\widehat{R}(i), \widehat{R}(\sigma(i))) \cong \mathbf{Ext}^2(\widehat{R}(i), \widehat{R}(i+1))$. Hence by (74.2) we have the following exact description of the sequencing permutation:

(74.9)
$$\sigma(i) = \begin{cases} i+1: & 1 \leq i \leq q-1 \\ 1: & i = q. \end{cases}$$

Consequently, for $1 \leq i \leq q-1$, $\mathcal{Z}(i)$ takes the form

$$\mathcal{Z}(i) = (0 \to R(i+1) \to P(i) \to \Lambda \to R(i) \to 0)$$

where $P(i)$ is a projective Λ-module of rank 1 whilst

$$\mathcal{Z}(q) = (0 \to R(1) \to \Lambda \to \Lambda \to R(q) \to 0)$$

is an alternative description of the basic sequence. Moreover $\bigoplus_{i=1}^{q-1} P(i) \cong \Lambda^{(q-1)}$. For future reference we write $\mathcal{Z}(i)$ in the form

(74.10)
$$\mathcal{Z}(i) \;=\; (0 \longrightarrow R(i+1) \longrightarrow P(i) \xrightarrow{\;\nearrow K(i) \searrow\;} \Lambda \longrightarrow R(i) \longrightarrow 0)$$

where $K(i) = \mathrm{Ker}(\Lambda \to R(i))$. Splicing together the sequences $\mathcal{Z}(i)$ we obtain:

(74.11) There exists a projective resolution with period $2q$ as follows:

$$\begin{array}{ccccccccccccccccc}
 & & & K(q) & & & K(q-1) & & K(2) & & & & K(1) & & & & \\
 & & & \nearrow\;\searrow & & & \nearrow\;\searrow & & \nearrow\;\searrow & & & & \nearrow\;\searrow & & & & \\
0 \to & R(1) & \to \Lambda & \longrightarrow & \Lambda & \longrightarrow & P(q-1) \longrightarrow & \cdots & \longrightarrow & \Lambda & \longrightarrow & P(1) & \longrightarrow & \Lambda & \longrightarrow R(1) & \to 0 \\
 & & & & \searrow\;\nearrow & & & & & \searrow\;\nearrow & & & & & & & \\
 & & & & R(q) & & & & & R(2) & & & & & & &
\end{array}$$

in which each $P(i)$ is projective of rank 1 over Λ and $\bigoplus_{i=1}^{q} P(i) \cong \Lambda^q$.
As portrayed, the above sequence is a resolution of $R(1)$. However, as it is periodic, by varying the starting point, it gives a resolution of each $R(i)$; for example we have a resolution for $R(2)$ as follows:

$$\begin{array}{ccccccccccccccccc}
 & & & K(1) & & & K(q) & & K(3) & & & & K(2) & & & & \\
 & & & \nearrow\;\searrow & & & \nearrow\;\searrow & & \nearrow\;\searrow & & & & \nearrow\;\searrow & & & & \\
0 \to & R(2) & \to P(1) & \longrightarrow & \Lambda & \longrightarrow & \Lambda \longrightarrow & \cdots & \longrightarrow & \Lambda & \longrightarrow & P(2) & \longrightarrow & \Lambda & \longrightarrow R(2) & \to 0 \\
 & & & & \searrow\;\nearrow & & & & & \searrow\;\nearrow & & & & & & & \\
 & & & & R(1) & & & & & R(3) & & & & & & &
\end{array}$$

In the case of the first two non-dihedral metacyclic groups, namely $G(5,4)$ and $G(7,3)$, the existence of complete diagonal resolutions was established by direct computation in the theses of Jamil Nadim [48] and Jonathan Remez [53] respectively. In both these cases, for each i, $P(i) = \Lambda$.

§75: An element of $D_{2k+1}(\mathbb{Z})$:

Consider the exact sequences $\{\mathcal{Z}(i)\}_{1\le i\le q}$ constructed above. Defining $\mathcal{Z}(n) = \mathcal{Z}(i)$ when $n \equiv i \bmod q$ we obtain exact sequences $\{\mathcal{Z}(n)\}_{n\in\mathbb{Z}}$. Splicing the sequences $\mathcal{Z}(n)$ together gives the following exact sequence

$$\mathcal{S}_+ = (\dots \xrightarrow{\partial^+_{2n+3}} P(n+1) \xrightarrow{\partial^+_{2n+2}} \Lambda \xrightarrow{\partial^+_{2n+1}} P(n) \xrightarrow{\partial^+_{2n}} \Lambda \xrightarrow{\partial^+_{2n-1}} P(n-1) \xrightarrow{\partial^+_{2n-2}} \dots)$$

where $\partial^+_{2n-1} = \iota_n \circ \pi_n$ and $\partial^+_{2n} = \alpha_n$. Taking $\partial^-_{2n-1} = (y-1)_*$ and $\partial^+_{2n} = (\Sigma_y)_*$ where $\Sigma_y = 1+y+\dots+y^{q-1}$ then the following sequence $\mathcal{S}_-$ is exact

$$\mathcal{S}_- = (\dots \to \Lambda \xrightarrow{\partial^-_{2n+3}} \Lambda \xrightarrow{\partial^-_{2n+2}} \Lambda \xrightarrow{\partial^-_{2n+1}} \Lambda \xrightarrow{\partial^-_{2n}} \Lambda \xrightarrow{\partial^-_{2n-1}} \Lambda \xrightarrow{\partial^-_{2n-2}} \dots).$$

Indeed, if $j : C_q \hookrightarrow G(p,q)$ is the inclusion then $\mathcal{S}_-$ is the induced resolution $\mathcal{S}_- = j_*(\mathcal{E})$ where $\mathcal{E}$ is the standard resolution of $\mathbb{Z}$ over $\mathbb{Z}[C_q]$

$$\mathcal{E} = (\dots \xrightarrow{y-1} \mathbb{Z}[C_q] \xrightarrow{\Sigma_y} \mathbb{Z}[C_q] \xrightarrow{y-1} \mathbb{Z}[C_q] \xrightarrow{\Sigma_y} \mathbb{Z}[C_q] \xrightarrow{y-1} \mathbb{Z}[C_q] \xrightarrow{\Sigma_y} \dots).$$

Taking direct sums we obtain the following exact sequence

$$\mathcal{S}_+ \oplus \mathcal{S}_- = (\dots \xrightarrow{\begin{pmatrix} \partial^+_{2n+3} & 0 \\ 0 & \partial^-_{2n+3} \end{pmatrix}} P(n+1)\oplus\Lambda \xrightarrow{\begin{pmatrix} \partial^+_{2n+2} & 0 \\ 0 & \partial^-_{2n+2} \end{pmatrix}} \Lambda\oplus\Lambda$$
$$\xrightarrow{\begin{pmatrix} \partial^+_{2n+1} & 0 \\ 0 & \partial^-_{2n+1} \end{pmatrix}} P(n)\oplus\Lambda \xrightarrow{\begin{pmatrix} \partial^+_{2n} & 0 \\ 0 & \partial^-_{2n} \end{pmatrix}} \dots).$$

Evidently $\mathcal{S}_+ \oplus \mathcal{S}_-$ is infinite in both directions and is periodic with period $2q$. Truncating at the third differential gives an exact sequence, infinite to the left:

$$\textbf{(75.1)} \qquad \dots \xrightarrow{\begin{pmatrix} \partial^+_5 & 0 \\ 0 & \partial^-_5 \end{pmatrix}} P(2)\oplus\Lambda \xrightarrow{\begin{pmatrix} \partial^+_4 & 0 \\ 0 & \partial^-_4 \end{pmatrix}} \Lambda\oplus\Lambda \xrightarrow{\begin{pmatrix} \partial^+_3 & 0 \\ 0 & \partial^-_3 \end{pmatrix}} P(1)\oplus\Lambda.$$

However, we also have an exact sequence

$$\textbf{(75.2)} \qquad P(1)\oplus\Lambda \xrightarrow{\begin{pmatrix} \partial^+_2 & 0 \\ 0 & \partial^-_2 \end{pmatrix}} \Lambda\oplus\Lambda \xrightarrow{\partial^+_1 + \partial^-_1} \Lambda \xrightarrow{\epsilon} \mathbb{Z} \to 0.$$
$$\searrow \quad \nearrow$$
$$\overline{I_C} \oplus [y-1]$$

Merging the two gives a complete resolution of $\mathbb{Z}$ which begins

$$\cdots \xrightarrow{\begin{pmatrix} \partial^+_{2n+3} & 0 \\ 0 & y-1 \end{pmatrix}} P(1) \oplus \Lambda \xrightarrow{\begin{pmatrix} \partial^+_2 & 0 \\ 0 & \partial^-_2 \end{pmatrix}} \Lambda \oplus \Lambda \xrightarrow{\partial^+_1 + \partial^-_1} \Lambda \xrightarrow{\epsilon} \mathbb{Z} \to 0$$

and continues

$$\ldots P(n+1) \oplus \Lambda \xrightarrow{\begin{pmatrix} \partial^+_{2n+2} & 0 \\ 0 & \Sigma_y \end{pmatrix}} \Lambda \oplus \Lambda \xrightarrow{\begin{pmatrix} \partial^+_{2n+1} & 0 \\ 0 & y-1 \end{pmatrix}} P(n) \oplus \Lambda \xrightarrow{\begin{pmatrix} \partial^+_{2n} & 0 \\ 0 & \Sigma_y \end{pmatrix}} \Lambda \oplus \Lambda \ldots$$

and where $\begin{cases} P(q) = \Lambda; & P(k+mq) = P(k) \\ \partial^+_{k+2mq} = \partial^+_k; & \partial^-_{k+2m} = \partial^-_k. \end{cases}$

We have constructed a diagonal resolution of $\mathbb{Z}$ with period $2q$. Moreover, as we have seen, $\bigoplus_{i=1}^{q-1} P(i) \cong \Lambda^{q-1}$. As is well known, the groups $G(p,q)$ have cohomological period $2q$. However, it follows immediately from the above that a somewhat stronger statement holds, namely:

(75.3) The groups $G(p,q)$ all have free period $2q$.

The above diagonal resolution shows that $R(k) \oplus [y-1)$ is a generalized syzygy of $R(1) \oplus [y-1)$. In fact we have:

(75.4) $R(k+1) \oplus [y-1) \in D_{2k}(R(1) \oplus [y-1))$ for $1 \le k \le q-1$.

A straightforward rank calculation shows that $\mathrm{rk}_{\mathbb{Z}}(R(k) \oplus [y-1)) = pq - 1$ so that $R(k) \oplus [y-1)$ is necessarily minimal in its projective equivalence class. As $R(1) \oplus [y-1) \cong I_G \in \Omega_1(\mathbb{Z})$ we obtain the following:

(75.5) $R(k+1) \oplus [y-1)$ is a minimal element of $D_{2k+1}(\mathbb{Z})$ for each k.

In view of this we may ask whether $R(k+1) \oplus [y-1)$ is a genuine syzygy of $R(1) \oplus [y-1)$. In Chapter Thirteen we shall show that this is false in general.

Chapter Ten

A cancellation theorem for extensions

The module cancellation problem asks whether, given modules X, X' and Y over a ring Λ, the existence of an isomorphism $X \oplus Y \cong X' \oplus Y$ implies that $X \cong X'$. In the next chapter we shall establish cancellation criteria for certain modules over $\mathbb{Z}[G(p,q)]$ defined by extensions of the form

$$0 \to K \to X \to Q \to 0$$

where K is a direct sum of row modules $R(i)$ and Q is either $\mathbb{Z}[C_q]$ or its augmentation ideal. The present chapter is a preliminary to this and establishes a general cancellation criterion for such extensions. The method of $\mathfrak{K}$-$\mathfrak{Q}$-modules, introduced in [37] and elaborated here, is, in principle, applicable in wider contexts than strictly necessary for $G(p,q)$ modules.

§76: $\mathfrak{K}$-$\mathfrak{Q}$-modules:

We work throughout in the category $\mathcal{M}od_\Lambda$ of right modules over a ring Λ. Let $\mathfrak{C}$ be a class of modules which is closed with respect to isomorphism. Recall that $\mathfrak{C}$ is a semigroup with respect to '$\oplus$' when if $C, C' \in \mathfrak{C}$ then $C \oplus C' \in \mathfrak{C}$. In that case we say that $\mathfrak{C}$ is a *cancellation semigroup* when

(a) If $C, C', C'' \in \mathfrak{C}$ and $C \oplus C' \cong C \oplus C''$ then $C' \cong C''$.

Furthermore we shall say that $\mathfrak{C}$ is a *strong cancellation semigroup* when in addition

(b) If C, S are nonzero Λ-modules such that $C \oplus S \in \mathfrak{C}$ then

$$C \in \mathfrak{C} \Longrightarrow S \in \mathfrak{C}.$$

Let $\mathfrak{K}$ and $\mathfrak{Q}$ be classes of modules which, without further mention, we assume to be closed with respect to isomorphism. We say that $(\mathfrak{K}, \mathfrak{Q})$ is an *admissible pair* when the following properties **(I)**–**(V)** are satisfied:

(I) $\mathfrak{K}$ is a cancellation semigroup;
(II) If $K \in \mathfrak{K}$ and $K' \subset K$ is a Λ-submodule then $K' \in \mathfrak{K}$;
(III) $\mathfrak{Q}$ is a strong cancellation semigroup;
(IV) $\mathrm{Hom}_\Lambda(K, Q) = 0$ for all $K \in \mathfrak{K}$ and all $Q \in \mathfrak{Q}$;
(V) If $Q \in \mathfrak{Q}$ then Q is 1-coprojective.

For the rest of this section we shall assume without further mention that $(\mathfrak{K}, \mathfrak{Q})$ is an admissible pair. If $K \in \mathfrak{K}$ and $Q \in \mathfrak{Q}$ we say that a Λ-module X is a $\mathfrak{K}$-$\mathfrak{Q}$-module *of type* (K, Q) when there exists an exact sequence of the form

$$\mathcal{E} = (0 \to K \xrightarrow{i} X \xrightarrow{p} Q \to 0).$$

$\mathcal{E}$ is then said to be an *admissible* $\mathfrak{K}$-$\mathfrak{Q}$ *sequence* for X. Given another exact sequence $\mathcal{E}' = (0 \to K' \xrightarrow{i'} X' \xrightarrow{p'} Q' \to 0)$ we write $\mathcal{E} \cong \mathcal{E}'$ when there exists a commutative diagram

$$\begin{array}{ccccccccc} 0 \to & K & \xrightarrow{i} & X & \xrightarrow{p} & Q & \to 0 \\ & \downarrow h_- & & \downarrow h & & \downarrow h_+ & \\ 0 \to & K' & \xrightarrow{i'} & X' & \xrightarrow{p'} & Q' & \to 0 \end{array}$$

in which h_-, h and h_+ are all isomorphisms. We note the following general statement:

(76.1) Let $\mathcal{E} = (0 \to K \xrightarrow{i} X \xrightarrow{p} Q \to 0)$; $\mathcal{E}' = (0 \to K' \xrightarrow{i'} X' \xrightarrow{p'} Q' \to 0)$ be admissible exact sequences; if $X \cong X'$ then $\mathcal{E} \cong \mathcal{E}'$; in particular, $K \cong K'$ and $Q \cong Q'$.

Proof. Suppose given an isomorphism $h : X \xrightarrow{\simeq} X'$. As $\mathrm{Hom}_\Lambda(K, Q') = 0$ then $p' \circ h \circ i = 0$ and so $h(\mathrm{Im}(i)) \subset \mathrm{Ker}(p')$. However, $\mathrm{Im}(i) = \mathrm{Ker}(p)$. Hence h induces a homomorphism $h_* : X/\mathrm{Ker}(p) \to X'/\mathrm{Ker}(p')$. From the Noether isomorphisms $Q \cong X/\mathrm{Ker}(p)$ and $Q' \cong X'/\mathrm{Ker}(p')$ we obtain a commutative diagram

$$(*)\qquad \begin{array}{c}\mathcal{E}\\ \downarrow\\ \mathcal{E}'\end{array} = \begin{pmatrix} 0\to & K & \xrightarrow{i} & X & \xrightarrow{p} & Q & \to 0\\ & \downarrow h_- & & \downarrow h & & \downarrow h_+ & \\ 0\to & K' & \xrightarrow{i'} & X' & \xrightarrow{p'} & Q' & \to 0\end{pmatrix}$$

in which h_+ is necessarily surjective. Likewise, taking an isomorphism $g : X' \xrightarrow{\simeq} X$ we get a commutative diagram

$$(**)\qquad \begin{array}{ccccccc} 0\to & K' & \xrightarrow{i'} & X' & \xrightarrow{p'} & Q' & \to 0\\ & \downarrow g_- & & \downarrow g & & \downarrow g_+ & \\ 0\to & K & \xrightarrow{i} & X & \xrightarrow{p} & Q & \to 0\end{array}$$

Taking $g = h^{-1}$ and composing (*) and (**) we get a commutative diagram

$$\begin{array}{ccccccc} 0\to & K & \xrightarrow{i} & X & \xrightarrow{p} & Q & \to 0\\ & \downarrow g_-\circ h_- & & \downarrow \mathrm{Id}_X & & \downarrow g_+\circ h_+ & \\ 0\to & K & \xrightarrow{i} & X & \xrightarrow{p} & Q & \to 0\end{array}$$

from which it follows that $g_+ \circ h_+ = \mathrm{Id}_Q$. Thus h_+ is also injective and hence we have an isomorphism $h_+ : Q \xrightarrow{\simeq} Q'$. Extending (*) to the left by adding zeroes it follows from the Five Lemma that $h_- : K \xrightarrow{\simeq} K'$ is also an isomorphism. Thus (*) gives the required isomorphism $h : \mathcal{E} \xrightarrow{\simeq} \mathcal{E}'$. □

It follows that the type of a $\mathfrak{K}$-$\mathfrak{Q}$ module is invariant under isomorphism; that is:

Corollary 76.2: *Let X, X' be $\mathfrak{K}$-$\mathfrak{Q}$-modules of types (K, Q) and (K', Q') respectively; if $X' \cong X$ then $K' \cong K$ and $Q' \cong Q$.*

The above is a form of rigidity for $\mathfrak{K}$-$\mathfrak{Q}$-modules. There is a slightly less obvious but quite powerful form of rigidity for extensions. To pursue this we first consider the notion of separating an extension as a direct sum. Thus given an extension $\mathcal{E} = (0 \to K \xrightarrow{i} X \xrightarrow{p} Q \to 0)$ where X is the internal direct sum $X = X_1 \dot{+} X_2$ of submodules X_1, X_2 we put $K_i = X_i \cap \mathrm{Ker}(p)$

so that we have extensions

$$\mathcal{E}_r = (0 \to K_r \xrightarrow{i_r} X_r \xrightarrow{p_r} p(X_r) \to 0).$$

The direct sum extension $\mathcal{E}_1 \oplus \mathcal{E}_2$ is then defined by

$$\mathcal{E}_1 \oplus \mathcal{E}_2 = (0 \to K_1 \oplus K_2 \xrightarrow{\binom{i_1\ 0}{0\ i_2}} X_1 \oplus X_2 \xrightarrow{\binom{p_1\ 0}{0\ p_2}} p(X_1) \oplus p(X_2) \to 0)$$

and there is a mapping of extensions $\mu : \mathcal{E}_1 \oplus \mathcal{E}_2 \to \mathcal{E}$

$$\begin{matrix} \mathcal{E}_1 \oplus \mathcal{E}_2 \\ \downarrow \mu \\ \mathcal{E} \end{matrix} = \left(\begin{matrix} 0 \to & K_1 \oplus K_2 & \xrightarrow{\binom{i_1\ 0}{0\ i_2}} & X_1 \oplus X_2 & \xrightarrow{\binom{p_1\ 0}{0\ p_2}} & p(X_1) \oplus p(X_2) & \to 0 \\ & \downarrow \mu_K & & \downarrow \mu_X & & \downarrow \mu_Q & \\ 0 \to & K & \xrightarrow{i} & X & \xrightarrow{p} & Q & \to 0 \end{matrix} \right)$$

where μ_K, μ_X, μ_Q are given by the appropriate additions $\begin{pmatrix} x_1 \\ x_2 \end{pmatrix} \mapsto x_1 + x_2$. On extending to the left by zeroes it follows from the Five Lemma that:

(76.3) $\mu : \mathcal{E}_1 \oplus \mathcal{E}_2 \to \mathcal{E}$ is an isomorphism $\Longleftrightarrow$ $p(X_1) \cap p(X_2) = 0$.

Proposition 76.4: *Let $\mathcal{E} = (0 \to K \xrightarrow{i} X \xrightarrow{p} Q \to 0)$ be an admissible $\mathfrak{K}$-$\mathfrak{Q}$ exact sequence in which X is the internal direct sum $X = X_1 \dot{+} X_2$ of submodules X_1, X_2. Suppose also that $p(X_1) \cap p(X_2) = 0$ and that X_1 is a $\mathfrak{K}$-$\mathfrak{Q}$ module of type (K_1, Q_1); then there exist $K_2 \in \mathfrak{K}$ and $Q_2 \in \mathfrak{Q}$ such that X_2 is a $\mathfrak{K}$-$\mathfrak{Q}$ of type (K_2, Q_2) where $K \cong K_1 \oplus K_2$ and $Q \cong Q_1 \oplus Q_2$.*

Proof. By hypothesis, X_1 is defined by a $\mathfrak{K}$-$\mathfrak{Q}$ exact sequence

$$\mathcal{E}_1 = (0 \to K_1 \xrightarrow{i_1} X_1 \xrightarrow{p_1} Q_1 \to 0).$$

We compare this with the exact sequence

$$\mathcal{E}' = (0 \to K' \xrightarrow{i'} X_1 \xrightarrow{p} p(X_1) \to 0)$$

where $K' = \mathrm{Ker}(p) \cap X_1$ via the diagram

$$\mathcal{D}_+ \qquad \left\{ \begin{matrix} 0 \to & K' & \xrightarrow{i'} & X_1 & \xrightarrow{p} & p(X_1) & \to 0 \\ & & & \downarrow \mathrm{Id} & & & \\ 0 \to & K_1 & \xrightarrow{i_1} & X_1 & \xrightarrow{p_1} & Q_1 & \to 0 \end{matrix} \right.$$

As K' is a submodule of $K \in \mathfrak{K}$ then $K' \in \mathfrak{K}$ by property **(II)**. As $Q_1 \in \mathfrak{Q}$ then $p_1 \circ i' = 0$ by property **(IV)**. Thus we may complete $\mathcal{D}_+$ to a commutative diagram

$$\widetilde{\mathcal{D}}_+ \quad \begin{cases} 0 \to & K' & \stackrel{i'}{\to} & X_1 & \stackrel{p}{\to} & p(X_1) & \to 0 \\ & \downarrow h_- & & \downarrow \mathrm{Id} & & \downarrow h_+ & \\ 0 \to & K_1 & \stackrel{i_1}{\to} & X_1 & \stackrel{p_1}{\to} & Q_1 & \to 0 \end{cases}$$

in which h_+ is necessarily surjective. In similar fashion, consider the diagram

$$\mathcal{D}_- \quad \begin{cases} 0 \to & K_1 & \stackrel{i_1}{\to} & X_1 & \stackrel{p_1}{\to} & Q_1 & \to 0 \\ & & & \downarrow \mathrm{Id} & & & \\ 0 \to & K' & \stackrel{i'}{\to} & X_1 & \stackrel{p}{\to} & p(X_1) & \to 0 \end{cases}$$

As $K_1 \in \mathfrak{K}$ and $Q \in \mathfrak{Q}$ then $\mathrm{Hom}_\Lambda(K_1, Q) = 0$ by property **(IV)**. As $p(X_1) \subset Q$ then $\mathrm{Hom}_\Lambda(K_1, p(X_1)) = 0$. In particular, $p \circ i_1 = 0$ so we can complete $\mathcal{D}_-$ to a commutative diagram

$$\widetilde{\mathcal{D}}_- \quad \begin{cases} 0 \to & K_1 & \stackrel{i_1}{\to} & X_1 & \stackrel{p_1}{\to} & Q_1 & \to 0 \\ & \downarrow g_- & & \downarrow \mathrm{Id} & & \downarrow g_+ & \\ 0 \to & K' & \stackrel{i'}{\to} & X_1 & \stackrel{p}{\to} & p(X_1) & \to 0 \end{cases}$$

Composing $\widetilde{\mathcal{D}}_- \circ \widetilde{\mathcal{D}}_+$ we obtain a commutative diagram

$$\begin{cases} 0 \to & K' & \stackrel{i'}{\to} & X_1 & \stackrel{p}{\to} & p(X_1) & \to 0 \\ & \downarrow g_- \circ h_- & & \downarrow \mathrm{Id} & & \downarrow g_+ \circ h_+ & \\ 0 \to & K' & \stackrel{i'}{\to} & X_1 & \stackrel{p}{\to} & p(X_1) & \to 0 \end{cases}$$

from which it follows that $g_+ \circ h_+ = \mathrm{Id}$. Thus h_+ is also injective and so gives an isomorphism $h_+ : p(X_1) \xrightarrow{\simeq} Q_1$; thus $p(X_1) \in \mathfrak{Q}$. As $p(X_1) \cap p(X_2) = 0$ then $Q \cong Q_1 \oplus p(X_2)$. As $\mathfrak{Q}$ is a strong cancellation semigroup then $p(X_2) \in \mathfrak{Q}$. Thus putting $Q_2 = p(X_2)$ we see that X_2 is given by an exact sequence

$$(*) \qquad (0 \to K_2 \xrightarrow{i_2} X_2 \xrightarrow{p} Q_2 \to 0)$$

where $K_2 = \mathrm{Ker}(p) \cap X_2$. As K_2 is a submodule of $K \in \mathfrak{K}$ then $K_2 \in \mathfrak{K}$ so that the exact sequence (*) is admissible and defines X_2 as a $\mathfrak{K}$-$\mathfrak{Q}$ module of type (K_2, Q_2) where $K \cong K_1 \oplus K_2$ and $Q \cong Q_1 \oplus Q_2$. □

For $r = 1, 2$ let $\mathcal{E}_r = (0 \to K_r \xrightarrow{i_r} Y_r \xrightarrow{p_r} Q_r \to 0)$ be admissible exact sequences and let S be a Λ-module such that $Y_1 \oplus S \cong Y_1 \oplus Y_2$. We say that the Λ module S is *well positioned with respect to* $(\mathcal{E}_1, \mathcal{E}_2)$ when there exists an isomorphism $\varphi : Y_1 \oplus S \xrightarrow{\simeq} Y_1 \oplus Y_2$ such that $p \circ \varphi(Y_1) \bigcap p \circ \varphi(S) = 0$ where $p = (p_1, p_2)$;

Theorem 76.5: *For $r = 1, 2$ let Y_r be a $\mathfrak{K}$-$\mathfrak{Q}$ module given by an admissible exact sequence $\mathcal{E}_r = (0 \to K_r \xrightarrow{i_r} Y_r \xrightarrow{i_r} Q_r \to 0)$ and let S be a Λ-module which is well positioned with respect to $(\mathcal{E}_1, \mathcal{E}_2)$; then S is a $\mathfrak{K}$-$\mathfrak{Q}$ module of the same type as Y_2; that is, S is described by an exact sequence of the form $0 \to K_2 \xrightarrow{j} S \xrightarrow{\pi} Q_2 \to 0$.*

Proof. Put $X = Y_1 \oplus Y_2$, $K = K_1 \oplus K_2$ and $Q = Q_1 \oplus Q_2$ and $i = \begin{pmatrix} i_1 & 0 \\ 0 & i_2 \end{pmatrix}$. Then $K \in \mathfrak{K}$ and, by hypothesis, $Q \in \mathfrak{Q}$ so that the exact sequence

$$\mathcal{E} = (0 \to K \xrightarrow{i} X \xrightarrow{p} Q \to 0)$$

is admissible and defines X as a $\mathfrak{K}$-$\mathfrak{Q}$ module. Let $\varphi : Y_1 \oplus S \xrightarrow{\simeq} Y_1 \oplus Y_2$ be an isomorphism such that $p \circ \varphi(Y_1) \bigcap p \circ \varphi(S) = 0$ where $p = (p_1, p_2)$. Put $X_1 = \varphi(Y_1)$ and $X_2 = \varphi(S)$. Then X is the internal direct sum $X = X_1 \dot{+} X_2$ and, by hypothesis, $p(X_1) \bigcap p(X_2) = 0$. Moreover, as $X_1 \cong Y_1$ then X_1 is a $\mathfrak{K}$-$\mathfrak{Q}$ module. It follows from (76.4) that X_2 is also a $\mathfrak{K}$-$\mathfrak{Q}$ module with a defining sequence of the form

$$0 \to K' \to X_2 \to Q' \to 0$$

for some $K' \in \mathfrak{K}$ and some $Q' \in \mathfrak{Q}$ such that $Q_1 \oplus Q' \cong Q_1 \oplus Q_2$. As $\mathfrak{Q}$ is a strong cancellation semigroup then $Q' \cong Q_2$. As $S \cong X_2$ then, collecting the above remarks, we see that S is $\mathfrak{K}$-$\mathfrak{Q}$ module with a defining sequence of the form

$$0 \to K' \xrightarrow{j} S \xrightarrow{\pi} Q_2 \to 0.$$

To conclude we must show that $K' \cong K_2$. In this direction we note that $Y_1 \oplus S$ is described by an exact sequence of the form

$$\mathcal{F} = (0 \to K_1 \oplus K' \to Y_1 \oplus S \to Q_1 \oplus Q_2 \to 0)$$

whilst $Y_1 \oplus Y_2$ is described by the exact sequence

$$\mathcal{E} = (0 \to K_1 \oplus K_2 \to Y_1 \oplus Y_2 \to Q_1 \oplus Q_2 \to 0)$$

already considered. Comparing these by means of φ and φ^{-1} it follows from (76.1) that $K_1 \oplus K_2 \cong K_1 \oplus K'$ so that, by property **(I)**, $K_2 \cong K'$. □

§77: A strong cancellation semigroup:

We denote by $\Lambda = \mathbb{Z}[G]$ the integral group ring of a finite group G and by I the augmentation ideal $I = \mathrm{Ker}(\epsilon : \Lambda \to \mathbf{Z})$. We define $Q(a,b) = \Lambda^{(a)} \oplus I^{(b)}$ for integers a, $b \geq 0$ and put

$$\mathfrak{Q} = \{Q(a,b) | a + b > 0\}.$$

Clearly $Q(a,b) \oplus Q(c,d) \cong Q(a+c, b+d)$ so that, up to isomorphism, $\mathfrak{Q}$ forms an additive semigroup under '$\oplus$'. For the remainder of this chapter, we will use the term 'Swan module' in its original context (cf §44); namely, a Swan module S will mean an extension module of the form

$$0 \to I \to S \to \mathbb{Z} \to 0.$$

We will show that $\mathfrak{Q}$ is a strong cancellation semigroup provided G satisfies the Eichler condition and each projective Swan module over Λ is free. When this is the case, then every Λ lattice satisfies the Eichler condition and we may restate Jacobinski's cancellation theorem (4.10) as follows;

(77.1) Let X, Y, M, N be Λ-lattices such that $X \oplus M \cong X \oplus N$ and $X \oplus Y \cong M^{(m)}$ for some $m \geq 1$; if G satisfies the Eichler condition then $M \cong N$.

Denote by $\mathcal{P}(a,b)$ the following statement:

$$\mathcal{P}(a,b): \qquad Q(a,b) \oplus S \cong Q(a,b) \oplus Q(c,d) \implies S \cong Q(c,d)$$

(77.2) $\mathcal{P}(1,0)$ holds provided G satisfies the Eichler condition.

Proof. Suppose $\Lambda \oplus Q(c,d) \cong \Lambda \oplus S$ and first consider the case where $c = 0$ so that $\Lambda \oplus I^{(d)} \cong \Lambda \oplus S$. Let $h : \Lambda \oplus S \xrightarrow{\cong} \Lambda \oplus I^{(d)}$ be an isomorphism and let $\eta = (\epsilon, 0) : \Lambda \oplus I^{(d)} \to \mathbb{Z}$. Then is it straightforward to see that h restricts to an isomorphism $h : I \oplus S \xrightarrow{\cong} I \oplus I^{(d)}$. Applying

(77.1) $X = I$, $M = I^{(d)}$ and $N = S$ it now follows that $S \cong I^{(d)} = Q(0, d)$ as required. In the general case where $c \neq 0$ we may write

$$\Lambda \oplus (\Lambda \oplus Q(c-1, d)) \;\cong\; \Lambda \oplus S.$$

We apply (77.1) with $X = \Lambda$, $M = \Lambda \oplus Q(c-1, d)$, $N = S$ to conclude that $\Lambda \oplus Q(c-1, d) \cong S$. Thus $S \cong Q(c, d)$. □

(77.3) $\mathcal{P}(0, 1)$ holds provided G satisfies the Eichler condition and every projective Swan module over Λ is isomorphic to Λ.

Proof. Suppose $I \oplus Q(c, d) \cong I \oplus S$ and first consider the case where $d = 0$ so that $I \oplus \Lambda^c \cong I \oplus S$. Now $\mathrm{Ext}^1(\Lambda, N) = 0$ for any Λ module N and hence

$$\mathrm{Ext}^1(I, N) \oplus \mathrm{Ext}^1(S, N) \cong \mathrm{Ext}^1(I, N).$$

As $\mathrm{Ext}^1(I, N)$ is finite then $\mathrm{Ext}^1(S, N) = 0$ so that, by (15.7), S is projective. Now suppose that $c = 1$. Let $h : I \oplus \Lambda \xrightarrow{\simeq} I \oplus S$ be an isomorphism. We claim that $S \cong \Lambda$. To see this, put $\eta = (0, \epsilon, 0) \circ h^{-1} : I \oplus S \to \mathbb{Z}$. We have a diagram of Λ-homomorphisms in which the rows are exact

$$\begin{array}{ccccccc} 0 \to & I \oplus I & \stackrel{i}{\to} & I \oplus \Lambda & \stackrel{(\epsilon,0)}{\to} & \mathbb{Z} & \to 0 \\ & & & \downarrow h & & & \\ 0 \to & \mathrm{Ker}(\eta) & \stackrel{j}{\to} & I \oplus S & \stackrel{\eta}{\to} & \mathbb{Z} & \to 0. \end{array}$$

As $\mathrm{Hom}_\Lambda(I, \mathbb{Z}) = 0$ then $\eta \circ h \circ i : I \oplus I \to \mathbb{Z}$ is identically zero and we may complete the above to the following commutative diagram

$$\begin{array}{ccccccc} 0 \to & I \oplus I & \stackrel{i}{\to} & I \oplus \Lambda & \stackrel{(\epsilon,0)}{\to} & \mathbb{Z} & \to 0 \\ & \downarrow h_- & & \downarrow h & & \downarrow h_+ & \\ 0 \to & \mathrm{Ker}(\eta) & \stackrel{j}{\to} & I \oplus S & \stackrel{\eta}{\to} & \mathbb{Z} & \to 0. \end{array}$$

As h is an isomorphism then $h_+ : \mathbb{Z} \to \mathbb{Z}$ is certainly surjective and therefore also an isomorphism. By extending one place to the left by zeroes, it follows from the Five Lemma that $h_- : I \oplus I \to \mathrm{Ker}(\eta)$ is an isomorphism. Let η_0 be the restriction $\eta_0 = \eta_{|S}$ and put $S_0 = \mathrm{Ker}(\eta_0)$. As $\mathrm{Hom}_\Lambda(I, \mathbb{Z}) = 0$ then $\eta = (0, \eta_0)$ and so

$$\mathrm{Ker}(\eta) = I \oplus S_0.$$

Thus $h_- : I \oplus I \xrightarrow{\simeq} I \oplus S_0$ is an isomorphism. Applying (77.1) with $X = Y = M = I$ it follows that $S_0 \cong I$. Moreover, again appealing to the fact that $\mathrm{Hom}_\Lambda(I, \mathbb{Z}) = 0$ we see that η_0 is surjective and so S occurs in an exact sequence of the form

$$0 \to I \to S \to \mathbb{Z} \to 0.$$

Thus S is a Swan module. As S is also projective then, by hypothesis, $S \cong \Lambda$ as claimed. This disposes of the case where $c = 1$ and $d = 0$.

Now consider the cases where $c \geq 2$ and $d = 0$ and suppose that

$$(*) \qquad I \oplus \Lambda^c \cong I \oplus S.$$

As S is projective it follows from the theorem of Swan (4.15) that $S \cong \Lambda^{c-1} \oplus P$ where P is a projective of rank 1. Hence we can re-write (*) as

$$\Lambda^{c-1} \oplus I \oplus \Lambda \cong \Lambda^{c-1} \oplus I \oplus P.$$

Applying (77.1) with $X = \Lambda^{(c-1)}$ and $M = I \oplus \Lambda$ then $I \oplus \Lambda \cong I \oplus P$ so that, by the case $c = 1$ above it follows that $P \cong \Lambda$ and hence $S \cong \Lambda^c$. This disposes of all cases where $d = 0$. Finally suppose that $I \oplus Q(c, d) \cong I \oplus S$ where $d \geq 1$. Then I is a direct summand of $Q(c, d)$ so that $Q(c, d) \cong S$ by Jacobinski's Theorem. This completes the proof of $\mathcal{P}(0, 1)$. □

(77.4) $\mathcal{P}(1, 0) \wedge \mathcal{P}(a, b) \implies \mathcal{P}(a + 1, b)$

Proof. Suppose that $Q(a + 1, b) \oplus Q(c, d) \cong Q(a + 1, b) \oplus S$. We must show that $Q(c, d) \cong S$. Re-writing we see that $\Lambda \oplus Q(c + a, b + d) \cong \Lambda \oplus Q(a, b) \oplus S$. By $\mathcal{P}(1, 0)$ it follows that $Q(c + a, b + d) \cong Q(a, b) \oplus S$ and hence

$$Q(a, b) \oplus Q(c, d) \cong Q(a, b) \oplus S.$$

By $\mathcal{P}(a, b)$ it now follows, as desired, that $Q(c, d) \cong S$. □

(77.5) $\mathcal{P}(0, 1) \wedge \mathcal{P}(a, b) \implies \mathcal{P}(a, b + 1)$

Proof. Suppose that $Q(a, b+1) \oplus Q(c, d) \cong Q(a, b+1) \oplus S$. Re-writing we see that $I \oplus Q(c + a, b + d) \cong I \oplus Q(a, b) \oplus S$. By $\mathcal{P}(0, 1)$ it follows that $Q(c + a, b + d) \cong Q(a, b) \oplus S$ and hence

$$Q(a, b) \oplus Q(c, d) \cong Q(a, b) \oplus S.$$

The desired conclusion that $Q(c, d) \cong S$ now follows from $\mathcal{P}(a, b)$. □

It follows easily from (77.2)–(77.5) that $\mathcal{P}(a,b)$ is true for all $a, b \geq 0$ such that $a+b \neq 0$. Thus we have shown that if G satisfies the Eichler condition and each projective Swan module is free then $\mathfrak{Q}$ is a cancellation semigroup. Finally, suppose that $Q \in \mathfrak{Q}$ and that S is a nonzero Λ-module such that $Q \oplus S \in \mathfrak{Q}$. Writing $Q = Q(a,b)$ and $Q \oplus S = Q(e,f)$ then consideration of rational Wedderburn decompositions shows that $a \leq e$ and $b \leq f$. Writing $c = e - a$ and $d = f - b$ we see that $Q(a,b) \oplus S \cong Q(a,b) \oplus Q(c,d)$. It follows from (77.5) that $S \cong Q(c,d)$ and hence $S \in \mathfrak{Q}$. Thus as claimed we have shown:

(77.6) If G satisfies the Eichler condition and each projective Swan module is free then $\mathfrak{Q}$ is a strong cancellation semigroup.

§78: $\mathfrak{K}$-$\mathfrak{Q}$-modules for metacyclic groups:

Taking $\Lambda = \mathbb{Z}[G(p,q)]$ we define classes $\mathfrak{K}$, $\mathfrak{Q}$ of Λ-modules as follows:

$\mathfrak{K}$: $K \in \mathfrak{K} \iff K$ is isomorphic to a submodule of $\underbrace{\mathcal{T}_q \oplus \cdots \oplus \mathcal{T}_q}_{m}$ for some $m \geq 1$. and we take $\mathfrak{Q}$ to be the class of modules of §77, but now considered as Λ-modules via the homomorphism $\Lambda \to \mathbb{Z}[C_q]$; that is:

$\mathfrak{Q} : Q \in \mathfrak{Q} \iff Q \cong \widetilde{\mathbb{Z}}[C_q]^{(a)} \oplus \widetilde{I_q^*}^{(b)}$ for some integers a, b with $a \geq 0$, $b \geq 0$ and $a + b \neq 0$.

We claim that $(\mathfrak{K}, \mathfrak{Q})$ satisfies properties **(I)**–**(V)** of §76. To verify property **(II)** we first define

$$\mathcal{P} = \{\text{isomorphism classes of finitely generated projective modules over } \mathcal{T}_q\}.$$

Clearly $\mathcal{P} \subset \mathfrak{K}$. To show that $\mathfrak{K} \subset \mathcal{P}$ we first recall from (67.10) that $\mathcal{T}_q$ is an hereditary ring. It now follows, for example from Kaplansky's Theorem (67.1), that if K is isomorphic to a submodule of $\underbrace{\mathcal{T}_q \oplus \cdots \oplus \mathcal{T}_q}_{m}$ then $K \in \mathcal{P}$. Thus $\mathfrak{K} \subset \mathcal{P}$ and hence $\mathfrak{K} = \mathcal{P}$. This establishes property **(II)**. Moreover, as $\mathfrak{K} = \mathcal{P}$ then property **(I)** is also established as, by (51.10), $\mathcal{P}$ is a cancellation semigroup. Property **(III)** for $\mathfrak{Q}$ holds by (77.6) and (45.7). Moreover, as any Λ-lattice is 1-coprojective then property **(V)** also holds. Thus to show that $(\mathfrak{K}$-$\mathfrak{Q})$ is an admissible pair, it suffices to show that property **(IV)** holds. As each $Q \in \mathfrak{Q}$ imbeds as a submodule of some

$\widetilde{\mathbb{Z}}[C_q]^{(m)}$ it follows easily from (69.10) that:

(78.1) $\mathrm{Hom}_\Lambda(\mathcal{T}_q, Q) = 0$ for all $Q \in \mathfrak{Q}$.

Theorem 78.2: $\mathrm{Hom}_\Lambda(K, Q) = 0$ *for all* $K \in \mathfrak{K}$ *and all* $Q \in \mathfrak{Q}$.

Proof. As we have already observed, $\mathcal{T}_q$ is an hereditary ring. Thus every $\mathcal{T}_q$-submodule of $\mathcal{T}_q^{(m)} = \underbrace{\mathcal{T}_q \oplus \cdots \oplus \mathcal{T}_q}_{m}$ is projective over $\mathcal{T}_q$. In particular, if $L \subset \mathcal{T}_q^{(m)}$ is a $\mathcal{T}_q$-submodule then there exists a $\mathcal{T}_q$-module L' such that $L \oplus L' \cong_{\mathcal{T}_q} \mathcal{T}_q^{(m)}$. Now suppose that $K \in \mathfrak{K}$ is Λ-submodule of $\mathcal{T}_q^{(m)}$. Then there exists a $\mathcal{T}_q$-submodule L of $\mathcal{T}_q^{(m)}$ such that $K \cong_\Lambda (\pi_+)^*(L)$ and hence also a Λ-module K' of the form $K' = (\pi_+)^*(L')$ such that

$$K \oplus K' \cong_\Lambda \mathcal{T}_q^{(m)}.$$

Taking $Q \in \mathfrak{Q}$ we see that $\mathrm{Hom}_\Lambda(K, Q) \oplus \mathrm{Hom}_\Lambda(K', Q) \cong \mathrm{Hom}_\Lambda(\mathcal{T}_q, Q)^{(m)}$. However, $\mathrm{Hom}_\Lambda(\mathcal{T}_q, Q) = 0$ by (78.1) so that $\mathrm{Hom}_\Lambda(K, Q) = 0$. □

Collecting our results we now conclude that:

(78.3) $(\mathfrak{K}, \mathfrak{Q})$ is an admissible pair.

If M is a Λ-lattice we put $M_\mathbb{Q} = M \otimes_\mathbb{Z} \mathbb{Q}$ so that $M_\mathbb{Q}$ is a semisimple module over the semisimple ring $\Lambda_\mathbb{Q}$. Similar proofs to (69.10) and (78.2), using the fact that $\mathrm{Hom}_{\mathbb{Q}[C_p]}(I[C_p]_\mathbb{Q}, \mathbb{Q}) = 0$ show the following, which we note for future reference:

(78.4) $\mathrm{Hom}_{\Lambda_\mathbb{Q}}(K_\mathbb{Q}, Q_\mathbb{Q}) = 0$ for all $K \in \mathfrak{K}$ and all $Q \in \mathfrak{Q}$.

We note there is a dual statement to (78.2), namely:

(78.5) $\mathrm{Hom}_\Lambda(Q, K) = 0$ for all $K \in \mathfrak{K}$ and all $Q \in \mathfrak{Q}$.

§79: A cancellation theorem for extensions:

In this section we continue with the notation from §78. In addition, we will adhere to the following notation:

Q' : a module in $\mathfrak{Q}$;

L : a module of type $(K_1, Q(a, b))$;

X : a module of type $(K_2, Q(c, d))$.

We define $r(a,b) = aq + b(q-1) = \mathrm{rk}_{\mathbb{Z}}(Q(a,b)$ so that

$$r(a+c, b+d) = r(a,b) + r(c,d).$$

Moreover, we take L and X to be defined by admissible sequences

$$\mathcal{L} \;=\; (0 \to K_1 \to L \to Q(a,b) \to 0)$$

$$\mathcal{X} \;=\; (0 \to K_2 \to X \to Q(c,d) \to 0$$

With this notation we have:

Proposition 79.1: *If $f : L \to Q'$ is a Λ-homomorphism then*

$$\mathrm{rk}_{\mathbb{Z}}(\mathrm{Im}(f)) \;\leq\; r(a,b).$$

Proof. It follows from (78.2) that the composition $f \circ i : K_2 \to Q'$ is identically zero. Hence f induces a Λ-homomorphism $f_* : X/K_2 \cong \mathbb{Z}[C_q]^{(a)} \oplus I_q^{(b)} \to Q'$. The conclusion follows as $\mathrm{rk}_{\mathbb{Z}}(\mathrm{Im}(f_*)) \;\leq\; \mathrm{rk}_{\mathbb{Z}}(I_q^{(a)} \oplus \mathbb{Z}[C_q]^{(b)}) = r(a,b)$. □

Likewise we have:

Proposition 79.2: *If $f : X \to Q'$ is a Λ-homomorphism then*

$$\mathrm{rk}_{\mathbb{Z}}(\mathrm{Im}(f)) \leq r(c,d).$$

In what follows S will denote a Λ-module *which we do not assume, in advance, is a* $\mathfrak{K}$-$\mathfrak{Q}$ module. With this notation we shall prove:

Theorem 79.3: *If $L \oplus S \cong L \oplus X$ then S is well positioned with respect to $(\mathcal{L}, \mathcal{X})$.*

Proof. We suppose L is given by an exact sequence of the form

$$0 \to K_1 \xrightarrow{i_1} L \xrightarrow{p_1} Q(a,b) \to 0$$

and that X is given by an exact sequence

$$0 \to K_2 \xrightarrow{i_2} X \xrightarrow{p_2} Q(c,d) \to 0$$

Let $h : L \oplus S \xrightarrow{\simeq} L \oplus X$ be an isomorphism and put $j = h^{-1} \circ (i_1 \times i_2)$ and $\pi = (p_1 \times p_2) \circ h$ so that $L \oplus S$ is a $\mathfrak{K}$-$\mathfrak{Q}$-module defined by the exact sequence

$$0 \to K_1 \oplus K_2 \xrightarrow{j} L \oplus S \xrightarrow{\pi} Q(a,b) \oplus Q(c,d) \to 0.$$

It suffices to show that

$$(*) \qquad \pi(L) \cap \pi(S) = 0.$$

To see this, observe that $L_{\mathbb{Q}} \oplus S_{\mathbb{Q}} \cong L_{\mathbb{Q}} \oplus X_{\mathbb{Q}}$. By Wedderburn's Theorem it follows that $S_{\mathbb{Q}} \cong X_{\mathbb{Q}}$. Thus by (79.2), the restriction π_S of $\pi : S_{\mathbb{Q}} \to (Q(a,b) \oplus Q(c,d))_{\mathbb{Q}}$ has $\dim_{\mathbb{Q}}(\mathrm{Im}(\pi_S)) \leq r(c,d)$. Likewise the restriction $\pi_L : L_{\mathbb{Q}} \to (Q(a,b) \oplus Q(c,d))_{\mathbb{Q}}$ has $\dim_{\mathbb{Q}}(\mathrm{Im}(\pi_L)) \leq r(a,b)$. As $\pi : L \oplus S \xrightarrow{\pi} Q(a,b) \oplus Q(c,d) = Q(a+c,b+d)$ is surjective then

$$\begin{aligned} r(a+c,b+d) &= \mathrm{rk}_{\mathbb{Z}}(\pi(L) + \pi(S)) \\ &\leq \mathrm{rk}_{\mathbb{Z}}(\pi(L)) + \mathrm{rk}_{\mathbb{Z}}(\pi(S)) \\ &\leq r(a,b) + r(c,d). \end{aligned}$$

As $r(a+c,b+d) = r(a,b) + r(c,d)$ it follows that

$$\mathrm{rk}_{\mathbb{Z}}[\pi(L) + \pi(S)] = \mathrm{rk}_{\mathbb{Z}}(\pi(L)) + \mathrm{rk}_{\mathbb{Z}}(\pi(S)).$$

However, $\mathrm{rk}_{\mathbb{Z}}[\pi(L) \cap \pi(S)] = \mathrm{rk}_{\mathbb{Z}}(\pi(L)) + \mathrm{rk}_{\mathbb{Z}}(\pi(S)) - \mathrm{rk}_{\mathbb{Z}}[\pi(L) + \pi(S)]$. Hence $\mathrm{rk}_{\mathbb{Z}}[\pi(L) \cap \pi(S)] = 0$ and $\pi(L) \cap \pi(S) = 0$ as required. □

It now follows from (76.5) that:

Theorem 79.4: *Suppose $L \oplus S \cong L \oplus X$ where L, X are $\mathfrak{K}$-$\mathfrak{Q}$ modules of types (K_1, Q_1) and (K_2, Q_2) respectively; then S is a $\mathfrak{K}$-$\mathfrak{Q}$ module of the same type as X; that is, S is described by an exact sequence*

$$0 \to K_2 \to S \to Q_2 \to 0.$$

Chapter Eleven

Cancellation of quasi-Swan modules

In this chapter we apply the theory of $\mathfrak{K}$-$\mathfrak{Q}$ modules developed in Chapter Ten to study a particular class of extensions, the *quasi-Swan* modules. These take the form

$$0 \longrightarrow \bigoplus_{j \in J} R(j)^{(a_j)} \longrightarrow X \longrightarrow Q \longrightarrow 0$$

where J is a subset of $\{1, \ldots, q\}$ and Q is either $\mathbb{Z}[C_q]$ or I_q, here considered as Λ-modules via coinduction. We show that if the cohomology class defining such an extension is suitably *degenerate* then X possesses the property of cancelling copies of Λ; that is,

$$X \oplus \Lambda \cong X' \oplus \Lambda \implies X \cong X'.$$

In particular, the stable class $[X]$ of such a module is straight. This cancellation result is a generalization of [37]. It can be contrasted with the non-cancellation results of Swan [64] and Klingler [41]. In this context one should regard degeneracy as defining a stable range in which further non-cancellation cannot occur.

§80: The $\mathrm{End}_{\mathcal{D}\mathrm{er}}(K)$-module structure on $\mathrm{Ext}^1(Q, K)$

Let $\mathcal{E} = (0 \to K \stackrel{i}{\to} X \stackrel{p}{\to} Q \to 0)$ be an extension of Λ-modules. If $f \in \mathrm{End}_\Lambda(K)$ form the pushout extension $f_*(\mathcal{E}) = (0 \to K \to \varinjlim(f, i) \to Q \to 0)$; then the pairing

$$\begin{array}{ccc} \mathrm{End}_\Lambda(K) \times \mathrm{Ext}^1(Q, K) & \to & \mathrm{Ext}^1(Q, K) \\ (f, \mathcal{E}) & \mapsto & f_*(\mathcal{E}) \end{array}$$

defines the structure of a left $\mathrm{End}_\Lambda(K)$-module on $\mathrm{Ext}^1(Q, K)$. In general we expect the ring $\mathrm{End}_\Lambda(K)$ to be complicated. However by imposing the restriction that Q be 1-coprojective, the action of $\mathrm{End}_\Lambda(K)$ on $\mathrm{Ext}^1(Q, K)$ is reducible to an action of $\mathrm{End}_{\mathcal{D}\mathrm{er}}(K)$ which is usually much more straightforward. Thus suppose that $f \in \mathrm{End}_\Lambda(K)$ factors through a projective module P:

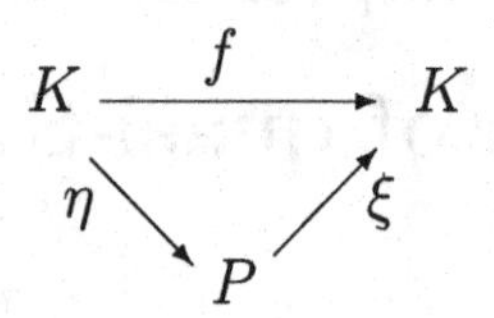

Then $f_*(\mathcal{E}) = \xi_*\eta_*(\mathcal{E})$ where $\eta_*(\mathcal{E}) = (0 \to P \to \varinjlim(\eta, i) \to Q \to 0)$. If Q is 1-coprojective then $\eta_*(\mathcal{E})$ splits and hence $f_*(\mathcal{E}) = \xi_*\eta_*(\mathcal{E})$ also splits. More generally, if $f - f'$ factors through a projective module then $f_*(\mathcal{E})$ and $f'_*(\mathcal{E})$ define the same class in $\mathrm{Ext}^1(Q, K)$; that is:

(80.1) If Q is 1-coprojective then $\mathrm{Ext}^1(Q, K)$ is a left module over $\mathrm{End}_{\mathcal{D}\mathrm{er}}(K)$.

In addition for $g \in \mathrm{End}_\Lambda(Q)$ then we may likewise form the pullback extension $g^*(\mathcal{E}) = (0 \to K \to \varprojlim(p, g) \to Q \to 0)$; the pairing

$$\begin{array}{ccc} \mathrm{Ext}^1(Q, K) \times \mathrm{End}_\Lambda(Q) & \to & \mathrm{Ext}^1(Q, K) \\ (\mathcal{E}, g) & \mapsto & g^*(\mathcal{E}) \end{array}$$

then defines the structure of a right $\mathrm{End}_\Lambda(Q)$-module on $\mathrm{Ext}^1(Q, K)$. Moreover, as $\mathrm{Ext}^1(P, K) = 0$ when P is projective this action descends to one of $\mathrm{End}_{\mathcal{D}\mathrm{er}}(Q)$ with no restriction on K. Taking the two together we obtain:

(80.2) If Q is 1-coprojective then $\mathrm{Ext}^1(Q, K)$ has the natural structure of an $\mathrm{End}_{\mathcal{D}\mathrm{er}}(K)$-$\mathrm{End}_{\mathcal{D}\mathrm{er}}(Q)$-bimodule.

Although we have noted it for the record, the right action of $\mathrm{End}_\Lambda(Q)$ on $\mathrm{Ext}^1(Q, K)$ plays no further part in our calculations. To elaborate on this, in this chapter and the next, our principal focus will be with modules X defined by means of extensions

$$\mathcal{E} = (0 \to K \to X \to Q \to 0)$$

and the extent to which homomorphisms between such modules are realized by homomorphisms between such extensions; that is, via commutative diagrams

$$\begin{array}{ccc} \mathcal{E} & & \\ \downarrow & = & \\ \mathcal{E}' & & \end{array} \left(\begin{array}{ccccccc} 0 \to & K & \stackrel{i}{\to} & X & \stackrel{p}{\to} & Q & \to 0 \\ & \downarrow h_- & & \downarrow h & & \downarrow h_+ & \\ 0 \to & K' & \stackrel{i'}{\to} & X' & \stackrel{p'}{\to} & Q' & \to 0 \end{array} \right)$$

Usually we will take $Q' = Q$ and $h_+ = \mathrm{Id}_Q$. In that case, h is an isomorphism if and only if h_- is an isomorphism. This situation occurs so frequently that we reserve the notation

(80.3) $$X \cong_{\mathrm{Id}} X'$$

to imply the existence of a commutative diagram of the form

$$\begin{array}{ccc} \mathcal{E} & & \\ \downarrow & = & \\ \mathcal{E}' & & \end{array} \left(\begin{array}{ccccccc} 0 \to & K & \stackrel{i}{\to} & X & \stackrel{p}{\to} & Q & \to 0 \\ & \downarrow h_- & & \downarrow h & & \downarrow \mathrm{Id} & \\ 0 \to & K' & \stackrel{i'}{\to} & X' & \stackrel{p'}{\to} & Q & \to 0 \end{array} \right)$$

in which both h and h_- are isomorphisms.

In addition, the following circumstance occurs frequently enough to merit special attention; suppose Q is 1-coprojective, let $f : K \to K$ be a Λ-homomorphism and consider the pushout extension

$$\begin{array}{c} \mathcal{E} \\ \downarrow \\ f_*(\mathcal{E}) \end{array} = \left(\begin{array}{ccccccc} 0 \to & K & \stackrel{i}{\to} & X & \stackrel{p}{\to} & Q & \to 0 \\ & \downarrow f & & \downarrow \nu & & \downarrow \mathrm{Id} & \\ 0 \to & K & \stackrel{i'}{\to} & \varinjlim(f,i) & \stackrel{p'}{\to} & Q & \to 0 \end{array} \right)$$

In view of the $\mathrm{End}_{\mathcal{D}\mathrm{er}}(K)$-module structure on $\mathrm{Ext}^1(Q,K)$ the congruence class of $f_*(\mathcal{E})$, and hence the isomorphism class of $\varinjlim(f,i)$, depends only on the class $\overline{f}$ of f in $\mathcal{D}\mathrm{er}(\Lambda)$. Fix a specific element f' of $\mathrm{End}_{\mathcal{D}\mathrm{er}}(K)$ to represent $\overline{f}$ and write $X(\overline{f}) = \varinjlim(f',i)$. Then we can rewrite the above as a commutative diagram thus

$$\begin{array}{ccccccc} 0 \to & K & \stackrel{i}{\to} & X & \stackrel{p}{\to} & Q & \to 0 \\ & \downarrow f & & \downarrow \nu(f) & & \downarrow \mathrm{Id} & \\ 0 \to & K & \stackrel{i'}{\to} & X(\overline{f}) & \stackrel{p'}{\to} & Q & \to 0 \end{array}$$

where the dependence of the natural map ν on f is now manifest. The question arises as to whether f, and thereby $\nu(f)$, can be chosen to be a Λ-isomorphism. A necessary condition for this to occur is that $\overline{f}$ should define an element of $\mathrm{Aut}_{\mathcal{D}\mathrm{er}}(K)$. However, although the natural ring homomorphism $\mathrm{End}_\Lambda(K) \to \mathrm{End}_{\mathcal{D}\mathrm{er}}(K)$ is surjective, in general the induced map on unit groups $\natural : \mathrm{Aut}_\Lambda(K) \to \mathrm{Aut}_{\mathcal{D}\mathrm{er}}(K)$ fails to be surjective. Thus one may choose ν to be an isomorphism precisely when $\overline{f}$ lifts to an element of $\mathrm{Aut}_\Lambda(K)$. We will encounter two situations where this can be achieved and which are expressed in their most basic terms as follows:

Example 1: Suppose $\mathrm{End}_{\mathcal{D}\mathrm{er}}(R)$ is a field $\mathbb{F}$ and put $K \cong R \oplus R$. In this case a sufficient condition for $\overline{f} \in \mathrm{Aut}_{\mathcal{D}\mathrm{er}}(K) \cong GL_2(\mathbb{F})$ is that $\det(\overline{f}) = 1$. For then $\overline{f}$ is a product of elementary transvections $\overline{f} = E_1 \dots E_N$, each of which is liftable to $\mathrm{Aut}_\Lambda(K)$ as $\mathrm{End}_\Lambda(R) \to \mathrm{End}_{\mathcal{D}\mathrm{er}}(R)$ is surjective. We shall elaborate on this example in §83 and §84 below.

Example 2: Suppose $K \cong R_1 \oplus R_2$ where each $\mathrm{End}_{\mathcal{D}\mathrm{er}}(R_i)$ is the field $\mathbb{F}_p$ with p elements. Suppose further that $R_2 \subset R_1$ is a submodule such that $p^k R_1 \subset R_2$ and that $\mathrm{Hom}_{\mathcal{D}\mathrm{er}}(R_1, R_2) = \mathrm{Hom}_{\mathcal{D}\mathrm{er}}(R_2, R_1) = 0$. Let $\delta \in \mathbb{F}_p^* \cong \mathrm{Aut}_{\mathcal{D}\mathrm{er}}(R_i)$ and choose integers a, b, d such that $ad + bp^k = 1$ where d represents δ mod p. Then

$$\begin{pmatrix} a & b \\ -p^k & d \end{pmatrix}$$

defines an Λ-automophism of K which lifts

$$\begin{pmatrix} \delta^{-1} & 0 \\ 0 & \delta \end{pmatrix}.$$

We shall elaborate on this example in §84 below.

§81: Nondegenerate modules of type $\mathfrak{R}$-$\mathbb{Z}[C_q]$:

In this section $\mathbf{a} = (a_1, \dots, a_N)$ will denote a sequence of integers satisfying

$$1 \leq a_1 < a_2 < \cdots < a_N \leq q.$$

An extension of the form $\mathcal{X} = (0 \to R(\mathbf{a}) \to X \to \mathbb{Z}[C_q] \to 0)$ is classified up to congruence by a sequence $\mathbf{c} = (c_1, \dots, c_N)$ where $c_i \in \mathrm{Ext}^1(\mathbb{Z}, R(i))$.

We say that the extension $\mathcal{X}$ is *nondegenerate* when each $c_i \neq 0$. As $\mathrm{Ext}^1(\mathbb{Z}[C_q], R(i)) \cong \mathbb{F}_p$ for all i then we can assume that, for a nondegenerate extension, we have $c_i \in \mathbb{F}_p^*$ for all i. For the rest of this sectiom, without further mention, $\mathbf{c}$ will denote an N-tuple $\mathbf{c} = (c_1, \ldots, c_N) \in \underbrace{\mathbb{F}_p^* \times \cdots \times \mathbb{F}_p^*}_{N}$.

Given $\mathbf{c}$ of this form we denote by $T(\mathbf{a}, \mathbf{c})$ the module defined by the extension $\mathcal{T} = (0 \to R(\mathbf{a}) \to T(\mathbf{a}, \mathbf{c}) \to \mathbb{Z}[C_q] \to 0)$ classified by $\mathbf{c} = (c_1, \ldots, c_N)$. We first consider a special case:

Special case $N = q$:

In this case $\mathbf{a}$ orders the entire set $\{1, 2, \ldots, q\}$ so further mention of the parameter $\mathbf{a}$ is unnecessary and we consider $T(c_1, c_2, \ldots c_q)$ defined by the sequence

$$0 \to \bigoplus_{j=1}^{q} R(j) \to T(c_1, c_2, \ldots c_q) \to \mathbb{Z}[C_q] \to 0)$$

which is classified by $(c_1, \ldots, c_q) \in \underbrace{\mathbb{F}_p^* \times \cdots \times \mathbb{F}_p^*}_{q}$. We first observe that:

Proposition 81.1: *$T(c_1, c_2, \ldots c_q)$ is projective for each $(c_1, \ldots, c_q) \in \underbrace{\mathbb{F}_p^* \times \cdots \times \mathbb{F}_p^*}_{q}$.*

Proof. We recall Swan's Theorem that the integral group ring of any finite group is indecomposable. Thus Λ is indecomposable and occurs in an exact sequence

$$0 \to \bigoplus_{j=1}^{q} R(j) \to \Lambda \to \mathbb{Z}[C_q] \to 0$$

classified by a sequence $\mathbf{l} = (l_1, l_2, \ldots, l_q)$ where $l_i \in \mathrm{Ext}^1(\mathbb{Z}[C_q], R(i)) \cong \mathbb{F}_p$. We claim that each $l_i \neq 0$. Otherwise if $l_j = 0$ then Λ would split as a direct sum $\Lambda \cong R(j) \oplus Y$ for some Y, thereby contradicting Swan's Theorem. Via the Corepresentation Theorem we now make the identifications

$$\underbrace{\mathbb{F}_p^* \times \cdots \times \mathbb{F}_p^*}_{q} \overset{\simeq}{\longleftrightarrow} \mathrm{Ext}^1(\mathbb{Z}[C_q], \bigoplus_{j=1}^{q} R(j)) \overset{\simeq}{\longleftrightarrow} \mathrm{End}_{\mathcal{D}\mathrm{er}}(\bigoplus_{j=1}^{q} R(j)).$$

Thus we may regard $(c_1 l_1^{-1}, c_2 l_2^{-1}, \ldots, c_q l_q^{-1}) \in \underbrace{\mathbb{F}_p^* \times \cdots \times \mathbb{F}_p^*}_{q}$ as an element of $\mathrm{Aut}_{\mathcal{D}\mathrm{er}}(\bigoplus_{j=1}^{q} R(j))$. Consequently there exists a Λ-endomorphism

α of $\bigoplus_{j=1}^{q} R(j)$ whose image in $\mathcal{D}er(\Lambda)$ is $(c_1 l_1^{-1}, c_2 l_2^{-1}, \dots, c_q l_q^{-1}) \in \mathrm{Aut}_{\mathcal{D}\mathrm{er}}(\bigoplus_{j=1}^{q} R(j))$. Consider the pushout extension

$$\begin{array}{ccccccc} 0 \to & \bigoplus_{j=1}^{q} R(j) & \stackrel{\iota}{\to} & \Lambda & \to & \mathbb{Z}[C_q] & \to 0 \\ & \downarrow \alpha & & \downarrow \nu & & \downarrow \mathrm{Id} & \\ 0 \to & \bigoplus_{j=1}^{q} R(j) & \to & \underrightarrow{\lim}(\alpha, \iota) & \to & \mathbb{Z}[C_q] & \to 0. \end{array}$$

One sees easily that $\underrightarrow{\lim}(\alpha, \iota) \cong T(c_1, c_2, \dots, c_q)$. However, as the image of α lies in $\mathrm{Aut}_{\mathcal{D}\mathrm{er}}(\bigoplus_{j=1}^{q} R(j))$ then $\underrightarrow{\lim}(\alpha, \iota)$ is projective by Swan's criterion (36.2). □

It follows immediately that:

Corollary 81.2: $\mathrm{Hom}_{\mathcal{D}\mathrm{er}}(T(c_1, c_2, \dots, c_q), R(k)) = 0$ *for each* k.

Generic case $N < q$: Assume the extension $0 \to R(\mathbf{a}) \stackrel{\widetilde{i}}{\to} T(\mathbf{a}, \mathbf{c}) \stackrel{\widetilde{p}}{\to} \mathbb{Z}[C_q] \to 0$ is classified by $\mathbf{c} = (c_1, \dots, c_N) \in \underbrace{\mathbb{F}_p^* \times \cdots \times \mathbf{F}_p^*}_{N}$ with $N < q$; we claim:

Theorem 81.3: $\mathrm{Hom}_{\mathcal{D}\mathrm{er}}(T(\mathbf{a}, \mathbf{c}), R(k)) = 0$ *for all* k.

Proof. We have an exact sequence

$$\mathrm{Hom}_{\mathcal{D}\mathrm{er}}(\mathbb{Z}[C_q], R(k)) \stackrel{\widetilde{p}^*}{\to} \mathrm{Hom}_{\mathcal{D}\mathrm{er}}(T(\mathbf{a}, \mathbf{c}), R(k)) \stackrel{\widetilde{i}^*}{\to} \mathrm{Hom}_{\mathcal{D}\mathrm{er}}(R(\mathbf{a}), R(k))$$
$$\stackrel{\widetilde{\partial}}{\to} \mathrm{Ext}^1(\mathbb{Z}[C_q], R(k)).$$

Now $\mathrm{Hom}_\Lambda(\mathbb{Z}[C_q], R(k)) = 0$ by (78.5) and hence $\mathrm{Hom}_{\mathcal{D}\mathrm{er}}(\mathbb{Z}[C_q], R(k)) = 0$. Thus it suffices to show that $\widetilde{\partial} : \mathrm{Hom}_{\mathcal{D}\mathrm{er}}(R(\mathbf{a}), R(k)) \stackrel{\widetilde{\partial}}{\to} \mathrm{Ext}^1(\mathbb{Z}[C_q], R(k))$ is injective. If $k \notin \{a_1, \dots, a_N\}$ then $\widetilde{\partial}$ is trivially injective as $\mathrm{Hom}_{\mathcal{D}\mathrm{er}}(R(\mathbf{a}), R(k)) = 0$. Thus suppose $k \in \{a_1, \dots, a_N\}$ and define $\mathbf{c}' = (c'_1, \dots, c'_q)$ by

$$c'_i = \begin{cases} c_j & \text{if } i = a_j \\ 1 & \text{if } i \notin \{a_1, \dots . a_N\} \end{cases}$$

Then we have an inclusion of exact sequences

$$\begin{array}{ccccccccc}
0 \to & R(\mathbf{a}) & \stackrel{\widetilde{i}}{\to} & T(\mathbf{a},\mathbf{c}) & \stackrel{\widetilde{p}}{\to} & \mathbb{Z}[C_q] & \to 0 \\
& \cap & & \cap & & || & \\
0 \to & \bigoplus_{j=1}^q R(j) & \stackrel{i}{\to} & T(c_1', c_2', \dots c_q') & \stackrel{p}{\to} & \mathbb{Z}[C_q] & \to 0
\end{array}$$

and hence a commutative diagram with exact rows as follows:

$$\begin{array}{ccccc}
\mathrm{Hom}_{\mathcal{D}\mathrm{er}}(T(c_1', c_2', \dots, c_q'), R(k)) & \stackrel{i^*}{\to} & \mathrm{Hom}_{\mathcal{D}\mathrm{er}}(\bigoplus_{j=1}^q R(j), R(k)) & \stackrel{\partial}{\to} & \mathrm{Ext}^1(\mathbb{Z}[C_q], R(k)) \\
\downarrow & & \downarrow \nu & & \downarrow \mathrm{Id} \\
\mathrm{Hom}_{\mathcal{D}\mathrm{er}}(T(\mathbf{a},\mathbf{c}), R(k)) & \stackrel{\widetilde{i}^*}{\to} & \mathrm{Hom}_{\mathcal{D}\mathrm{er}}(R(\mathbf{a}), R(k)) & \stackrel{\widetilde{\partial}}{\to} & \mathrm{Ext}^1(\mathbb{Z}[C_q], R(k))
\end{array}$$

where ν is induced from the inclusion $R(\mathbf{a}) \subset \bigoplus_{j=1}^q R(j)$. As $k \in \{a_1, \dots, a_N\}$ then $\nu : \mathrm{Hom}_{\mathcal{D}\mathrm{er}}(\bigoplus_{j=1}^q R(j), R(k)) \to \mathrm{Hom}_{\mathcal{D}\mathrm{er}}(R(\mathbf{a}), R(k))$ is an isomorphism. However, $\mathrm{Hom}_{\mathcal{D}\mathrm{er}}(T(c_1', c_2', \dots, c_q'), R(k)) = 0$ by (81.2). Consequently the connecting homomorphism $\partial : \mathrm{Hom}_{\mathcal{D}\mathrm{er}}(\bigoplus_{j=1}^q R(j), R(k)) \to \mathrm{Ext}^1(\mathbb{Z}[C_q], R(k))$ is injective and so, as claimed, $\widetilde{\partial} = \partial \circ \nu^{-1}$ is also injective. □

Suppose the module $T = T(\mathbf{a}, \mathbf{c})$ also occurs in an exact sequence

$$\mathcal{E} = (0 \to R(\mathbf{b}) \to T \to \mathbb{Z}[C_q] \to 0).$$

We claim that the extension $\mathcal{E}$ is nondegenerate. For $R(\mathbf{b}) \cong R(\mathbf{a})$ by (78.3) and (76.2) so that $\mathcal{E}$ is also classified by $\gamma = (\gamma_1, \dots, \gamma_N)$ with $\gamma_i \in \mathrm{Ext}(\mathbb{Z}[C_q], R(a_i))$. If $\gamma_k = 0$ for some k then T splits as a direct sum $T \cong R(a_k) \oplus T'$ where T' occurs in the extension

$$0 \to \bigoplus_{j \neq k} R(a_j) \to T' \to \mathbb{Z}[C_q] \to 0$$

classified by $(\gamma_j)_{j \neq k}$. As $\mathrm{Hom}_{\mathcal{D}\mathrm{er}}(R(a_k),\ R(a_k)) \cong \mathbb{F}_p$ then $\mathrm{Hom}_{\mathcal{D}\mathrm{er}}(T, R(a_k)) \neq 0$, contradicting (81.3). Thus we see that:

(81.4) Nondegeneracy is an isomorphism invariant for modules of type $\mathfrak{R}$-$\mathbb{Z}[C_q]$.

§82: Nondegenerate modules of type $\mathfrak{R}$-I_q :

We now consider the case of modules defined by extensions of the form

$$\mathcal{X} = (0 \to R(\mathbf{a}) \to X \to I_q \to 0).$$

Such an extension is again classified up to congruence by a sequence (c_i) where, in this case, $c_i \in \mathrm{Ext}^1(I_q, R(i))$. Nondegeneracy is again defined by the conditions that each $c_i \neq 0$. The difference from the previous section is that now $\mathrm{Ext}^1(I_q, R(1)) = 0$. Thus in considering nondegenerate extensions we must exclude the possibility that '$a_1 = 1$' and to this end we restrict ourselves to sequences of the form

$$2 \leq a_2 < \cdots < a_N \leq q.$$

For the rest of this section, without further mention, $\mathbf{c}$ will denote an $N-1$-tuple $\mathbf{c} = (c_2, \ldots, c_N) \in \underbrace{\mathbb{F}_p^* \times \cdots \times \mathbb{F}_p^*}_{N}$ and $S(\mathbf{a}, \mathbf{c})$ will denote the module defined by the extension

$$\mathcal{S} = (0 \to R(\mathbf{a}) \to S(\mathbf{a}, \mathbf{c}) \to I_q \to 0)$$

classified by $\mathbf{c} = (c_2, \ldots, c_N) \in \underbrace{\mathbb{F}_p^* \times \cdots \times \mathbb{F}_p^*}_{N-1}$.

Theorem 82.1: $\mathrm{Hom}_{\mathcal{D}\mathrm{er}}(S(\mathbf{a}, \mathbf{c}), R(k)) = 0$ *for all* k.

Proof. As $\mathbf{c} = (c_2, \ldots, c_N) \in \underbrace{\mathbb{F}_p^* \times \cdots \times \mathbb{F}_p^*}_{N-1}$ there is a corresponding extension of the form $0 \to R(\mathbf{a}) \to T(\mathbf{a}, \mathbf{c}) \to \mathbb{Z}[C_q] \to 0$ and an inclusion of exact sequences

$$\begin{array}{ccccccc} 0 \to & R(\mathbf{a}) & \stackrel{i}{\to} & S(\mathbf{a}, \mathbf{c}) & \stackrel{p}{\to} & I_q & \to 0 \\ & || & & \cap & & \cap & \\ 0 \to & R(\mathbf{a}) & \stackrel{\tilde{i}}{\to} & T(\mathbf{a}, \mathbf{c}) & \stackrel{\tilde{p}}{\to} & \mathbb{Z}[C_q] & \to 0 \end{array}$$

and hence a commutative diagram with exact rows as follows

$$\begin{array}{ccccccc} \mathrm{Hom}_{\mathcal{D}\mathrm{er}}(\mathbb{Z}[C_q], R(k)) & \stackrel{\tilde{p}^*}{\to} & \mathrm{Hom}_{\mathcal{D}\mathrm{er}}(T(\mathbf{a}, \mathbf{c}), R(k)) & \stackrel{\tilde{i}^*}{\to} & \mathrm{Hom}_{\mathcal{D}\mathrm{er}}(R(\mathbf{a}), R(k)) & \stackrel{\tilde{\partial}}{\to} & \mathrm{Ext}^1(\mathbb{Z}[C_q], R(k)) \\ \downarrow \nu_1 & & \downarrow \nu_2 & & \downarrow \mathrm{Id} & & \downarrow \nu_1 \\ \mathrm{Hom}_{\mathcal{D}\mathrm{er}}(I_q, R(k)) & \stackrel{p^*}{\to} & \mathrm{Hom}_{\mathcal{D}\mathrm{er}}(S(\mathbf{a}, \mathbf{c}), R(k)) & \stackrel{i^*}{\to} & \mathrm{Hom}_{\mathcal{D}\mathrm{er}}(R(\mathbf{a}), R(k)) & \stackrel{\partial}{\to} & \mathrm{Ext}^1(I_q, R(k)) \end{array}$$

where ν_1, ν_2 are induced from the appropriate inclusions. From (78.5) we again see that $\mathrm{Hom}_\Lambda(\mathbb{Z}[C_q], R(k)) = 0$. Hence it follows that $\mathrm{Hom}_{\mathcal{D}\mathrm{er}}(\mathbb{Z}[C_q], R(k)) = 0$. Likewise $\mathrm{Hom}_\Lambda(I_q, R(k)) = \mathrm{Hom}_{\mathcal{D}\mathrm{er}}(I_q, R(k)) = 0$. Thus the above diagram reduces to

$$
\begin{array}{ccccccc}
0 \to & \mathrm{Hom}_{\mathcal{D}\mathrm{er}}(T(\mathbf{a},\mathbf{c}), R(k)) & \stackrel{\widetilde{i^*}}{\to} & \mathrm{Hom}_{\mathcal{D}\mathrm{er}}(R(\mathbf{a}), R(k)) & \stackrel{\widetilde{\partial}}{\to} & \mathrm{Ext}^1(\mathbb{Z}[C_q], R(k)) \\
& \downarrow \nu_2 & & \downarrow \mathrm{Id} & & \downarrow \nu_1 \\
0 \to & \mathrm{Hom}_{\mathcal{D}\mathrm{er}}(S(\mathbf{a},\mathbf{c}), R(k)) & \stackrel{i^*}{\to} & \mathrm{Hom}_{\mathcal{D}\mathrm{er}}(R(\mathbf{a}), R(k)) & \stackrel{\partial}{\to} & \mathrm{Ext}^1(I_q, R(k))
\end{array}
$$

When $k \neq 1$ then $\nu_1 : \mathrm{Ext}^1(\mathbb{Z}[C_q], R(k)) \to \mathrm{Ext}^1(I_q, R(k))$ is an isomorphism when it follows that $\nu_2 : \mathrm{Hom}_{\mathcal{D}\mathrm{er}}(T(\mathbf{a},\mathbf{c}), R(k)) \to \mathrm{Hom}_{\mathcal{D}\mathrm{er}}(S(\mathbf{a},\mathbf{c}), R(k))$ is bijective. By (81.3), $\mathrm{Hom}_{\mathcal{D}\mathrm{er}}(T(\mathbf{a},\mathbf{c}), R(k)) = 0$. Hence $\mathrm{Hom}_{\mathcal{D}\mathrm{er}}(S(\mathbf{a},\mathbf{c}), R(k)) = 0$. If $k = 1$ then as $1 \in \{a_1, \ldots, a_N\}$ we see that $\mathrm{Hom}_{\mathcal{D}\mathrm{er}}(R(\mathbf{a}), R(k)) = 0$. The bottom row now reduces to $0 \to \mathrm{Hom}_{\mathcal{D}\mathrm{er}}(S(\mathbf{a},\mathbf{c}), R(1)) \stackrel{i^*}{\to} 0 \stackrel{\partial}{\to} \mathrm{Ext}^1(I_q, R(1))$ from which we again see that $\mathrm{Hom}_{\mathcal{D}\mathrm{er}}(S(\mathbf{a},\mathbf{c}), R(1)) = 0$. □

On substituting (82.1) for (81.3) a similar argument to (81.4) shows:

(82.2) Nondegeneracy is an isomorphism invariant for modules of type $\mathfrak{R}$-I_q.

§83: A reduction theorem:

In what follows, we denote by

R : a row module $R(i)$ where $1 \le i \le q$;
Q : a module isomorphic to either $\mathbb{Z}[C_q]$ or I_q;
Z : a Λ-module, otherwise unspecified.

If $m \ge 1$ we denote by $\langle R^{(m)}, Z, Q\rangle$ the class of modules X defined by extensions

$$0 \to R^m \oplus Z \to X \to Q \to 0.$$

Such extensions are classified up to congruence by sequences $(c_1, \ldots, c_m; c_Z)$ where $c_i \in \mathrm{Ext}^1(Q, R) \cong \mathbb{F}_p$ for $1 \le i \le m$ and $c_Z \in \mathrm{Ext}^1(Q, Z)$. We have the following elementary reduction theorem.

Proposition 83.1: *Let $X \in \langle R^{(m)}, Z, Q\rangle$; if $m \geq 2$ then*

$$X \cong_{\mathrm{Id}} R^{(m-1)} \oplus X'$$

where $X' \in \langle R, Z, Q\rangle$.

Proof. First consider the case where $m = 2$. Then we write

$$0 \to R_1 \oplus R_2 \oplus Z \to X \to Q \to 0$$

where $R_1 \cong R_2 \cong R(i)$ and where the sequence is classified by $(c_1, c_2; c_Z)$ with $c_j \in \mathrm{Ext}^1(Q, R_j)$. We need only consider two cases:

Case 1: $c_1 = 0$;

Case 2: $c_1 \neq 0$.

In Case 1, $X \cong_{\mathrm{Id}} R \oplus X'$ where $X' \in \langle R, Z, Q\rangle$ is classified by $(c_2; c_Z)$.

In Case 2 $\mathrm{Ext}^1(Q, R) \cong \mathbb{F}_p$. After identifying $\mathrm{Ext}^1(Q, R \oplus R) \cong \mathbb{F}_p \oplus \mathbb{F}_p$ put

$$E = \begin{pmatrix} c_2c_1^{-1} & -1 \\ 1 & 0 \end{pmatrix} = \begin{pmatrix} 0 & -1 \\ 1 & 0 \end{pmatrix} \begin{pmatrix} 1 & 0 \\ -c_2c^{-1} & 1 \end{pmatrix} \in E_2(\mathbb{F}_p).$$

Now lift E to an automorphism $\widetilde{E} \in \mathrm{Aut}_\Lambda(R \oplus R)$, extend to the automorphism $\widehat{E} = \widetilde{E} \oplus \mathrm{Id}_Z : R_1 \oplus R_2 \oplus Z \to R_1 \oplus R_2 \oplus Z$ and consider the pushout extension

$$\begin{array}{ccccccc} 0 \to & R \oplus R \oplus Z & \xrightarrow{i} & X & \xrightarrow{\pi} & Q & \to 0 \\ & \downarrow \widehat{E} & & \downarrow \nu & & \downarrow \mathrm{Id} & \\ 0 \to & R \oplus R \oplus Z & \xrightarrow{i'} & \varinjlim(\widehat{E}, i) & \xrightarrow{\pi'} & Q & \to 0. \end{array}$$

Moreover $\widehat{E}$ acts on the classifying sequence thus

$$\begin{pmatrix} c_2c_1^{-1} & -1 & 0 \\ 1 & 0 & 0 \\ 0 & 0 & 1 \end{pmatrix} \begin{pmatrix} c_1 \\ c_2 \\ c_Z \end{pmatrix} = \begin{pmatrix} 0 \\ c_1 \\ c_Z \end{pmatrix}.$$

Thus $\varinjlim(\widehat{E}, i)$ is in Case I, classified by $(0, c_1; c_Z)$. Hence $\varinjlim(\widehat{E}, i) \cong R \oplus X'$ for some $X' \in \langle R, Z; Q\rangle$. The conclusion for $m = 2$ follows as the canonical map ν is an isomorphism by the Five Lemma.

Now suppose $m \geq 3$ and assume the statement proved for $m-1$. We write X as an extension

$$0 \to R \oplus R \oplus R^{m-2} \oplus Z \to X \to Q \to 0.$$

Putting $Z' = R^{m-2} \oplus Z$ it follows by the above that $X \cong_{\mathrm{Id}} R \oplus X'$ where $X'' \in \langle R^{m-1}, Z, Q\rangle$. Inductively $X'' \cong_{\mathrm{Id}} R^{m-2} \oplus X'$ where $X' \in \langle R, Z, Q\rangle$. The conclusion now follows as $X \cong_{\mathrm{Id}} R \oplus R^{m-2} \oplus X' \cong_{\mathrm{Id}} R^{m-1} \oplus X'$. □

In the above we note that Case II excludes the possibility that $Q = I_q$ and $i \neq 1$ as $\mathrm{Ext}^1(I_q, R(1)) = 0$.

In (83.1) let $(c'; c_Z)$ be the classifying sequence for X' where $c' \in \mathrm{Ext}^1(Q, R)$. We note the following addendum to (83.1):

(83.2) If $(c'; c_Z)$ is the classifying sequence for X' then $c' \neq 0$ provided that some $c_i \neq 0$ in the original classifying sequence $(c_1, \ldots, c_m; c_Z)$ for X.

Now let $\mathbf{a} = (a_1, \ldots, a_N)$ be a sequence of positive integers such that

$$1 \leq a_1 < a_2 < \cdots < a_N \leq q$$

and let $\mathbf{e} = (e_1, \ldots, e_N)$ be a sequence of positive integers. We define

$$R(\mathbf{a}, \mathbf{e}) = \bigoplus_{i=1}^{n} R(a_i)^{e_i}$$

$$R(\mathbf{a}, \mathbf{e} - \mathbf{1}) = \bigoplus_{i=1}^{n} R(a_i)^{e_i - 1}.$$

In the latter case we adhere to the convention that $R(a_i)^0 = 0$. In the special case where each $e_i = 1$ we write simply $R(\mathbf{a}) = \bigoplus_{i=1}^{N} R(a_i)$.

We shall consider extensions of the form

$$0 \to R(\mathbf{a}) \oplus Z \to X \to Q \to 0.$$

Such an extension is classified by $(c_1, \ldots, c_N; c_Z)$ where $c_i \in \mathrm{Ext}^1(Q, R(a_i))$. We say that X is *relatively nondegenerate* when each $c_i \neq 0$. More generally, we consider extensions of the form

$$0 \to R(\mathbf{a}, \mathbf{e}) \oplus Z \to X \to Q \to 0.$$

Up to congruence, such an extension is classified by a sequence $(c(1), \ldots, c(N); c_Z)$ where $c(i) = (c(i)_1, \ldots, c(i)_{e_i}) \in \mathrm{Ext}^1(Q, R(a_i)^{e_i})$ and $c_Z \in \mathrm{Ext}^1(Q, Z)$. There are two exceptional cases; the first of these is the *totally degenerate* case in which each $c(i) \equiv 0$. Then X decomposes as a direct sum

$$X \cong R(\mathbf{a}, \mathbf{e}) \oplus X'$$

where X' is the extension $0 \to Z \to X' \to Q \to 0$ classified by $c_Z \in \mathrm{Ext}^1(Q, Z)$. This case plays no significant role in our investigations. More interesting is the case where each $c(i) \neq 0$. Then we have the following:

Theorem 83.3: *Let X be defined by an extension of the form*

$$0 \to R(\mathbf{a}, \mathbf{e}) \oplus Z \to X \to Q \to 0$$

where $c(i) \not\equiv 0$ for $1 \leq i \leq N$. Then $X \cong_{\mathrm{Id}} R(\mathbf{a}, \mathbf{e} - \mathbf{1}) \oplus X'$ where X' is a relatively nondegenerate extension of the form $0 \to R(\mathbf{a}) \oplus Z \to X' \to Q \to 0$.

Proof. The proof is by induction on N where $\mathbf{a} = (a_1, \ldots, a_N)$. In the case $N = 1$ we have $0 \to R(a_1)^{e_1} \oplus Z \to X \to Q \to 0$ and the conclusion follows from (83.1) and (83.2).

Suppose proved for $N - 1$ and that we are given an extension

$$0 \to R(\mathbf{a}, \mathbf{e}) \oplus Z \to X \to Q \to 0$$

where $\mathbf{a} = (a_1 < a_2 < \cdots < a_N)$. Write $\mathbf{a}' = (a_1 < a_2 < \cdots < a_{N-1})$ and $\mathbf{e}' = (e_1, \ldots, e_{N-1})$. Then X is an extension of the form

$$0 \to R(a_n)^{e_n} \oplus R(\mathbf{a}', \mathbf{e}') \oplus Z \to X \to Q \to 0.$$

Putting $Z' = R(\mathbf{a}', \mathbf{e}') \oplus Z$ it follows inductively that $X \cong_{\mathrm{Id}} R(a_N)^{e_N - 1} \oplus X''$ where X'' is an extension of the form $0 \to R(a_N) \oplus R(\mathbf{a}', \mathbf{e}') \oplus Z \to X'' \to Q \to 0$. Put $Z'' = R(a_n) \oplus Z$ and write X'' as an extension

$$0 \to R(\mathbf{a}', \mathbf{e}') \oplus Z'' \to X'' \to Q \to 0.$$

Then by the Induction hypothesis $X'' \cong_{\mathrm{Id}} R(\mathbf{a}', \mathbf{e}' - \mathbf{1}) \oplus X'$ where X' is an extension of the form $0 \to R(\mathbf{a}') \oplus Z'' \to X' \to Q \to 0$. However $R(\mathbf{a}') \oplus Z'' = R(\mathbf{a}) \oplus Z$ so that $X \cong_{\mathrm{Id}} R(a_N)^{e_N - 1} \oplus R(\mathbf{a}', \mathbf{e}' - \mathbf{1}) \oplus X'$ where X' is an extension of the form $0 \to R(\mathbf{a}) \oplus Z \to X' \to Q \to 0$. However $R(a_N)^{e_N - 1} \oplus R(\mathbf{a}', \mathbf{e}' - \mathbf{1}) \cong R(\mathbf{a}, \mathbf{e} - \mathbf{1})$ from which the conclusion now follows. We note that the relative nondegeneracy of X' is a consequence, at each stage, of (83.2). □

For a general extension of the form $0 \to R(\mathbf{a}, \mathbf{e}) \oplus Z \to X \to Q \to 0$ we may suppose that $c(i) \neq 0$ for some i and that $c(j) = 0$ for some j. In this case we first consider the subsequences $\mathbf{a}' = (a'_1 < \cdots < a'_{N'})$ and $\mathbf{b} = (b_1 < \cdots < b_M)$ of $\mathbf{a}$ which correspond respectively to those summands

$R(a_i)$ and $R(a_j)$ for which $c(i) \not\equiv 0$ and $c(j) \equiv 0$. Formally, we put

$$\mathbf{a}' = \{i \,|1 \le i \le N \quad \text{such that } c(i) \not\equiv 0\}$$

$$\mathbf{b} = \{j \,|1 \le j \le N \quad \text{such that } c(j) \equiv 0\}.$$

Put $N' = |\mathbf{a}'|$ and $M = |\mathbf{b}|$ and let $\mathbf{a}' = (a'_1 < \cdots < a'_{N'})$ and $\mathbf{b} = (b_1 < \cdots < b_M)$ be the natural orderings. For $1 \le i \le N'$ put $e'_i = e_k$ when $a'_i = a_k$ and for $1 \le j \le M$ put $d_j = e_k$ when $b_j = a_k$. If $0 \to R(\mathbf{a}, \mathbf{e}) \oplus Z \to X \to Q \to 0$ is such an extension we first decompose

$$X \cong_{\text{Id}} R(\mathbf{b}, \mathbf{d}) \oplus Y$$

where Y is the extension $0 \to R(\mathbf{a}', \mathbf{e}') \oplus Z \to Y \to Q \to 0$ classified by $(c'(1), \ldots, c'(N'); c_Z)$ where $c'(i) \in \text{Ext}^1(Q, R(a'_i)^{e'_i})$ and $c'(i) \neq 0$ for each i. Then appealing to (83.3) we further decompose

$$Y \cong_{\text{Id}} R(\mathbf{a}', \mathbf{e}' - \mathbf{1}) \oplus X'$$

where X' is a relatively nondegenerate extension of the form

$$(*) \qquad 0 \to R(\mathbf{a}') \oplus Z \to X' \to Q \to 0.$$

Consequently we may write $X \cong_{\text{Id}} R(\mathbf{a}', \mathbf{e}' - 1) \oplus R(\mathbf{b}, \mathbf{d})\, oplus\, X'$. As $\mathbf{a} = \mathbf{a}' \cup \mathbf{b}$ then writing $\mathbf{f} = (f_1, \ldots, f_N)$ where

$$f_i = \begin{cases} e_i - 1 & \text{if} \quad c(i) \not\equiv 0 \\ e_i & \text{if} \quad c(i) \equiv 0 \end{cases}$$

we now have $X \cong_{\text{Id}} R(\mathbf{a}, \mathbf{f}) \oplus X'$ where X' is the relatively nondegenerate extension at (*) above. Finally, taking $Z = 0$ we obtain:

Theorem 83.4: *Let* $0 \to R(\mathbf{a}, \mathbf{e}) \to X \to Q \to 0$ *be an extension which is not totally degenerate; then*

$$X \cong_{\text{Id}} R(\mathbf{a}, \mathbf{f}) \oplus X_{nd}$$

where X_{nd} is a nondegenerate extension of the form $0 \to R(\mathbf{a}') \to X_{nd} \to Q \to 0$ for some subsequence $\mathbf{a}'$ of $\mathbf{a}$ and f_i is either $e_i - 1$ or e_i according to whether $c(i) \not\equiv 0$ or $c(i) \equiv 0$.

We say that a quasi-Swan module X is *degenerate* when it is either totally degenerate or else $X \neq X_{nd}$.

§84: An isomorphism theorem:

In this section we adhere to the convention that when M is an Λ-module then $\overline{\alpha} \in \mathrm{End}_{\mathcal{D}\mathrm{er}}(M)$ will denote the canonical image of $\alpha \in \mathrm{End}_\Lambda(M)$. Moreover, without further mention, for any j in the range $1 \le j \le q$, we will make the identifications

(84.1) $$\mathrm{End}_{\mathcal{D}\mathrm{er}}(R(j) \oplus R(j)) \longleftrightarrow M_2(\mathbb{F}_p)$$

and hence

(84.2) $$\mathrm{Aut}_{\mathcal{D}\mathrm{er}}(R(j) \oplus R(j)) \longleftrightarrow GL_2(\mathbb{F}_p).$$

Let Z denote some otherwise unspecified Λ-module and let Q denote either $\mathbb{Z}[C_q]$ or I_q. Fix integers j, k such that $1 \le j \le j+k \le q$. Choose $c, c' \in \mathbb{F}_p$, $c_Z \in \mathrm{Ext}^1(Q, Z)$ and let $X(c', c; c_Z)$ denote the extension module

(84.3) $$0 \to R(j) \oplus R(j+k) \oplus Z \xrightarrow{i} X(c', c; c_Z) \xrightarrow{\pi} Q \to 0$$

classified by $(c', c; c_Z)$. We shall prove that

(84.4) $$X(c', c;\ c_Z) \cong_{\mathrm{Id}} X(\delta c', \delta^{-1} c;\ c_Z) \ \text{ for each } \delta \in \mathbb{F}_p^*.$$

There are two cases to consider, according to whether $k = 0$ or $k > 0$. To establish the case $k = 0$ we first prove:

Proposition 84.5: *If $\delta \in \mathbb{F}_p^*$ there exists $\alpha \in \mathrm{Aut}_\Lambda(R(j) \oplus R(j))$ such that*

$$\overline{\alpha} = \begin{pmatrix} \delta & 0 \\ 0 & \delta^{-1} \end{pmatrix}.$$

Proof. Choose $d, e \in \mathrm{End}_\Lambda(R(j))$ such that $\overline{d} = \delta$ and $\overline{e} = \delta^{-1}$ and define $\alpha \in \mathrm{End}_\Lambda(R(j) \oplus R(j))$ by

$$\alpha = \begin{pmatrix} d & d \circ e - \mathrm{Id} \\ \mathrm{Id} - e \circ d & 2e - e \circ d \circ e \end{pmatrix}.$$

As $\overline{d}\,\overline{e} = 1 = \overline{e}\,\overline{d}$ it follows that

$$\overline{\alpha} = \begin{pmatrix} \delta & 0 \\ 0 & \delta^{-1} \end{pmatrix}.$$

However, as α is a product of elementary automorphisms thus

$$\alpha = \begin{pmatrix} 0 & -\mathrm{Id} \\ \mathrm{Id} & 0 \end{pmatrix} \begin{pmatrix} \mathrm{Id} & e \\ 0 & \mathrm{Id} \end{pmatrix} \begin{pmatrix} \mathrm{Id} & 0 \\ -d & \mathrm{Id} \end{pmatrix} \begin{pmatrix} \mathrm{Id} & e \\ 0 & \mathrm{Id} \end{pmatrix}$$

it follows that $\alpha \in \mathrm{Aut}_\Lambda(R(j) \oplus R(j))$. □

Proposition 84.6: *When $k = 0$ then $X(c', c; c_Z) \cong_{\mathrm{Id}} X(\delta c', \delta^{-1} c c_Z)$ for each $\delta \in \mathbb{F}_p^*$,*

Proof. We are given $\delta \in \mathbf{F}_p^*$ and an extension

$$0 \to R(j) \oplus R(j) \oplus Z \xrightarrow{i} X(c', c; c_Z) \xrightarrow{\pi} Q \to 0$$

Construct $\alpha \in \mathrm{Aut}_\Lambda(R(j) \oplus R(j))$ as in (84.5) and extend to an automorphism A of $R(j) \oplus R(j) \oplus Z$ via the identity on Z thus:

$$A = \alpha \oplus \mathrm{Id}_Z.$$

Now consider the pushout extension

$$\begin{array}{ccccccc} 0 \to & R(j) \oplus R(j) \oplus Z & \xrightarrow{i} & X(c, c'; c_Z) & \xrightarrow{\pi} & Q & \to 0 \\ & \downarrow A & & \downarrow \nu & & \downarrow \mathrm{Id} & \\ 0 \to & R(j) \oplus R(j) \oplus Z & \xrightarrow{i'} & \varinjlim(A, i) & \xrightarrow{\pi'} & Q & \to 0 \end{array}$$

By (84.5) we see that $\overline{A}$ has the form

$$\overline{A} = \begin{pmatrix} \delta & 0 & 0 \\ 0 & \delta^{-1} & 0 \\ 0 & 0 & \mathrm{Id} \end{pmatrix}$$

and so acts diagonally on the extension parameters thus

$$\overline{A} \begin{pmatrix} c' \\ c \\ c_Z \end{pmatrix} = \begin{pmatrix} \delta & 0 & 0 \\ 0 & \delta^{-1} & 0 \\ 0 & 0 & \mathrm{Id} \end{pmatrix} \begin{pmatrix} c' \\ c \\ c_Z \end{pmatrix} = \begin{pmatrix} \delta c' \\ \delta^{-1} c \\ c_Z \end{pmatrix}.$$

Hence $\varinjlim(A, i) \cong X(\delta c', \delta^{-1} c; c_Z)$. The conclusion follows as the canonical map ν is an isomorphism by the Five Lemma. □

Next we consider the general case where $k > 0$.

Proposition 84.7: *When $k > 0$ then $X(c', c;\ c_Z) \cong_{\mathrm{Id}} X(\delta c', \delta^{-1} c;\ c_Z)$ for each $\delta \in \mathbb{F}_p^*$.*

Proof. We are given $\delta \in \mathbf{F}_p^*$ and an extension

$$0 \to R(j) \oplus R(j+k) \oplus Z \xrightarrow{i} X(c', c; c_Z) \xrightarrow{\pi} Q \to 0.$$

Let $d \in \{1, \ldots, p-1\}$ be the unique integer whose image in $\mathbb{F}_p^*$ is δ. Then $(d, p^k) = 1$ so we may choose integers a, b such that $a\widehat{\delta} + bp^k = 1$. Observe that the image of a in $\mathbb{F}_p^*$ is δ^{-1}. Put

$$\alpha = \begin{pmatrix} d & -b \\ p^k & a \end{pmatrix} \in SL_2(\mathbb{Z}).$$

As $R(j+k) \subset R(j)$ with index p^k it follows that α defines an automorphism of $R(j) \oplus R(j+k)$. Extend α to an automorphism A of $R(j) \oplus R(j+k) \oplus Z$ via the identity on Z thus: $A = \alpha \oplus \mathrm{Id}_Z$. Again consider the pushout extension

$$\begin{array}{ccccccc} 0 \to & R(j) \oplus R(j+k) \oplus Z & \xrightarrow{i} & X(c, c'; c_Z) & \xrightarrow{\pi} & Q & \to 0 \\ & \downarrow A & & \downarrow \nu & & \downarrow \mathrm{Id} & \\ 0 \to & R(j) \oplus R(j+k) \oplus Z & \xrightarrow{i'} & \varinjlim(A, i) & \xrightarrow{\pi'} & Q & \to 0 \end{array}$$

As $\mathrm{Hom}_{\mathcal{D}\mathrm{er}}(R(j), R(j+k)) = \mathrm{Hom}_{\mathcal{D}\mathrm{er}}(R(j+k), R(j)) = 0$ then $\overline{A}$ acts diagonally on the extension parameters as follows:

$$\begin{pmatrix} \delta & 0 & 0 \\ 0 & \delta^{-1} & 0 \\ 0 & 0 & 1 \end{pmatrix} \begin{pmatrix} c' \\ c \\ c_Z \end{pmatrix} = \begin{pmatrix} \delta c' \\ \delta^{-1} c \\ c_Z \end{pmatrix}.$$

Hence $\varinjlim(A, i) \cong X(\delta c', \delta^{-1} c; c_Z)$. The conclusion follows as the canonical map ν is an isomorphism by the Five Lemma. □

This completes the proof of (84.4). Maintaining the above notation we observe:

Corollary 84.8: *If* $c_1, c_2 \in \mathbb{F}_p^*$ *then* $X(0, c_1; c_Z) \cong_{\mathrm{Id}} X(0, c_2; c_Z)$.

Proof. Apply (84.4) with $c' = 0$, $c = c_1$ and $\delta = c_1 c_2^{-1}$. □

Interchanging the roles of c' and c we also obtain:

Corollary 84.9: *If* $c_1, c_2 \in \mathbb{F}_p^*$ *then* $X(c_1, 0; c_Z) \cong_{\mathrm{Id}} X(c_2, 0; c_Z)$.

§85: Rearranging the extension parameters:

We first consider modules of type $T(\mathbf{a}, \mathbf{c})$. Thus for

$$\mathbf{c} = (c_1, c_2, \dots, c_N) \in \underbrace{\mathbb{F}_p^* \times \cdots \times \mathbb{F}_p^*}_{N}$$

consider the sequences $\mathbf{c}^{(1)}, \mathbf{c}^{(2)}, \dots, \mathbf{c}^{(N)}$ obtained successively as follows;

$$\begin{aligned}
\mathbf{c}^{(1)} &= (c_1, c_2, \dots, c_{N-1}, c_N); \\
\mathbf{c}^{(2)} &= (c_1, c_2, \dots, c_{N-1}c_N, 1) \\
&\vdots \\
\mathbf{c}^{(N-1)} &= (c_1, c_2 \dots c_{N-1}c_N, \dots, 1, 1) \\
\mathbf{c}^{(N)} &= (c_1c_2, c_3 \dots, c_{N-1}c_N, 1, \dots, 1, 1).
\end{aligned}$$

Formally $\mathbf{c}^{(m)} = (c_1^{(m)}, c_2^{(m)}, \dots, c_{N-1}^{(m)}, c_N^{(m)})$ where

$$c_j^{(m)} = \begin{cases} c_j & j < N - m + 1 \\ \prod_{r=N-m+1}^{N} c_r & j = N - m + 1 \\ 1 & N - m + 1 < j. \end{cases}$$

Now fix $\mathbf{a} = (1 \leq a_1 < a_2 < \cdots < a_N \leq q)$. Then we have modules $T(\mathbf{a}, \mathbf{c}^{(m)})$ for $1 \leq m \leq N$ given by extensions

$$\mathcal{T}(m) = (0 \to R(\mathbf{a}) \to T(\mathbf{a}, \mathbf{c}^{(m)}) \to \mathbb{Z}[C_q] \to 0)$$

classified by $\mathbf{c}^{(m)} \in \underbrace{\mathbb{F}_p^* \times \cdots \times \mathbb{F}_p^*}_{N}$. With this notation we then have:

Theorem 85.1: $T(\mathbf{a}, \mathbf{c}^{(m)}) \cong_{\mathrm{Id}} T(\mathbf{a}, \mathbf{c})$ *for all* m.

Proof. It suffices to show that $T(\mathbf{a}, \mathbf{c}^{(m)}) \cong_{\mathrm{Id}} T(\mathbf{a}, \mathbf{c}^{(m-1)})$ for $2 \leq m \leq N$. To show that $T(\mathbf{a}, \mathbf{c}^{(2)}) \cong_{\mathrm{Id}} T(\mathbf{a}, \mathbf{c}^{(1)}) = T(\mathbf{a}, \mathbf{c})$ we apply (84.4) above with $j = a_{N-1}$, $k = a_N - a_{N-1}$, $\delta = c_N$, $Z = \bigoplus_{i=1}^{N-2} R(i)$ and

$c_Z = (c_1, \ldots, c_{N-2})$. To show that $T(\mathbf{a}, \mathbf{c}^{(m)}) \cong_{\mathrm{Id}} T(\mathbf{a}, \mathbf{c}^{(m-1)})$ we apply (84.4) with

$$\begin{aligned} j &= a_{N-m+1}; \\ k &= a_{N-m+2} - a_{N-m+1}; \\ \delta &= \prod_{i=N-m+2}^{N} c_i; \\ Z &= \bigoplus_{i=1}^{N-m} R(i) \oplus \bigoplus_{i=N-m+3}^{N} R(i) \end{aligned}$$

and where $c_Z = (c'_1, \ldots, c'_{N-m} | \cdots | \cdots | \; c'_{N-m+3}, \ldots, c'_N)$ is the sequence given by

$$c'_i = \begin{cases} c_i & i \leq N-m \\ 1 & N-m+3 \leq i. \end{cases}$$

Note at this stage we have $R(j) = R(a_{N-m+1})$ and $R(j+k) = R(a_{N-m+2})$. □

The above allows some convenient alternative descriptions for the modules $T(\mathbf{a}, \mathbf{c})$; thus if $\mathbf{c} = (c_1, c_2, \ldots, c_N) \in \underbrace{\mathbb{F}_p^* \times \cdots \times \mathbb{F}_p^*}_{N}$ put

$$d = d(\mathbf{c}) = \prod_{j=1}^{N} c_j.$$

Then (85.1) shows that :

(85.2) $$T(\mathbf{a}, \mathbf{c}) \cong_{\mathrm{Id}} T(\mathbf{a}, (d, 1, \ldots, 1)).$$

Moreover, the same proof also shows that

$$\begin{aligned} T(\mathbf{a}, (d, 1, 1, \ldots, 1)) &\cong_{\mathrm{Id}} T(\mathbf{a}, (1, d, 1, \ldots, 1)) \\ &\vdots \\ &\cong_{\mathrm{Id}} T(\mathbf{a}, (1, 1, \ldots, 1, d, 1)) \\ &\cong_{\mathrm{Id}} T(\mathbf{a}, (1, 1, \ldots, 1, , d)). \end{aligned}$$

Formally, if $d \in \mathbb{F}_p^*$ put $\gamma(d, r) = (\gamma_1, \ldots, \gamma_N) \in \underbrace{\mathbb{F}_p^* \times \cdots \times \mathbb{F}_p^*}_{N}$ where

$$\gamma_i = \begin{cases} d & i = r \\ 1 & i \neq r. \end{cases}$$

Then:

Theorem 85.3: $T(\mathbf{a},\mathbf{c}) \cong_{\mathrm{Id}} T(\mathbf{a},\gamma(\prod_{j=1}^{N} c_j, r))$ *for all* r.

Similar considerations apply to the modules $S(\mathbf{a},\mathbf{c})$ after the appropriate notational changes. Here we take sequences $\mathbf{c} = (c_2,\ldots,c_N) \in \underbrace{\mathbb{F}_p^* \times \cdots \times \mathbb{F}_p^*}_{N-1}$ and consider the sequences $\mathbf{c}^{(1)},\mathbf{c}^{(2)},\ldots,\mathbf{c}^{(N-1)}$ obtained successively as follows:

$$\begin{aligned}
\mathbf{c}^{(1)} &= (c_2,\ldots,c_{N-1},c_N);\\
\mathbf{c}^{(2)} &= (c_2,\ldots,c_{N-1}c_N,1)\\
&\vdots\\
\mathbf{c}^{(N-2)} &= (c_2,c_3\ldots,c_{N-1}c_N,\ldots,1,1)\\
\mathbf{c}^{(N-1)} &= (c_2c_2c_3\ldots c_{N-1}c_N,1,\ldots,1,1).
\end{aligned}$$

Formally $\mathbf{c}^{(m)} = (c_2^{(m)},\ldots,c_{N-1}^{(m)},c_N^{(m)})$ where

$$c_j^{(m)} = \begin{cases} c_j & j < N-m+1 \\ \prod_{r=N-m+1}^{N} c_r & j = N-m+1 \\ 1 & N-m+1 < j. \end{cases}$$

For $\mathbf{a} = (1 \le a_2 < \cdots < a_N \le q)$ we have modules $S(\mathbf{a},\mathbf{c}^{(m)})$ for $1 \le m \le N-1$ given by extensions

$$\mathcal{S}(m) = (0 \to R(\mathbf{a}) \to S(\mathbf{a},\mathbf{c}^{(m)}) \to I_q \to 0)$$

classified by $\mathbf{c}^{(m)} \in \underbrace{\mathbb{F}_p^* \times \cdots \times \mathbb{F}_p^*}_{N-1}$. With this notation the same proof as (85.1) then shows:

Theorem 85.4: $S(\mathbf{a},\mathbf{c}^{(m)}) \cong_{\mathrm{Id}} S(\mathbf{a},\mathbf{c})$ *for all* m.

As before, we obtain alternative descriptions for the modules $S(\mathbf{a},\mathbf{c})$; thus for $\mathbf{c} = (c_2, \dots, c_N) \in \underbrace{\mathbb{F}_p^* \times \cdots \times \mathbb{F}_p^*}_{N-1}$ put

$$\widehat{d} = \widehat{d}(\mathbf{c}) = \textstyle\prod_{j=2}^{N} c_j.$$

Then

$$\textbf{(85.5)} \qquad S(\mathbf{a},\mathbf{c}) \cong_{\mathrm{Id}} S(\mathbf{a},(\widehat{d},1,\dots,1)).$$

And also

$$\begin{aligned} S(\mathbf{a},(\widehat{d},1,1,\dots,1)) &\cong_{\mathrm{Id}} S(\mathbf{a},(1,\widehat{d},1,\dots,1)) \\ &\vdots \\ &\cong_{\mathrm{Id}} S(\mathbf{a},(1,1,\dots,1,\widehat{d},1)) \\ &\cong_{\mathrm{Id}} S(\mathbf{a},(1,1,\dots,1,\widehat{d})). \end{aligned}$$

Formally, if $\widehat{d} \in \mathbb{F}_p^*$ put $\widehat{\gamma}(d,r) = (\gamma_2, \dots, \gamma_N) \in \underbrace{\mathbb{F}_p^* \times \cdots \times \mathbb{F}_p^*}_{N-1}$ where

$$\gamma_i = \begin{cases} \widehat{d} & i = r \\ 1 & i \neq r. \end{cases}$$

Then:

Theorem 85.6: $S(\mathbf{a},\mathbf{c}) \cong_{\mathrm{Id}} S(\mathbf{a},\gamma(\prod_{j=2}^{N} c_j, r))$ *for all* r.

§86: A stability theorem:

In what follows, Q will denote either $\mathbb{Z}[C_q]$ or I_q and $\mathbf{a}$ will denote a sequence of integers of the form either

$$\mathbf{a} = (1 \le a_1 < a_2 < \cdots < a_N \le q)$$

or

$$\mathbf{a} = (2 \le a_2 < a_2 < \cdots < a_N \le q)$$

according to whether $Q = \mathbb{Z}[C_q]$ or I_q. We then denote by C the nondegenerate extension module

$$0 \to R(\mathbf{a}) \to C \to Q \to 0$$

classified by the constant sequence $(1,1,\dots,1)$. With this we have:

Theorem 86.1: *Let V be a nondegenerate extension module of the form*

$$0 \to R(\mathbf{a}) \to V \to Q \to 0$$

Then $R(j) \oplus V \cong_{\text{Id}} R(j) \oplus C$ for any row module $R(j)$.

Proof. It suffices to consider three cases:

Case I: $R(j) \cong R(a_s)$ for some s:

By means of (85.3) or (85.6) we may describe V as a nondegenerate extension of the form $0 \to R(\mathbf{a}) \to V \to Q \to 0$ classified by a sequence $(1, \dots, 1 \vdots d \vdots 1, \dots, 1)$ in which $d \in \mathbb{F}_p^*$ occurs in the sth place. In the notation of (84.3), we may describe $R(j) \oplus V = R(a_s) \oplus V$ as the extension module

$$0 \to R(j) \oplus R(j) \oplus Z \to X(0, d; c_Z) \to Q \to 0$$

where $Z = \bigoplus_{t \neq s} R(a_t)$ and $c_Z = (1, 1, \dots, 1)$ is the constant sequence of the appropriate length. However, $R(j) \oplus C$ is described as the extension

$$0 \to R(j) \oplus R(j) \oplus Z \to X(0, 1\,; c_Z) \to Q \to 0$$

and, by (84.8), $X(0, d; c_Z) \cong_{\text{Id}} X(0, 1; c_Z)$. Thus $R(j) \oplus V \cong_{\text{Id}} R(j) \oplus C$, so completing the proof in Case I.

Case II: $j \notin \mathbf{a}$ and $j < a_N$.

Describe V as a nondegenerate extension of the form $0 \to R(\mathbf{a}) \to V \to Q \to 0$ classified by a sequence $(1, \dots, 1, d)$. Then $R(j) \oplus V$ is described as the extension module

$$0 \to R(j) \oplus R(j+k) \oplus Z \to X(0, d; c_Z) \to Q \to 0$$

where $k = a_N - j$, $Z = \bigoplus_{t \neq N} R(a_t)$ and $c_Z = (1, 1, \dots, 1)$ is the constant sequence whilst $R(j) \oplus C$ is described as the extension module

$$0 \to R(j) \oplus R(j+k) \oplus Z \to X(0, 1; c_Z) \to Q \to 0.$$

Then $X(0, d; c_Z) \cong_{\text{Id}} X(0, 1; c_Z)$ by (84.8). Hence $R(j) \oplus V \cong_{\text{Id}} R(j) \oplus C$, so completing the proof in Case II.

Case III: $a_N < j$.

This is similar to Case II except that we now describe $R(j) \oplus V$ as the extension

$$0 \to R(a_N) \oplus R(a_N + k) \oplus Z \to X(d, 0; c_Z) \to Q \to 0$$

where $k = j - a_N$, $Z = \bigoplus_{t \neq N} R(a_t)$ and where $c_Z = (1, 1, \ldots, 1)$ is the constant sequence. In this case, $R(j) \oplus C$ is the extension module

$$0 \to R(a_N) \oplus R(a_N + k) \oplus Z \to X(1,\, 0;\, c_Z) \to Q \to 0.$$

It now follows from (84.9) $X(d,\, 0;\, c_Z) \cong_{\mathrm{Id}} X(1,\, 0;\, c_Z)$. Then again we have $R(j) \oplus V \cong_{\mathrm{Id}} R(j) \oplus C$, so completing the proof. □

As an immediate consequence we have:

Corollary 86.2: *Let V, V' nondegenerate extension modules of the same form*

$$0 \to R(\mathbf{a}) \to V \to Q \to 0 \quad ; \quad 0 \to R(\mathbf{a}) \to V' \to Q \to 0;$$

then $R(j) \oplus V \cong_{\mathrm{Id}} R(j) \oplus V'$ for any row module $R(j)$.

§87: Straightness of degenerate modules:

As $$\mathrm{Hom}_{\mathcal{D}\mathrm{er}}(R(i), R(k)) = \begin{cases} \mathbb{F}_p & j = k \\ 0 & j \neq k \end{cases}$$

it follows immediately that

(87.1) $R(\mathbf{a}, \mathbf{e}) \cong R(\mathbf{b}, \mathbf{f}) \iff \mathrm{Hom}_{\mathcal{D}\mathrm{er}}(R(\mathbf{a}, \mathbf{e}), R(k)) \cong \mathrm{Hom}_{\mathcal{D}\mathrm{er}}(R(\mathbf{b}, \mathbf{f}), R(k))$ for all k.

Theorem 87.2: *Let X, Y be nondegenerate extension modules of the form*

$$0 \to R(\mathbf{a}') \to X \to Q \to 0 \quad ; \quad 0 \to R(\mathbf{b}') \to Y \to Q \to 0.$$

If $R(\mathbf{a}, \mathbf{e}) \oplus X \oplus \Lambda \cong R(\mathbf{b}, \mathbf{f}) \oplus Y \oplus \Lambda$ then $R(\mathbf{a}, \mathbf{e}) \cong R(\mathbf{b}, \mathbf{f})$ and $R(\mathbf{a}') \cong R(\mathbf{b}')$.

Proof. As X is nondegenerate then $\mathrm{Hom}_{\mathcal{D}\mathrm{er}}(X \oplus \Lambda, R(k)) = 0$ and thus

$$\mathrm{Hom}_{\mathcal{D}\mathrm{er}}(R(\mathbf{a}, \mathbf{e}) \oplus X \oplus \Lambda, R(k)) = \mathrm{Hom}_{\mathcal{D}\mathrm{er}}(R(\mathbf{a}, \mathbf{e}), R(k)).$$

Likewise $\mathrm{Hom}_{\mathcal{D}\mathrm{er}}(R(\mathbf{b}, \mathbf{f}) \oplus Y \oplus \Lambda, R(k)) = \mathrm{Hom}_{\mathcal{D}\mathrm{er}}(R(\mathbf{b}, \mathbf{f}), R(k))$. Hence $R(\mathbf{a}, \mathbf{e}) \cong R(\mathbf{b}, \mathbf{f})$ by (87.1). Now $R(\mathbf{a}, \mathbf{e}) \oplus X$ is an extension of the form

$$0 \to R(\mathbf{a}, \mathbf{e}) \oplus R(\mathbf{a}) \to R(\mathbf{a}, \mathbf{e}) \oplus X \to Q \oplus \mathbb{Z}(C_q] \to 0.$$

Likewise $R(\mathbf{b}, \mathbf{f}) \oplus Y$ is an extension of the form

$$0 \to R(\mathbf{b}, \mathbf{f}) \oplus R(\mathbf{b}') \to R(\mathbf{a}, \mathbf{e}) \oplus Y \to Q \to 0.$$

As $R(\mathbf{a}, \mathbf{e}) \oplus X \oplus \Lambda \cong R(\mathbf{b}, \mathbf{f}) \oplus Y \oplus \Lambda$ it follows from (78.3) and (76.2) that

$$R(\mathbf{a}, \mathbf{e}) \oplus R(\mathbf{a}') \cong R(\mathbf{b}, \mathbf{f}) \oplus R(\mathbf{b}').$$

As $R(\mathbf{a}, \mathbf{e}) \cong R(\mathbf{b}, \mathbf{f})$ and $\mathfrak{K}$ is a cancellation semigroup, then $R(\mathbf{a}') \cong R(\mathbf{b}')$. □

We arrive at the following which is Theorem I of the Introduction.

Theorem 87.3: *Let X be a degenerate $\mathfrak{R}$-$\mathfrak{Q}$-module; then the stability class of X is straight.*

Proof. Write $X = R(\mathbf{a}, \mathbf{e}) \oplus X_{nd}$ where $R(\mathbf{a}, \mathbf{e}) \neq 0$ and X_{nd} is nondegenerate defined by an exact sequence of the form

$$(*) \qquad 0 \to R(\mathbf{c}) \to X_{nd} \to Q \to 0.$$

Now suppose $Y \oplus \Lambda \cong X \oplus \Lambda$. We claim that $Y \cong X$. Observe that X is defined by an extension of the form

$$0 \to R(\mathbf{a}, \mathbf{e}) \oplus R(\mathbf{c}) \to X \to Q \to 0.$$

As $Y \oplus \Lambda \cong X \oplus \Lambda$ then, by (79.4), Y is also given by a sequence of the form

$$0 \to R(\mathbf{a}, \mathbf{e}) \oplus R(\mathbf{c}) \to Y \to Q \to 0.$$

By the reduction theorem, we may also write $Y = R(\mathbf{b}, \mathbf{f}) \oplus Y_{nd}$ where Y_{nd} is nondegenerate defined by an exact sequence of the form

$$0 \to R(\mathbf{d}) \to Y_{nd} \to Q \to 0.$$

Consequently we have $R(\mathbf{a}, \mathbf{e}) \oplus X_{nd} \oplus \Lambda \cong R(\mathbf{b}, \mathbf{f}) \oplus Y_{nd} \oplus \Lambda$. It follows from (87.2) that $R(\mathbf{a}, \mathbf{e}) \cong R(\mathbf{b}, \mathbf{f})$ and $R(\mathbf{c}) \cong R(\mathbf{d})$. Hence we have $Y \cong R(\mathbf{a}, \mathbf{e}) \oplus Y_{nd}$ where Y_{nd} is nondegenerate defined by an exact sequence

$$(**) \qquad 0 \to R(\mathbf{c}) \to Y_{nd} \to Q \to 0.$$

Write $R(\mathbf{a}, \mathbf{e}) = Z \oplus R(a_1)$ where $Z = R(a_1)^{e_1 - 1} \oplus \bigoplus_{j=2}^{N} R(a_j)^{e_j}$. (Note that $Z = 0$ in the special case $N = 1$ and $e_1 = 0$.) Comparing (*) and (**)

it follows from (86.2) that $R(a_1) \oplus Y_{nd} \cong R(a_1) \oplus X_{nd}$. Hence $Z \oplus R(a_1) \oplus Y_{nd} \cong Z \oplus R(a_1) \oplus X_{nd}$. Hence $R(\mathbf{a}, \mathbf{e}) \oplus Y_{nd} \cong R(\mathbf{a}, \mathbf{e}) \oplus X_{nd}$ and so, as claimed, $Y \cong X$. More generally, if $Y \oplus \Lambda \cong X \oplus \Lambda^a$ for $a \geq 2$ then $Y \cong X \oplus \Lambda^{a-1}$ by the Swan-Jacobinski Theorem. □

Corollary 87.4: *For any $\mathfrak{R}$-$\mathfrak{Q}$-module X and any row module $R(k)$ the stability class of $R(k) \oplus X$ is straight.*

As $[y-1)$ is an $\mathfrak{R}$-$\mathfrak{Q}$-module of the form $0 \to \bigoplus_{j=2}^{q} R(j) \to [y-1) \to I_q \to 0$ we see the following, which is Corollary II of the Introduction:

Corollary 87.5: *The stability class of $R(k) \oplus [y-1)$ is straight for any k.*

Chapter Twelve

Swan homomorphisms for metacyclic groups

In Chapter Seven we described $\Lambda = \mathbb{Z}[G(p,q)]$ as a fibre product thus:

$$(\mathfrak{S}) \qquad \begin{array}{ccc} \Lambda & \xrightarrow{\pi_-} & \mathcal{T}_q(A,\pi) \\ \downarrow \pi_+ & & \downarrow \varphi_- \\ \mathbb{Z}[C_q] & \xrightarrow{\varphi_+} & \mathbb{F}_p[C_q]. \end{array}$$

Recall that $\mathcal{T}_q(A,\pi)$ decomposes as direct sum $\mathcal{T}_q(A,\pi) \cong R(1) \oplus R(2) \oplus \cdots \oplus R(q)$ where $R(k)$ is the Λ-module concentrated in the kth row of $\mathcal{T}_q(A,\pi)$. In this chapter we will prove the following which is Theorem III of the Introduction.

(*) $R(k) \oplus [y-1)$ is full.

All modules considered in this chapter will be lattices over either $\mathbb{Z}[C_q]$ or Λ. By (33.5), each Λ-lattice M occurs in an exact sequence $0 \to M \to \Lambda^m \to N \to 0$ where N is also a Λ-lattice. By (34.2), N is 1-coprojective so that M is tame and the Swan homomorphism $S_M : \mathrm{Aut}_{\mathcal{D}\mathrm{er}}(M) \to \widetilde{K}_0(\Lambda)$ is defined. Likewise the dual Swan homomorphism S^M is also defined. We shall also prove:

(**) $\mathrm{Im}(S_{R(k)}) = \mathcal{LF}(\Lambda)$ for $1 \le k \le q$

where $\mathcal{LF}(\Lambda) = \mathrm{Ker}(\widetilde{K}_0(\Lambda)) \xrightarrow{\pi_*} \widetilde{K}_0(\mathbb{Z}[C_q]) \oplus \widetilde{K}_0(\mathcal{T}_q(A,\pi))$.

The proofs of (*) and (**) involve a modification, in successive stages, of the Milnor sequence associated with $(\mathfrak{S})$ namely:

$$K_1(\mathbb{Z}[C_q]) \oplus K_1(\mathcal{T}_q(A,\pi)) \xrightarrow{\varphi_*} K_1(\mathbb{F}_p[C_q]) \xrightarrow{\partial} \widetilde{K}_0(\Lambda)$$
$$\xrightarrow{\pi_*} \widetilde{K}_0(\mathbb{Z}[C_q]) \oplus \widetilde{K}_0(\mathcal{T}_q(A,\pi)) \xrightarrow{\varphi_*} \widetilde{K}_0(\mathbb{F}_p[C_q]) \to 0.$$

§88: $\mathcal{T}_q(A, \pi)$ is full:

Given an exact sequence of Λ-lattices $0 \to M \xrightarrow{i} \Lambda \xrightarrow{p} N \to 0$ we recall that the Swan homomorphism $S_M : \mathrm{Aut}_{\mathcal{D}\mathrm{er}}(M) \to \widetilde{K}_0(\Lambda)$ is defined by

$$S_M([\alpha]) = [\varinjlim(\alpha, i)]$$

where $\varinjlim(\alpha, i)$ is the pushout in the following commutative diagram with exact rows

$$\begin{array}{ccccccccc} 0 & \longrightarrow & M & \xrightarrow{i} & \Lambda & \xrightarrow{p} & N & \longrightarrow & 0 \\ & & \downarrow \alpha & & \downarrow \widetilde{\alpha} & & \downarrow \mathrm{Id} & & \\ 0 & \longrightarrow & M & \xrightarrow{\overline{j}} & \varinjlim(\alpha, i) & \xrightarrow{\overline{p}} & N & \longrightarrow & 0. \end{array}$$

Let $\nu : \mathrm{Aut}_\Lambda(M) \to \mathrm{Aut}_{\mathcal{D}\mathrm{er}}(M)$ be the canonical homomorphism. If $[\alpha] = \nu(\widehat{\alpha})$ for some $\widehat{\alpha} \in \mathrm{Aut}_\Lambda(M)$ then, by the Five Lemma,

$$\widehat{\alpha} : \Lambda \to \varinjlim(\alpha, i)$$

is an isomorphism and $S_M([\alpha]) = 0$; that is $\mathrm{Im}(\nu) \subset \mathrm{Ker}(S_m)$. We say that M is *full* when $\mathrm{Im}(\nu) = \mathrm{Ker}(S_M)$; that is, when the sequence

$$\mathrm{Aut}_\Lambda(M) \xrightarrow{\nu} \mathrm{Aut}_{\mathcal{D}\mathrm{er}}(M) \xrightarrow{S_M} \widetilde{K}_0(\Lambda)$$

is exact. Now take $\mathcal{T}_q = \mathcal{T}_q(A, \pi)$ and consider the standard exact sequence

$$0 \to \mathcal{T}_q \xrightarrow{i} \Lambda \xrightarrow{p} \mathbb{Z}[C_q] \to 0$$

embedded in commutative diagrams of Λ-homomorphisms of the form

$$\mathfrak{D}(\alpha) \quad = \quad \left\{ \begin{array}{ccccccc} 0 \to & \mathcal{T}_q & \xrightarrow{i} & \Lambda & \xrightarrow{p} & \mathbb{Z}[C_q] & \to 0 \\ & \downarrow \alpha & & \downarrow \widetilde{\alpha} & & \downarrow \mathrm{Id} & \\ 0 \to & \mathcal{T}_q & \xrightarrow{\overline{i}} & P & \xrightarrow{\overline{p}} & \mathbb{Z}[C_q] & \to 0 \end{array} \right.$$

in which the second row is also exact.

Proposition 88.1: *Suppose given a commutative diagram with exact rows $\mathfrak{D}(\alpha)$ as above; if $P \cong \Lambda$ then there exists a commutative diagram with* *exact rows*

$$\mathfrak{D}(\beta) = \begin{cases} 0 \to & \mathcal{T}_q & \xrightarrow{i} & \Lambda & \xrightarrow{p} & \mathbb{Z}[C_q] & \to 0 \\ & \downarrow \beta & & \downarrow \widetilde{\beta} & & \downarrow \mathrm{Id} & \\ 0 \to & \mathcal{T}_q & \xrightarrow{\bar{i}} & P & \xrightarrow{\bar{p}} & \mathbb{Z}[C_q] & \to 0 \end{cases}$$

in which β and $\widetilde{\beta}$ are isomorphisms.

Proof. Choose an isomorphism $\psi : \Lambda \xrightarrow{\simeq} P$ and consider the following diagram

$$\begin{array}{ccccccc} 0 \to & \mathcal{T}_q & \xrightarrow{i} & \Lambda & \xrightarrow{p} & \mathbb{Z}[C_q] & \to 0 \\ & & & \downarrow \psi & & & \\ 0 \to & \mathcal{T}_q & \xrightarrow{\bar{i}} & P & \xrightarrow{\bar{p}} & \mathbb{Z}[C_q] & \to 0. \end{array}$$

As $\mathrm{Hom}_\Lambda(\mathcal{T}_q, \mathbb{Z}[C_q]) = 0$ then $\bar{p} \circ \psi \circ i = 0$ so the above can be completed to a commutative diagram as follows

$$\begin{array}{ccccccc} 0 \to & \mathcal{T}_q & \xrightarrow{i} & \Lambda & \xrightarrow{p} & \mathbb{Z}[C_q] & \to 0 \\ & \downarrow \psi_- & & \downarrow \psi & & \downarrow \psi_+ & \\ 0 \to & \mathcal{T}_q & \xrightarrow{\bar{i}} & P & \xrightarrow{\bar{p}} & \mathbb{Z}[C_q] & \to 0 \end{array}$$

in which $\psi_+ : \mathbb{Z}[C_q] \to \mathbb{Z}[C_q]$ is then necessarily surjective. As $\mathbb{Z}[C_q]$ is a free abelian group of finite rank then ψ_+ is necessarily bijective and defines an isomorphism of $\mathbb{Z}[C_q]$ modules $\psi_+ : \mathbb{Z}[C_q] \xrightarrow{\simeq} \mathbb{Z}[C_q]$. Hence there is a unit $\mathbf{u} \in \mathbb{Z}[C_q]^*$ such that

$$\psi_+(\mathbf{z}) = \mathbf{u} \cdot \mathbf{z}$$

for all $\mathbf{z} \in \mathbb{Z}[C_q]$. Let $j : \mathbb{Z}[C_q] \to \Lambda$ be the ring homomorphism induced from the inclusion $C_q \subset G(p,q)$ so that $p \circ j = \mathrm{Id}$. Put $\mathbf{v} = j(\mathbf{u}^{-1})$ and define $\widetilde{\beta} : \Lambda \to P$ by $\widetilde{\beta}(\mathbf{x}) = \psi(\mathbf{v} \cdot \mathbf{x})$. As ψ is a Λ-isomorphism and $\mathbf{v} \in \Lambda^*$ then $\widetilde{\beta}$ is also a Λ-isomorphism. Observing that $p : \Lambda \to \mathbb{Z}[C_q]$ is a ring homomorphism we compute

$$\begin{aligned}\overline{p}\circ\beta(\mathbf{x}) &= \overline{p}\circ\psi(\mathbf{v}\cdot\mathbf{x})\\ &= \psi_{+}p(\mathbf{v}\cdot\mathbf{x})\\ &= \mathbf{u}\cdot p(\mathbf{v})\cdot p(\mathbf{x})\\ &= \mathbf{u}\cdot\mathbf{u}^{-1}\cdot p(\mathbf{x})\\ &= p(x).\end{aligned}$$

Observing that $\overline{p}\circ\widetilde{\beta}\circ i = 0$ as again $\mathrm{Hom}_\Lambda(\mathcal{T}_q,\mathbb{Z}[C_q]) = 0$ we have a commutative diagram

$$\mathfrak{D}(\beta) = \begin{cases} 0\to & \mathcal{T}_q & \xrightarrow{i} & \Lambda & \xrightarrow{p} & \mathbb{Z}[C_q] & \to 0\\ & \downarrow\beta & & \downarrow\widetilde{\beta} & & \downarrow\mathrm{Id} & \\ 0\to & \mathcal{T}_q & \xrightarrow{\overline{i}} & P & \xrightarrow{\overline{p}} & \mathbb{Z}[C_q] & \to 0\end{cases}$$

where β is the restriction $\beta = \widetilde{\beta}_{|\mathcal{T}_q}$. By continuing the diagram one place to the left by zeroes it follows from the Five Lemma that β is also an isomorphism. □

As a consequence, we now see that:

Theorem 88.2: *$\mathcal{T}_q(A,\pi)$ is full as a module over Λ.*

Proof. Let $[\alpha]\in\mathrm{Aut}_{\mathcal{D}\mathrm{er}}(\mathcal{T}_q)$ be such that $S_{\mathcal{T}_q}([\alpha]) = 0$. We must construct $\beta\in\mathrm{Aut}_\Lambda(\mathcal{T}_q)$ such that $\nu(\beta) = [\alpha]$ where $\nu:\mathrm{Aut}_\Lambda(\mathcal{T}_q)\to\mathrm{Aut}_{\mathcal{D}\mathrm{er}}(\mathcal{T}_q)$ is the canonical homomorphism. We first represent $[\alpha]\in\mathrm{Aut}_{\mathcal{D}\mathrm{er}}(\mathcal{T}_q)$ by a Λ-homomorphism $\alpha:\mathcal{T}_q\to\mathcal{T}_q$ and construct the pushout extension

$$\mathfrak{D}(\alpha) = \begin{cases} 0\to & \mathcal{T}_q & \xrightarrow{i} & \Lambda & \xrightarrow{p} & \mathbb{Z}[C_q] & \to 0\\ & \downarrow\alpha & & \downarrow\widetilde{\alpha} & & \downarrow\mathrm{Id} & \\ 0\to & \mathcal{T}_q & \xrightarrow{\overline{i}} & \underrightarrow{\lim}(\alpha,i) & \xrightarrow{\overline{p}} & \mathbb{Z}[C_q] & \to 0.\end{cases}$$

As $S_{\mathcal{T}_q}([\alpha]) = 0$ it follows that $\underrightarrow{\lim}(\alpha,i)$ is stably free of rank 1. However Λ satisfies the Eichler condition so that $\underrightarrow{\lim}(\alpha,i)\cong\Lambda$. It now follows from

(88.1) that there exists a commutative diagram

$$\mathfrak{D}(\beta) = \begin{cases} 0 \to & \mathcal{T}_q & \xrightarrow{i} & \Lambda & \xrightarrow{p} & \mathbb{Z}[C_q] & \to 0 \\ & \downarrow \beta & & \downarrow \widetilde{\beta} & & \downarrow \mathrm{Id} & \\ 0 \to & \mathcal{T}_q & \xrightarrow{\overline{i}} & \varinjlim(\alpha, i) & \xrightarrow{\overline{p}} & \mathbb{Z}[C_q] & \to 0 \end{cases}$$

in which $\beta \in \mathrm{Aut}_\Lambda(\mathcal{T}_q)$. Thus we have a commutative diagram

$$\begin{array}{ccccccc} 0 \to & \mathcal{T}_q & \xrightarrow{i} & \Lambda & \xrightarrow{p} & \mathbb{Z}[C_q] & \to 0 \\ & \downarrow \beta - \alpha & & \downarrow \widetilde{\beta} - \widetilde{\alpha} & & \downarrow 0 & \\ 0 \to & \mathcal{T}_q & \xrightarrow{\overline{i}} & \varinjlim(\alpha, i) & \xrightarrow{\overline{p}} & \mathbb{Z}[C_q] & \to 0 \end{array}$$

in which $\mathrm{Im}(\widehat{\beta} - \widehat{\alpha}) \subset \mathrm{Ker}(\overline{p}) = \mathrm{Im}(\overline{i})$. We now have a factorization of $\beta - \alpha$ through Λ thus:

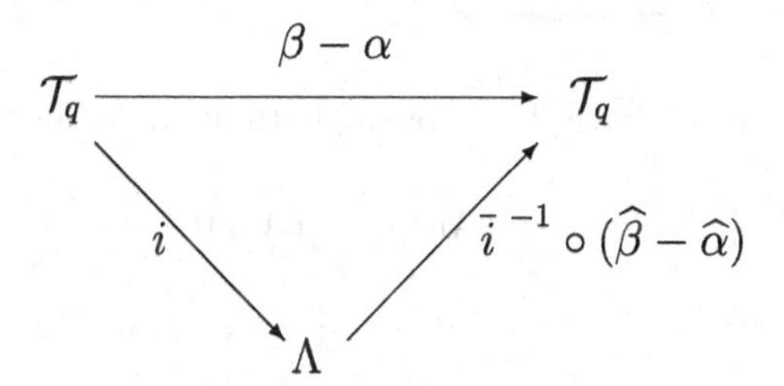

The conclusion follows as $[\alpha] = [\beta] = \nu(\beta)$ where $\beta \in \mathrm{Aut}_\Lambda(\mathcal{T}_q)$. □

§89: A model for $S_{\mathcal{T}_q}$:

As $\mathbb{F}_p^* \cong C_{p-1}$ it follows that, for any divisor e of $p-1$ there exists a unique subgroup $\mathcal{K}(e) \subset \mathbb{F}_p^*$ with the property that $\mathcal{K}(e) \cong C_e$. At this point, it is convenient to recall the following notation from Chapter Seven:

$$\hat{d} = \begin{cases} (p-1)/q & \text{if } q \text{ is odd} \\ 2(p-1)/q & \text{if } q \text{ is even.} \end{cases} \quad ; \quad \check{q} = \begin{cases} q & \text{if } q \text{ is odd} \\ q/2 & \text{if } q \text{ is even.} \end{cases}$$

Then both $\hat{d}$ and $\check{q}$ are divisors of $p-1$; in fact

(89.1) $$(p-1) = \hat{d} \cdot \check{q}.$$

Hence we have

(89.2) $$\mathbb{F}_p^*/\mathcal{K}(\hat{d}) \cong C_{\check{q}}.$$

Recall that $N : \underbrace{\mathbb{F}_p^* \times \cdots \times \mathbb{F}_p^*}_{q} \to \mathbb{F}_p^*$ is defined by $N(a_1, \ldots, a_q) = \prod_{r=1}^{q} a_r$.

Now consider again the canonical mapping $\natural : \mathcal{T}_q(A,\pi)^* \to \underbrace{\mathbb{F}_p^* \times \cdots \times \mathbb{F}_p^*}_{q}$.

An element $\mathbf{a} = (a_1, \ldots, a_q)$ is said to be *liftable to* $\mathcal{T}_q(A,\pi)^*$ when $\mathbf{a} \in \mathrm{Im}(\natural)$. By (62.20),

(89.3) $$\mathbf{a} \in \underbrace{\mathbb{F}_p^* \times \cdots \times \mathbb{F}_p^*}_{q} \text{ is liftable} \iff \mathbf{a} \in N^{-1}(\mathcal{K}(\hat{d})).$$

As $N : \underbrace{\mathbb{F}_p^* \times \cdots \times \mathbb{F}_p^*}_{q} \to \mathbb{F}_p^*$ is surjective and $\mathrm{Im}(\natural) = N^{-1}(\mathcal{K}(\hat{d}))$ then

(89.4) $$\underbrace{\mathbb{F}_p^* \times \cdots \times \mathbb{F}_p^*}_{q}/\mathrm{Im}(\natural) \cong \mathbb{F}_p^*/\mathcal{K}(\hat{d}) \cong C_{\check{q}}.$$

As $\mathcal{T}_q(A,\pi)$ is full we have the following faithful model for $S_{\mathcal{T}_q}$.

(89.5) $$\begin{array}{ccccccc} \mathrm{Aut}_\Lambda(\mathcal{T}_q(A,\pi)) & \stackrel{\nu}{\longrightarrow} & \mathrm{Aut}_{\mathcal{D}\mathrm{er}}(\mathcal{T}_q(A,\pi)) & \stackrel{S_{\mathcal{T}_q}}{\longrightarrow} & \mathrm{Im}(S_{\mathcal{T}_q}) & \to & 0 \\ \updownarrow & & \updownarrow & & \updownarrow & & \\ \mathcal{T}_q(A,\pi)^* & \stackrel{\natural}{\longrightarrow} & \underbrace{\mathbb{F}_p^* \times \cdots \times \mathbb{F}_p^*}_{q} & \stackrel{\mu\circ N}{\longrightarrow} & C_{\check{q}} & \to & 0 \end{array}$$

where both rows are exact and $\mu : \mathbb{F}_p^* \to \mathbb{F}_p^*/\mathcal{K}(\hat{d}) \cong C_{\check{q}}$ is the identification mapping. In particular, we see that:

(89.6) $$\mathrm{Im}(S_{\mathcal{T}_q}) \cong C_{\check{q}}.$$

§90: The identity $\mathrm{Im}(S_{R(k)}) = \mathrm{Im}(S_{\mathcal{T}_q})$:

In this section we will prove:

Theorem 90.1: $\mathrm{Im}(S_{R(k)}) = \mathrm{Im}(S_{\mathcal{T}_q})$ *for* $1 \le k \le q$.

First consider the case $k = q$. We begin by recalling that $\Lambda = \mathbb{Z}[G(p,q)]$ is described as the extension

$$\mathbf{L} = (0 \to \mathcal{T}_q \xrightarrow{i} \Lambda \xrightarrow{\eta} \mathbb{Z}[C_q] \to 0)$$

with extension class $(1, \dots, 1) \in \underbrace{\mathbb{F}_p \times \cdots \times \mathbb{F}_p}_{q} \cong \mathrm{Ext}^1(\mathbb{Z}[C_q], \mathcal{T}_q(A, \pi))$.

Define $R(\overline{q}) = \bigoplus_{i \neq q} R(i)$. Then from the filtration $R(q) \subset R(\overline{q}) \oplus R(q) = \mathcal{T}_q \subset \Lambda$ we obtain an exact sequence $\mathbf{X} = (0 \to R(q) \to \Lambda \xrightarrow{p} \Lambda/R(q) \to 0)$. Clearly $\Lambda/R(q)$ is an extension $\mathbf{Y} = (0 \to R(\overline{q}) \to \Lambda/R(q) \xrightarrow{\nu} \mathbb{Z}[C_q] \to 0)$. From $\mathbf{X}$ and $\mathbf{Y}$ we have the tautological identities

(90.2) $$\mathcal{T}_q = p^{-1}(R(\overline{q})) \text{ and } \nu \circ p = \eta$$

and the following alternative description of $\mathbf{L}$:

(90.3) $$\mathbf{L} = (0 \to p^{-1}(R(\overline{q})) \to \Lambda \xrightarrow{\eta} \mathbb{Z}[C_q] \to 0).$$

If $\alpha : R(q) \to R(q)$ is a Λ-homomorphism we define $\widehat{\alpha} : p^{-1}(R(\overline{q})) \to p^{-1}(R(\overline{q}))$ by

$$\widehat{\alpha} = \begin{pmatrix} I_{q-1} & 0 \\ 0 & \alpha \end{pmatrix}$$

where the matrix is taken relative to the decomposition $R(\overline{q}) \oplus R(q) = p^{-1}(R(\overline{q}))$. Now form the pushout extension

(90.4) $$\begin{matrix} \mathbf{L} \\ \downarrow \natural \\ \widehat{\alpha}_*(\mathbf{L}) \end{matrix} = \begin{pmatrix} 0 & \to & p^{-1}(R(\overline{q})) & \xrightarrow{i} & \Lambda & \xrightarrow{\eta} & \mathbb{Z}[C_q] & \to & 0 \\ & & \downarrow \widehat{\alpha} & & \downarrow \natural & & \downarrow \mathrm{Id} & & \\ 0 & \to & p^{-1}(R(\overline{q})) & \xrightarrow{\overline{i}} & \varinjlim(\widehat{\alpha}, i) & \xrightarrow{\overline{\eta}} & \mathbb{Z}[C_q] & \to & 0 \end{pmatrix}.$$

From $\mathbf{X}$ we form the pushout extension

(90.5) $$\begin{matrix} \mathbf{X} \\ \downarrow \natural \\ \widehat{\alpha}_*(\mathbf{X}) \end{matrix} = \begin{pmatrix} 0 & \to & R(q) & \xrightarrow{j} & \Lambda & \xrightarrow{\mu} & \Lambda/R(q) & \to & 0 \\ & & \downarrow \alpha & & \downarrow \natural & & \downarrow \mathrm{Id} & & \\ 0 & \to & R(q) & \xrightarrow{\overline{j}} & \varinjlim(\alpha, j) & \xrightarrow{\overline{\mu}} & \Lambda/R(q) & \to & 0 \end{pmatrix}.$$

In similar fashion to the above from $\widehat{\alpha}_*(\mathbf{X})$ and $\mathbf{Y}$ we have the tautological identities

(90.6) $$\mathcal{T}_q = \overline{p}^{-1}(R(\overline{q})) \quad \text{and} \quad \overline{\nu} \circ \overline{p} = \overline{\eta}$$

and the following alternative description of $\widehat{\alpha}_*(\mathbf{L})$

(90.7) $$\widehat{\alpha}_*(\mathbf{L}) = (0 \to \overline{p}^{-1}(R(\overline{q})) \to \varinjlim(\alpha, j) \xrightarrow{\eta} \mathbb{Z}[C_q] \to 0).$$

Comparing (90.4) and (90.7) we see that:

(90.8) $$\varinjlim(\alpha, j) \cong \varinjlim(\widehat{\alpha}, i).$$

Observe that if α represents an element of $\mathrm{Aut}_{\mathcal{D}\mathrm{er}}(R(q))$ then $\widehat{\alpha}$ represents an element of $\mathrm{Aut}_{\mathcal{D}\mathrm{er}}(p^{-1}(R(\overline{q}))) = \mathrm{Aut}_{\mathcal{D}\mathrm{er}}(\mathcal{T}_q)$ and hence $[\varinjlim(\widehat{\alpha}, i)] \in \mathrm{Im}(S_{\mathcal{T}_q})$. Thus it follows from (90.8) that:

(90.9) If $[\alpha] \in \mathrm{Aut}_{\mathcal{D}\mathrm{er}}(R(q))$ then $[\varinjlim(\alpha, j)] \in \mathrm{Im}(S_{\mathcal{T}_q})$.
However, $\mathrm{Im}(S_{R(q)})$ consists of all projective classes of the form $[\varinjlim(\alpha, j)]$ where $[\alpha]$ runs through the elements of $\mathrm{Aut}_{\mathcal{D}\mathrm{er}}(R(q))$. Thus we see that:

(90.10) $$\mathrm{Im}(S_{R(q)}) \subset \mathrm{Im}(S_{\mathcal{T}_q}).$$

As $\mathrm{Aut}_{\mathcal{D}\mathrm{er}}(R(q)) \cong \mathbb{F}_p^* \cong C_{p-1}$ and $\mathrm{Ker}(S_{R(q)}) \subset \mathrm{Aut}_{\mathcal{D}\mathrm{er}}(R(q))$ it follows that $\mathrm{Ker}(S_{R(q)}) \cong C_e$ where e divides $p-1$. As $\mathrm{Im}(S_{R(q)}) \cong C_{p-1}/\mathrm{Ker}(S_{R(q)})$ then

(90.11) $$|\mathrm{Im}(S_{R(q)})| = \frac{(p-1)}{e}.$$

However

(90.12) $$|\mathrm{Im}(S_{\mathcal{T}_q})| = \frac{(p-1)}{\hat{d}}.$$

It now follows from (90.10) that

(90.13) $$\hat{d} \leq e.$$

We now make the identification $\mathrm{Aut}_{\mathcal{D}\mathrm{er}}(S_{R(q)}) \longleftrightarrow \mathbb{F}_p^*$. Choosing $\alpha \in \mathrm{Ker}(S_{R(q)}) \subset \mathbb{F}_p^*$ such that $\mathrm{ord}(\alpha) = e$ consider again the modules $\varinjlim(\alpha, j)$ and $\varinjlim(\widehat{\alpha}, i)$ as defined in (90.5) and (90.4). Then by (90.8), $\varinjlim(\alpha, j) \cong \varinjlim(\widehat{\alpha}, i)$. As $\alpha \in \mathrm{Ker}(S_{R(q)})$ then $[\varinjlim(\alpha, j)] = 0$. Hence $[\varinjlim(\widehat{\alpha}, i)] = 0$ and $[\varinjlim(\widehat{\alpha}, i)] \in \mathrm{Ker}(S_{\mathcal{T}_q})$. As $\mathcal{T}_q(A, \pi)$ is full the extension class $(1, \dots, 1, \alpha) \in \mathrm{Ext}^1(\mathbb{Z}[C_q], \mathcal{T}_q(A, \pi)) \cong \underbrace{\mathbb{F}_p^* \times \cdots \times \mathbb{F}_p^*}_{q}$ is liftable to $\mathcal{T}_q(A, \pi)^*$. It now follows that $\alpha = N(1, \dots, 1, \alpha) \in \mathcal{K}(\hat{d})$ and so:

(90.14) $$e = \mathrm{ord}(\alpha) \leq \hat{d}.$$

Comparing (90.13) and (90.14) we see that $e = \hat{d}$. Comparing (90.11) and (90.12) then $|\mathrm{Im}(S_{R(q)})| = |\mathrm{Im}(S_{\mathcal{T}_q})|$. As $\mathrm{Im}(S_{R(q)}) \subset \mathrm{Im}(S_{\mathcal{T}_q})$ it follows that:

(90.15) $$\mathrm{Im}(S_{R(q)}) = \mathrm{Im}(S_{\mathcal{T}_q}).$$

This completes the proof of (90.1) when $k = q$. From the periodic resolution (74.11) we see that $R(k)$ is a generalized syzygy of $R(q)$. Hence $\mathrm{Im}(S_{R(k)}) = \mathrm{Im}(S_{R(q)})$ by (40.4), thereby completing the proof of (90.1) in the general case.

§91: A simplification:

In dealing with the segment

(91.1) $$K_1(\mathbb{F}_p[C_q]) \xrightarrow{\partial} K_0(\Lambda) \xrightarrow{\pi_*} K_0(\mathbb{Z}[C_q]) \oplus K_0(\mathcal{T}_q(A,\pi)) \xrightarrow{\varphi_*} K_0(\mathbb{F}_p[C_q]) \to 0$$

one is faced with the inconvenience that both $\mathcal{T}_q(A,\pi)$ and $\mathbb{F}_p[C_q]$ have projective modules which are smaller than the free module of rank 1. One can deal with this at the formal level by observing that, by the conclusions of Chapter Six, the surjective homomorphism

$$K_0(\mathcal{T}_q(A,\pi)) \overset{\varphi_*}{\twoheadrightarrow} K_0(\mathbb{F}_p[C_q])$$

is split with kernel equal to $Cl(A)$. Consequently (91.1) gives rise to an exact sequence

(91.2) $$K_1(\mathbb{F}_p[C_q]) \xrightarrow{\partial} K_0(\Lambda) \overset{\pi_*}{\twoheadrightarrow} K_0(\mathbb{Z}[C_q]) \oplus Cl(A) \to 0.$$

There is, however, a more intrinsic way to consider this: say that a projective module P over $\mathcal{T}_q(A,\pi)$ is *almost free of rank* m when

$$P \otimes_\Lambda \widehat{\Lambda} \cong \mathcal{T}_q(\widehat{A},\pi)^{(m)}.$$

Let $\mathcal{AF}(\mathcal{T}_q(A,\pi))$ be the set of isomorphism classes of such modules. Clearly $\mathcal{AF}(\mathcal{T}_q(A,\pi))$ is an additive semigroup under '$\oplus$'. Our previous description (51.10) of projective modules over $\mathcal{T}_q(A,\pi)$ shows that $\mathcal{AF}(\mathcal{T}_q(A,\pi))$ is a cancellation semigroup. To describe it explicitly, we note that, as before, addition in $Cl(A) \oplus \mathbb{N}$ is given by

$$(\mathbf{a}, m) \oplus (\mathbf{b}, n) = (\mathbf{ab}, m + n + 1).$$

If $\mathbf{a}$ is an ideal in A we define $[\mathbf{a}] = R(1) \oplus \cdots \oplus R(q-1) \oplus (R(q) \otimes \mathbf{a})$. With the convention that $\mathcal{T}_q(A,\pi)^{(0)} = 0$ it follows from (51.9) that the mapping

$$Cl(A) \oplus \mathbb{N} \longrightarrow \mathcal{AF}(\mathcal{T}_q(A,\pi)) \quad ; \quad (\mathbf{a}, n) \mapsto [\mathbf{a}] \oplus \mathcal{T}_q(A,\pi)^{(n)}$$

is an additive isomorphism; that is:

Theorem 91.3: $\mathcal{AF}(\mathcal{T}_q(A,\pi)) \cong Cl(A) \oplus \mathbb{N}$.

Denoting by $K_0(\mathcal{AF})$ the Grothendieck group of $\mathcal{AF}(\mathcal{T}_q(A,\pi))$ we then see that:

Corollary 91.4: $K_0(\mathcal{AF}) \cong Cl(A) \oplus \mathbb{Z}$.

We denote by $\mathcal{F}(\mathbb{F}_p[C_q])$ the isomorphism classes of (non zero) finitely generated free modules over $\mathbb{F}_p[C_q]$ so that $\mathcal{F}(\mathbb{F}_p[C_q]) \cong \mathbb{Z}_+$. We denote its Grothendieck group by $K_0(\mathcal{F})$ so that $K_0(\mathcal{F}) \cong \mathbb{Z}$. Swan's local freeness theorem for projective modules over Λ now shows that we can replace (91.1) by

(91.5) $K_1(\mathbb{F}_p[C_q]) \xrightarrow{\partial} K_0(\Lambda) \xrightarrow{\pi_*} K_0(\mathbb{Z}[C_q]) \oplus K_0(\mathcal{AF}) \xrightarrow{\varphi_*} K_0(\mathcal{F}) \to 0$

Again the surjection $K_0(\mathcal{AF}) \xrightarrow{\varphi_*} K_0(\mathcal{F})$ splits with kernel $Cl(A)$ from which we regain (91.2), although in a more intrinsic manner.

§92: The first modification:

In this section we will show that, for the fibre square

$$(\mathfrak{S}) \qquad \begin{cases} \mathbb{Z}[G(p,q)] & \xrightarrow{\pi_-} & \mathcal{T}_q(A,\pi) \\ \pi_+ \downarrow & & \downarrow \varphi_- \\ \mathbb{Z}[C_q] & \xrightarrow{\varphi_+} & \mathbb{F}_p[C_q] \end{cases}$$

the sequence (91.2) can be modified to an exact sequnece

(92.1) $$0 \to \overline{GL_1}(\mathfrak{S}) \xrightarrow{\partial} \widetilde{K}_0(\Lambda) \xrightarrow{\pi_*} \widetilde{K}_0(\mathbb{Z}[C_q]) \oplus Cl(A) \to 0$$

where, in accordance with the conventions of §7,

$$\overline{GL_1}(\mathfrak{S}) = GL_1(\mathcal{T}_q(A,\pi)) \backslash GL_1(\mathbb{F}_p[C_q]) / GL_1(\mathbb{Z}[C_q]).$$

In passing, we note that, for a general fibre square $(\mathfrak{F})$, the set $\overline{GL_1}(\mathfrak{F})$ admits no natural group structure. However, in the present case, as

$GL_1(\mathbb{F}_p[C_q])$ is an abelian group then so also is $\overline{GL_1}(\mathfrak{S})$. Moreover, it is finite as $GL_1(\mathbb{F}_p[C_q])$ is finite. Given a Milnor square of ring homomorphisms as in §7

$$(\mathfrak{F}) = \begin{cases} \Lambda & \stackrel{\eta_-}{\to} & \Lambda_- \\ \downarrow \eta_+ & & \downarrow \varphi_- \\ \Lambda_+ & \stackrel{\varphi_+}{\to} & \Lambda_0 \end{cases}$$

we recall that a projective module L over Λ is *locally free of rank n* when it is obtained by glueing a square in which the corners are free of rank n thus:

$$\begin{array}{ccc} L & \to & \Lambda_-^{(n)} \\ \downarrow \eta_+ & & \downarrow \varphi_- \\ \Lambda_+^{(n)} & \stackrel{\varphi_+}{\to} & \Lambda_0^{(n)}. \end{array}$$

Denoting by $\mathcal{LF}_n(\mathfrak{F})$ the isomorphism classes of such modules we have a bijection

$$\mathcal{LF}_n(\mathfrak{F}) \stackrel{\simeq}{\longleftrightarrow} GL_n(\Lambda_+)\backslash GL_n(\Lambda_0)/GL_n(\Lambda_-).$$

Observe that a finitely generated projective module $P \in \mathcal{P}(\Lambda)$ is locally free of rank n with respect to $(\mathfrak{F})$ if and only if $\eta_-(P)$ is free over Λ_- and $\eta_+(P)$ is free over Λ_+. Expressed differently, the stable class $[P]$ lies in $\mathrm{Ker}\left(\eta_* : \widetilde{K}_0(\Lambda) \to \widetilde{K}_0(\Lambda_+) \oplus \widetilde{K}_0(\Lambda_-)\right)$ precisely when $P \oplus \Lambda^n$ is locally free for some n. Stabilization gives a mapping $\mathcal{LF}_n(\mathfrak{F}) \to \mathcal{LF}_{n+1}(\mathfrak{F})$, thereby giving the colimit

$$\mathcal{LF}(\mathfrak{F}) = \varinjlim \mathcal{LF}_n(\mathfrak{F}).$$

By the above, $\mathcal{LF}(\mathfrak{F}) = \mathrm{Ker}(\eta_* : \widetilde{K}_0(\Lambda) \to \widetilde{K}_0(\Lambda_+) \oplus \widetilde{K}_0(\Lambda_-))$ so giving an exact sequence

(92.2) $$0 \to \mathcal{LF}(\mathfrak{F}) \stackrel{\partial}{\to} \widetilde{K}_0(\Lambda) \stackrel{\pi_*}{\to} \widetilde{K}_0(\mathbb{Z}[C_q]) \oplus Cl(A) \to 0.$$

Under reasonable hypotheses on $(\mathfrak{F})$ we are able to compute $\mathcal{LF}(\mathfrak{F})$ exactly. From (7.5) we have:

Proposition 92.3: *If $\mathcal{P}(\Lambda)$ is a cancellation semigroup and Λ_0 is commutative and weakly Euclidean then the stabilization mapping $s_n : \overline{GL_1}(\mathfrak{F}) \to \overline{GL_{n+1}}(\mathfrak{F})$ is bijective for all $n \geq 1$.*

Now specialize to the above fibre square ($\mathfrak{S}$). In this case $\mathbb{Z}[G(p,q)]$ satisfies the Eichler condition and so has projective cancellation by the Swan-Jacobinski Theorem. Moreover, $\mathbb{F}_p[C_q]$ is a product of fields

$$\mathbb{F}_p[C_q] \cong \underbrace{\mathbb{F}_p \times \cdots \times \mathbb{F}_p}_{q}$$

and so is commutative and weakly Euclidean. Thus ($\mathfrak{S}$) satisfies the hypotheses of (92.3). Stabilization now gives a bijection

$$\overline{GL_1}(\mathfrak{S}) \xrightarrow{\simeq} \varinjlim \overline{GL_{n+1}}(\mathfrak{S}) = \mathcal{LF}(\mathfrak{S}).$$

Substituting $\overline{GL_1}(\mathfrak{S})$ for $\mathcal{LF}(\mathfrak{S})$ in (92.2) gives the exact sequence

$$0 \to \overline{GL_1}(\mathfrak{S}) \xrightarrow{\partial} \widetilde{K}_0(\Lambda) \xrightarrow{\pi_*} \widetilde{K}_0(\mathbb{Z}[C_q]) \oplus Cl(A) \to 0$$

required to verify the statement (92.1) above.

§93: The second modification:

In this section we modify the sequence of (92.1) to one of the form

$$(\spadesuit) \qquad 0 \to \mathrm{Im}(S_{R(k)}) \to \widetilde{K}_0(\Lambda) \xrightarrow{\pi_*} \widetilde{K}_0(\mathbb{Z}[C_q]) \oplus Cl(A) \to 0.$$

As before, we put $C_q = \langle y | y^q = 1 \rangle$ and we maintain our previous choice of a in the range $2 \leq a \leq p-1$ with the property that the residue class $[a]$ has order q in the multiplicative group $\mathbb{F}_p^*$. For each r in the range $0 \leq r \leq q-1$ let $\chi_r : \mathbb{F}_p[C_q] \to \mathbb{F}_p$ be the ring homomorphism determined by

$$\textbf{(93.1)} \qquad \chi_r(y) = [a]^r.$$

Then there is a ring homomorphism $\chi : \mathbb{F}_p[C_q] \longrightarrow \underbrace{\mathbb{F}_p \times \cdots \times \mathbb{F}_p}_{q}$ given by

$$\chi(\alpha) = (\chi_{q-1}(\alpha), \chi_{q-1}(\alpha), \ldots, \chi_1(\alpha), \chi_0(\alpha)).$$

It is straightforward to verify that $\chi : \mathbb{F}_p[C_q] \xrightarrow{\simeq} \underbrace{\mathbb{F}_p \times \cdots \times \mathbb{F}_p}_{q}$ is a ring isomorphism and that the following diagram commutes

(93.2)
$$\begin{array}{ccc} & \mathcal{T}_q(A,\pi) & \\ \varphi \swarrow & & \searrow \natural \\ \mathbb{F}_p[C_q] & \xrightarrow{\quad \chi \quad} & \underbrace{\mathbb{F}_p \times \ldots \times \mathbb{F}_p}_{q} \end{array}$$

where $\varphi : \mathbb{Z}[C_q] \to \mathbb{F}_p[C_q]$ is the homomorphism which occurs in $(\mathfrak{S})$. We note, in particular, that χ induces an isomorphism of unit groups

(93.3)
$$\chi : \mathbb{F}_p[C_q]^* \xrightarrow{\simeq} \underbrace{\mathbb{F}_p^* \times \cdots \times \mathbb{F}_p^*}_{q}.$$

One sees easily that

(93.4)
$$\det(\rho_{reg}(\alpha)) = N(\alpha)$$

where $\rho_{reg} : \mathbb{F}_p[C_q] \hookrightarrow M_q(\mathbb{F}_p)$ is induced by the regular representation of C_q and $N : \mathbb{F}_p[C_q]^* \to \mathbb{F}_p^*$ is the norm homomorphism of §61. The natural surjective ring homomorphism $\natural : \mathbb{Z}[C_q] \twoheadrightarrow \mathbb{F}_p[C_q]$ induces a homomorphism of unit groups $\natural : \mathbb{Z}[C_q]^* \twoheadrightarrow \mathbb{F}_p[C_q]^*$ and we note:

Proposition 93.5: *If $\alpha \in \mathbb{Z}[C_q]^*$ then $\natural(\alpha) \in \mathbb{F}_p[C_q]^*$ admits a lifting to $\mathcal{T}_q(A,\pi)^*$.*

Proof. Observe that the following diagram commutes:

(93.6)
$$\begin{array}{ccccc} \mathbb{Z}[C_q]^* & \xrightarrow{\rho_{reg}} & GL_q(\mathbb{Z}) & \xrightarrow{\det} & \mathbb{Z}^* \\ \downarrow \natural & & \downarrow \natural & & \downarrow \natural \\ \mathbf{F}_p[C_q]^* & \xrightarrow{\rho_{reg}} & GL_q(\mathbb{F}_p) & \xrightarrow{\det} & \mathbb{F}_p^* \end{array}$$

As $\alpha \in \mathbb{Z}[C_q]^*$ then $\det(\rho_{reg}(\alpha)) \in \mathbb{Z}^*$. As $\mathbb{Z}^* = \{\pm 1\}$ it follows by commutativity of the above that $N(\alpha) = \pm 1$. The conclusion now follows from (62.21). □

As an immediate consequence we have a canonical bijection:

(93.7)
$$\mathcal{T}_q(A,\pi)^* \backslash \mathbb{F}_p[C_q]^* / \mathbb{Z}[C_q]^* \xleftrightarrow{\simeq} \mathcal{T}_q(A,\pi)^* \backslash \mathbb{F}_p[C_q]^*.$$

Recalling that $\overline{GL_1}(\mathfrak{S}) = \mathcal{T}_q(A,\pi)^*\backslash\mathbb{F}_p[C_q]^*/\mathbb{Z}[C_q]^*$ we can now modify (92.1) to an exact sequence

$$0 \to \mathcal{T}_q(A,\pi)^*\backslash\mathbb{F}_p[C_q]^* \xrightarrow{\partial} \widetilde{K}_0(\Lambda) \xrightarrow{\pi_*} \widetilde{K}_0(\mathbb{Z}[C_q]) \oplus Cl(A) \to 0$$

which, in turn, can be elongated to an exact sequence

(93.8) $$\mathcal{T}_q(A,\pi)^* \to \mathbb{F}_p[C_q]^* \xrightarrow{\partial} \widetilde{K}_0(\Lambda) \xrightarrow{\pi_*} \widetilde{K}_0(\mathbb{Z}[C_q]) \oplus Cl(A) \to 0.$$

The canonical mapping $\mathrm{Aut}_\Lambda(\mathcal{T}_q(A,\pi)) \to \mathrm{Aut}_{\mathcal{D}\mathrm{er}}(\mathcal{T}_q(A,\pi))$ with cokernel $\mathrm{Im}(S_{\mathcal{T}_q})$ is identical to mapping $\mathcal{T}_q(A,\pi)^* \to \mathbb{F}_p[C_q]^*$ so that (93.8) can be truncated to

(93.9) $$0 \to \mathrm{Im}(S_{\mathcal{T}_q}) \to \widetilde{K}_0(\Lambda) \xrightarrow{\pi_*} \widetilde{K}_0(\mathbb{Z}[C_q]) \oplus Cl(A) \to 0.$$

By (90.1) we may substitute $\mathrm{Im}(S_{R(k)})$ for $\mathrm{Im}(S_{\mathcal{T}_q})$ to obtain an exact sequence

(93.10) $$0 \to \mathrm{Im}(S_{R(k)}) \to \widetilde{K}_0(\Lambda) \xrightarrow{\pi_*} \widetilde{K}_0(\mathbb{Z}[C_q]) \oplus Cl(A) \to 0$$

which satisfies the description (♠) given at the start of this section. Moreover, as $\mathcal{LF}(\Lambda) = \mathrm{Ker}(\pi_*)$ it follows tautologically that:

Theorem 93.11: $\mathrm{Im}(S_{R(k)}) = \mathcal{LF}(\Lambda)$ *for* $1 \le k \le q$.

§94: Fullness of $R(k)$:

From (66.8) and (66.11) we see that $R(q) \cong \overline{I^*_C}$ and $\mathrm{End}_\Lambda(\overline{I^*_C}) \cong A$; hence:

(94.1) $$\mathrm{End}_\Lambda(R(q)) \cong A^*.$$

Elsewhere we have seen that

(94.2) $$\mathrm{End}_{\mathcal{D}\mathrm{er}}(R(q)) \cong \mathbb{F}_p.$$

There is a unique ring homomorphism $A \to \mathbb{F}_p$. It follows that the canonical ring homomorphism $\nu : \mathrm{End}_\Lambda(R(q)) \to \mathrm{End}_{\mathcal{D}\mathrm{er}}(R(q))$ can be identified with the ring homomorphism $A \to \mathbb{F}_p$. Moreover, we saw in (61.3) that:

(94.3) $$\mathrm{Im}(A^* \to \mathbb{F}_p^*) = \mathcal{K} = \begin{cases} C_d & q \text{ odd} \\ C_{2d} & q \text{ even.} \end{cases}$$

Let $S : \mathrm{Aut}_{\mathcal{D}\mathrm{er}}(\overline{I^*_C}) \to \widetilde{K}_0(\Lambda)$ denote the Swan homomorphism; then

(94.4) $$\mathrm{Im}(S) \cong \begin{cases} C_q & q \text{ odd} \\ C_{q/2} & q \text{ even.} \end{cases}$$

As $\mathrm{Im}(S_{R(q)}) \cong C_{\check{q}}$ we have the following model for $S_{R(q)}$

$$\begin{array}{ccccccc}
\mathrm{Aut}_\Lambda(R(q)) & \stackrel{\nu}{\longrightarrow} & \mathrm{Aut}_{\mathcal{D}\mathrm{er}}(R(q)) & \stackrel{S_{R(q)}}{\longrightarrow} & \mathrm{Im}(S_{R(q)}) & \rightarrow & 0 \\
\updownarrow & & \updownarrow & & \updownarrow & & \\
A^* & \stackrel{\natural}{\longrightarrow} & \mathbb{F}_p^* & \stackrel{\mu\circ N}{\longrightarrow} & C_{\check{q}} & \rightarrow & 0
\end{array}$$

and hence a commutative diagram with exact rows

$$\begin{array}{ccccccccc}
0 & \rightarrow & \mathrm{Im}(\nu) & \hookrightarrow & \mathrm{Aut}_{\mathcal{D}\mathrm{er}}(R(q) & \stackrel{S_{R(q)}}{\longrightarrow} & \mathrm{Im}(S_{R(q)}) & \rightarrow & 0 \\
& & \updownarrow & & \updownarrow & & \updownarrow & & \\
1 & \rightarrow & \mathcal{K} & \stackrel{\natural}{\longrightarrow} & \mathbb{F}_p^* & \stackrel{\mu\circ N}{\longrightarrow} & \mathbb{F}_p^*/\mathcal{K} & \rightarrow & 1
\end{array}$$

Thus $\mathrm{Ker}(S_{R(q)}) = \mathrm{Im}(\nu : \mathrm{Aut}_\Lambda(R(q)) \stackrel{\nu}{\longrightarrow} \mathrm{Aut}_{\mathcal{D}\mathrm{er}}(R(q)))$; that is:

(94.5) $R(q)$ is full.

Now suppose that $k \neq q$ and let $\nu : \mathrm{Aut}_\Lambda(R(k)) \stackrel{\nu}{\longrightarrow} \mathrm{Aut}_{\mathcal{D}\mathrm{er}}(R(k))$ denote the natural mapping. As $\mathrm{Im}(S_{R(k)}) = \mathrm{Im}(S_{\mathcal{T}_q}) = C_{\check{q}}$ then it follows that we have a commutative diagram

$$\begin{array}{ccccccccc}
0 & \rightarrow & \mathrm{Im}(\nu) & \hookrightarrow & \mathrm{Aut}_{\mathcal{D}\mathrm{er}}(R(k) & \stackrel{S_{R(k)}}{\longrightarrow} & \mathrm{Im}(S_{R(k)}) & \rightarrow & 0 \\
& & \cap & & \updownarrow & & \updownarrow & & \\
1 & \rightarrow & \mathcal{K}(\hat{d}) & \stackrel{\natural}{\longrightarrow} & \mathbb{F}_p^* & \stackrel{\mu\circ N}{\longrightarrow} & \mathbb{F}_p^*/\mathcal{K}(\hat{d}) & \rightarrow & 1
\end{array}$$

As the bottom sequence is exact, to show the top sequence is exact, and hence to show that $R(k)$ is full, it is enough to show that there exists a Λ-automorphism $\gamma : R(k) \rightarrow R(k)$ such that $\mathrm{ord}(\nu(\gamma)) = \hat{d}$. However, as $R(q)$ is full, there exists a Λ-automorphism $\alpha : R(q) \rightarrow R(q)$ such that $\mathrm{ord}([\alpha]) = \hat{d}$ where $[\alpha]$ is the image of α in $\mathrm{Aut}_{\mathcal{D}\mathrm{er}}(R(q))$. By (68.8) we can represent $R(k)$ in the form

$$R(k) = \pi^k R(q).$$

As π is the unique prime in A over p then $\alpha(\pi^k \mathbf{x}) = \pi^k \alpha(\mathbf{x})$ so that α restricts to a Λ-automorphism $\gamma : R(k) \rightarrow R(k)$ such that $\mathrm{ord}(\nu(\gamma)) = \hat{d}$. Hence we have:

(94.6) $R(k)$ is full.

§95: Changing rings from $\mathbb{Z}[C_q]$ to $\mathbb{Z}[G(p,q)]$:

In Chapter Four in the context of the derived module category we studied the process of changing rings from a subring Ω to a containing ring Λ.

We now specialize to the case where $\Lambda = \mathbb{Z}[G(p,q)]$ and $\Omega = \mathbb{Z}[C_q]$. We write $\mathbb{Z}$ for the trivial module over either Λ or $\mathbb{Z}[C_q]$. As in Chapter Eight the integral group ring $\mathbb{Z}[C_p]$ and its augmentation ideal $I(C_p)$ are invariant under the Galois action of y giving an exact sequence of Galois modules

(95.1) $$0 \longrightarrow \overline{I(C_p)} \longrightarrow \overline{\mathbb{Z}[C_p]} \stackrel{\epsilon}{\longrightarrow} \mathbb{Z} \longrightarrow 0.$$

Under the Galois action, C_q acts freely on the $\mathbb{Z}$-basis $\{x^k - 1\}_{1 \le k \le p-1}$ of $I(C_p)$ with $(p-1)/q$ orbits. It follows that:

$$j^*(\overline{I(C_p)}) \cong \mathbb{Z}[C_q]^{(p-1)/q}.$$

Applying j^* to (95.1) gives an exact sequence of $\mathbb{Z}[C_q]$-modules

$$0 \longrightarrow j^*(\overline{I(C_p)}) \longrightarrow j^*(\overline{\mathbb{Z}[C_p]}) \longrightarrow j^*(\mathbb{Z}) \longrightarrow 0$$

which splits as $j^*(\overline{I(C_p)})$ is free and $j^*(\mathbb{Z}) = \mathbb{Z}$ is torsion free; hence

$$j^*(\overline{\mathbb{Z}[C_p]}) \cong \mathbb{Z} \oplus \mathbb{Z}[C_q]^{(p-1)/q}.$$

However, we also have $j_*(\mathbb{Z}) = \overline{\mathbb{Z}[C_p]}$ and hence

(95.2) $$j^*j_*(\mathbb{Z}) \cong \mathbb{Z} \oplus \mathbb{Z}[C_q]^{(p-1)/q}.$$

As $\Lambda \cong \mathbb{Z}[C_q] \otimes_{\delta(\Lambda)} \overline{\mathbb{Z}[C_p]}$ then for any $\mathbb{Z}[C_q]$-lattice M it follows that

$$\begin{aligned} j^*(M \otimes_{\mathbb{Z}[C_q]} \Lambda) &= M \otimes_{\delta(C_q)} j^*(\overline{\mathbb{Z}[C_p]}) \\ &\cong M \otimes_{\delta(C_q)} (\mathbb{Z} \oplus \mathbb{Z}[C_q]^{(p-1)/q}) \\ &\cong M \oplus M \otimes_{\delta(C_q)} \mathbb{Z}[C_q]^{(p-1)/q}. \end{aligned}$$

Clearly $M \otimes_{\delta(C_q)} \mathbb{Z}[C_q]^{(p-1)/q} \cong \mathbb{Z}[C_q]^{\mathrm{r}(M)(p-1)/q}$ so that, for any $\mathbb{Z}[C_q]$-lattice M;

(95.3) $$j^*j_*(M) \cong M \oplus \mathbb{Z}[C_q]^{r(M)(p-1)/q}.$$

Hence $j^*j_*(M)$ and M are isomorphic in the derived module category of $\mathbb{Z}[C_q]$. Moreover, from the identity $j^*(M \otimes_{\mathbb{Z}[C_q]} \Lambda) = M \otimes_{\delta(C_q)} j^*(\overline{\mathbb{Z}[C_p]})$ we obtain a canonical $\mathbb{Z}[C_q]$-homomorphism $\pi_M = \mathrm{Id}_M \otimes \epsilon : j^*j_*(M) \longrightarrow M$ which defines an explicit isomorphism within $\mathcal{D}\mathrm{er}(\mathbb{Z}[C_q])$. In fact we have an exact sequence of $\mathbb{Z}[C_q]$-modules

(95.4) $$0 \longrightarrow \mathrm{Ker}(\pi_M) \longrightarrow j^*(j_*(M)) \xrightarrow{\pi_M} M \longrightarrow 0$$

which splits as $\mathrm{Ker}(\pi_M) \cong M \otimes_{\delta(C_q)} j^*(\overline{I(C_p)}) \cong \mathbb{Z}[C_q]^{r(M)(p-1)/q}$ is free and M is coprojective. If $i_M : M \to j^*j_*(M)$ is a right splitting of (95.4) we obtain a homomorphism of abelian groups $\nu : \mathrm{End}_{\mathbb{Z}[C_q]}(j^*j_*(M)) \to \mathrm{End}_{\mathbb{Z}[C_q]}(M)$ by

$$\nu(\alpha) = \pi_M \circ \alpha \circ i_M.$$

In general ν is *not a ring homomorphism*; however ν induces a ring isomorphism

$$\nu_* : \mathrm{End}_{\mathcal{D}\mathrm{er}}(j^*j_*(M)) \xrightarrow{\simeq} \mathrm{End}_{\mathcal{D}\mathrm{er}}(M)$$

which is independent of the particular splitting i_M chosen. It is now straightforward to see that we have a commutative square of ring homomorphisms

$$\begin{array}{ccc} \mathrm{End}_{\mathcal{D}\mathrm{er}(\Lambda)}(j_*(M)) & \xrightarrow{\ j^*\ } & \mathrm{End}_{\mathcal{D}\mathrm{er}(C_q)}(j^*j_*(M)) \\ \uparrow j_* & & \downarrow \nu_* \\ \mathrm{End}_{\mathcal{D}\mathrm{er}(C_q)}(M) & \xrightarrow{\ \mathrm{Id}\ } & \mathrm{End}_{\mathcal{D}\mathrm{er}(C_q)}(M). \end{array}$$

Hence we see that:

(95.5) For any $\mathbb{Z}[C_q]$-lattice M, $j_* : \mathrm{End}_{\mathcal{D}\mathrm{er}}(M) \to \mathrm{End}_{\mathcal{D}\mathrm{er}}(j_*(M))$ is an injective ring homomorphism with left inverse $\nu_* \circ j^* : \mathrm{End}_{\mathcal{D}\mathrm{er}}(j_*(M)) \to \mathrm{End}_{\mathcal{D}\mathrm{er}}(M)$.

§96: Fullness of $[y-1)$:

In this section and the next we use boldface symbols **Hom**, $\mathbf{Ext}^k$ when describing homomorphisms and extensions of Λ-modules and italics *Hom* and Ext^k, when referring to homomorphisms and extensions of modules over $\mathbb{Z}[C_q]$. Let I_Q denote the augmentation ideal of $\mathbb{Z}[C_q]$. Then I_Q is the right ideal of $\mathbb{Z}[C_q]$ generated by $y-1$. As in §17 we have

$$Ext^1(\mathbb{Z}, I_Q) \cong \mathbb{Z}/q.$$

Analogously, we denote by $[y-1)$ denote the right ideal of Λ generated by $y-1$. It is straightforward to show that:

(96.1) $$[y-1) = j_*(I_Q).$$

Applying j_* to the exact sequence $(0 \to I_Q \to \mathbb{Z}[C_q] \to \mathbb{Z} \to 0)$ gives an exact sequence $(0 \to j_*(I_Q) \to \Lambda \to j_*(\mathbb{Z}) \to 0)$ from which we see that

(96.2) $$[y-1) = \Omega_1(j_*(\mathbb{Z})).$$

By the Eckmann-Shapiro Theorem

$$\mathbf{Ext}^1(j_*(\mathbb{Z}), [y-1)) \cong \mathbf{Ext}^1(j_*(\mathbb{Z}), j_*(I_Q)) \cong Ext^1(j^*j_*(\mathbb{Z}), I_Q).$$

However, by (95.2), $j^*j_*(\mathbb{Z}) \cong \mathbb{Z} \oplus \mathbb{Z}[C_q]^{(p-1)/q}$ and so

$$Ext^1(j^*j_*(\mathbb{Z}), I_Q) \cong Ext^1(\mathbb{Z}, I_Q) \cong \mathbb{Z}/q.$$

Moreover by (96.1), (96.2) and the Corepresentation Theorem it follows that

$$\mathbf{End}_{\mathcal{D}\mathrm{er}}([y-1)) \cong \mathbf{Ext}^1(j_*(\mathbb{Z}), j_*(I_Q)).$$

Collecting our observations together we see that:

(96.3) $$\mathbf{End}_{\mathcal{D}\mathrm{er}}([y-1)) \cong \mathbb{Z}/q.$$

Again writing $[y-1) = j_*(I_Q)$, then as in (95.5) we have an injective ring homomorphism $j_* : \mathrm{End}_{\mathcal{D}\mathrm{er}(C_q)}(I_Q) \to \mathrm{End}_{\mathcal{D}\mathrm{er}(\Lambda)}([y-1))$. Moreover, noting that $[I_Q] = \Omega_1(\mathbb{Z})$ then, by coprojectivity of $\mathbb{Z}[C_q]$-lattices we see that

$$End_{\mathcal{D}\mathrm{er}}(I_Q) \cong End_{\mathcal{D}\mathrm{er}}(\mathbb{Z}) \cong \mathbb{Z}/q.$$

It follows that $j_* : End_{\mathcal{D}\mathrm{er}}(I_Q) \to \mathbf{End}_{\mathcal{D}\mathrm{er}}([y-1))$ is a ring isomorphism as it is injective and both sets have cardinal q. We now have a commutative diagram of group homomorphisms

$$\begin{array}{ccc}
\mathbf{Aut}_\Lambda([y-1)) & \xrightarrow{\ \nu\ } & \mathbf{Aut}_{\mathcal{D}\mathrm{er}}([y-1)) \\
\uparrow j_* & & \uparrow j_* \ \cong \\
Aut_{\mathbb{Z}[C_q]}(I_Q) & \xrightarrow{\ \nu\ } & Aut_{\mathcal{D}\mathrm{er}}(I_Q).
\end{array}$$

in which $j_* : Aut_{\mathcal{D}\text{er}}(I_Q) \to \mathbf{Aut}_{\mathcal{D}\text{er}}([y-1))$ is an isomorphism. The mapping $\nu : Aut_{\mathbb{Z}[C_q]}(I_Q) \longrightarrow Aut_{\mathcal{D}\text{er}}(I_Q)$ is surjective by (45.6). It now follows that

(96.4) $\nu : \mathrm{Aut}_\Lambda([y-1)) \to \mathrm{Aut}_{\mathcal{D}\text{er}(\Lambda)}([y-1))$ is surjective.

We know that $\mathrm{Im}(\nu) \subset \mathrm{Ker}(S_{[y-1)}) \subset \mathrm{Aut}_{\mathcal{D}\text{er}(\Lambda)}([y-1))$ where $S_{[y-1)}$ is the Swan homomorphism. It follows from (96.4) that $\mathrm{Im}(\nu) = \mathrm{Ker}(S_{[y-1)})$; that is:

(96.5) $[y-1)$ is full.

It follows also from (96.4) that $\mathrm{Ker}(S_{[y-1)}) = \mathrm{Aut}_{\mathcal{D}\text{er}(\Lambda)}([y-1))$; hence:

(96.6) The Swan homomorphism $S_{[y-1)} : \mathrm{Aut}_{\mathcal{D}\text{er}(\Lambda)}([y-1)) \to \widetilde{K}_0(\Lambda)$ is identically zero.

§97: Fullness of $R(k) \oplus [y-1)$:

If $0 \to N \to P_n \to P_{n-1} \to \cdots \to P_0 \to M \to 0$ is an exact sequence of Λ-lattices in which each P_i is finitely generated projective then as Λ-lattices are all coprojective it follows by iterating (24.7) that $\mathrm{End}_{\mathcal{D}\text{er}}(N) \cong \mathrm{End}_{\mathcal{D}\text{er}}(M)$. In particular from the resolution of (74.11), each $R(k)$ is a generalized syzygy of $R(1)$; thus, for all k:

(97.1) $\mathbf{End}_{\mathcal{D}\text{er}}(R(k) \oplus [y-1)) \cong \mathbf{End}_{\mathcal{D}\text{er}}(R(1) \oplus [y-1)).$

In (71.6) we decomposed the augmentation ideal I_G of G as a direct sum

$$I_G \cong R(1) \oplus [y-1)$$

and from the exact sequence $0 \to I_G \to \Lambda \to \mathbb{Z} \to 0$ it follows by repeating the above argument that $\mathrm{End}_{\mathcal{D}\text{er}}(R(1) \oplus [y-1)) \cong \mathrm{End}_{\mathcal{D}\text{er}}(\mathbb{Z}) \cong \mathbb{Z}/pq$. Hence for all k:

(97.2) $\mathbf{End}_{\mathcal{D}\text{er}}(R(k) \oplus [y-1)) \cong \mathbb{Z}/pq.$

We may represent $\alpha \in \mathbf{End}_{\mathcal{D}\text{er}}(R(k) \oplus [y-1))$ as a matrix

$$\alpha = \begin{pmatrix} \alpha_{11} & \alpha_{12} \\ \alpha_{21} & \alpha_{21} \end{pmatrix} \in \begin{pmatrix} \mathbf{End}_{\mathcal{D}\text{er}}(R(k)) & \mathbf{Hom}_{\mathcal{D}\text{er}}([y-1), R(k)) \\ \mathbf{Hom}_{\mathcal{D}\text{er}}(R(k), [y-1)) & \mathbf{End}_{\mathcal{D}\text{er}}([y-1)) \end{pmatrix}$$

from which we see that the diagonal terms give an injective ring homomorphism

$$\mathbf{End}_{\mathcal{D}\mathrm{er}}(R(k)) \times \mathbf{End}_{\mathcal{D}\mathrm{er}}([y-1)) \hookrightarrow \mathbf{End}_{\mathcal{D}\mathrm{er}}(R(k) \oplus [y-1)).$$

However, $\mathbf{End}_{\mathcal{D}\mathrm{er}}(R(k)) \cong \mathbb{Z}/p$ from (70.12) whilst $\mathbf{End}_{\mathcal{D}\mathrm{er}}([y-1)) \cong \mathbb{Z}/q$ from (96.3). As domain and codomain have the same cardinality we see that

(97.4) $$\mathbf{End}_{\mathcal{D}\mathrm{er}}(R(k)) \times \mathbf{End}_{\mathcal{D}\mathrm{er}}([y-1)) \cong \mathbf{End}_{\mathcal{D}\mathrm{er}}(R(k) \oplus [y-1)).$$

Moreover, as a corollary we have

(97.5) $$\mathbf{Hom}_{\mathcal{D}\mathrm{er}}([y-1), R(k)) = \mathbf{Hom}_{\mathcal{D}\mathrm{er}}(R(k), [y-1)) = 0.$$

It follows that a class $[\alpha] \in \mathbf{Aut}_{\mathcal{D}\mathrm{er}}(R(k) \oplus [y-1))$ takes the form

$$[\alpha] = \begin{pmatrix} [\alpha_1] & 0 \\ 0 & [\alpha_2] \end{pmatrix}$$

where $[\alpha_1] \in \mathbf{Aut}_{\mathcal{D}\mathrm{er}}(R(k))$ and $[\alpha_2] \in \mathbf{Aut}_{\mathcal{D}\mathrm{er}}([y-1))$ so that, by (39.3)

$$S_{R(k)\oplus[y-1)}[\alpha] = S_{R(k)}[\alpha_1] + S_{[y-1)}[\alpha_2].$$

However, by (96.6), $S_{[y-1)} \equiv 0$ and so $S_{R(k)\oplus[y-1)}[\alpha] = S_{R(k)}[\alpha_1]$; hence:

(97.6) $$\mathrm{Im}(S_{R(k)\oplus[y-1)}) = \mathrm{Im}(S_{R(k)}).$$

Furthermore, $\mathrm{Ker}(S_{R(k)\oplus[y-1)}) = \mathrm{Ker}(S_{R(k)}) \times \mathbf{Aut}_{\mathcal{D}\mathrm{er}}([y-1))$. The canonical mapping $\nu : \mathbf{Aut}_{\Lambda}(R(k) \oplus [y-1)) \to \mathbf{Aut}_{\mathcal{D}\mathrm{er}}(R(k) \oplus [y-1))$ has

$$\mathrm{Im}(\nu) \subset \mathrm{Ker}(S_{R(k)\oplus[y-1)}) = \mathrm{Ker}(S_{R(k)}) \times \mathbf{Aut}_{\mathcal{D}\mathrm{er}}([y-1)).$$

The mappings $\mathbf{Aut}_{\Lambda}(R(k)) \twoheadrightarrow \mathrm{Ker}(S_{R(k)})$ and $\mathbf{Aut}_{\Lambda}([y-1)) \twoheadrightarrow \mathbf{Aut}_{\mathcal{D}\mathrm{er}}([y-1))$ are surjective by (94.6) and (96.5) respectively. Thus

$$\nu : \mathbf{Aut}_{\Lambda}(R(k) \oplus [y-1)) \twoheadrightarrow \mathrm{Ker}(S_{R(k)\oplus[y-1)})$$

is surjective; we obtain the following which is Theorem III of the Introduction:

(97.7) $R(k) \oplus [y-1)$ is full.

As $I_G \cong R(1) \oplus [y-1)$ represents $\Omega_1(\mathbb{Z})$, then from (40.4) and (96.6) it follows that $\mathrm{Im}(S_{R(1)}) = \mathrm{Im}(S_{\mathbb{Z}})$; that is, the image of the original Swan homomorphism $S_{\mathbb{Z}} : (\mathbb{Z}/|G|)^* \to \widetilde{K}_0(\Lambda)$. For the sake of completeness, we record it thus:

(97.8) $$\mathrm{Im}(S_{\mathbb{Z}}) = \mathcal{LF}(\Lambda).$$

We note that this expression together with (92.2) reproduces the computation of $\widetilde{K}_0(\Lambda)$ given by Galovich, Reiner and Ullom in [22].

Chapter Thirteen

An obstruction to monogenicity

In Chapter Nine we showed the existence, for each $G(p, q)$, of a basic exact sequence $0 \to R(1) \to \Lambda \to \Lambda \to R(q) \to 0$. From this we derived exact sequences

$$\mathcal{Z}(k) = (0 \to R(k+1) \to P(k) \to \Lambda \to R(k) \to 0)$$

where $P(k)$ is projective of rank 1 and $\bigoplus_{k=1}^{q-1} P(k) \cong \Lambda^{(q-1)}$. It is natural to ask whether, in these derived sequences, we can arrange that $P(k) \cong \Lambda$. If this were always the case then we could modify the resolution of (74.11) to one of the form

$$\begin{array}{ccccccccccccc} & & K(q) & & & & K(q-1) & & K(2) & & K(1) & & \\ & & \nearrow\searrow & & & & \nearrow\searrow & & \nearrow\searrow & & \nearrow\searrow & & \\ 0 \to R(1) \to \Lambda & \longrightarrow & & \Lambda \longrightarrow & & \Lambda & \longrightarrow & \cdots & \longrightarrow \Lambda & \longrightarrow & \Lambda \longrightarrow \Lambda & \longrightarrow R(1) \longrightarrow 0 \\ & & & \searrow\nearrow & & & & & & \searrow\nearrow & & & \\ & & & R(q) & & & & & & R(2) & & & \end{array}$$

Such a resolution, if it exists, we will term *strongly monogenic*. In Chapter Fourteen we will give conditions which are sufficient to guarantee the existence of such strongly monogenic resolutions. The first such question to be answered is therefore:

Does there exist an exact sequence $0 \to R(2) \to \Lambda \to \Lambda \to R(1) \to 0$?

As we shall also see, a positive answer to this question is enough to guarantee that the $D(2)$-property holds for $G(p, q)$. In this chapter, however, we will show that no such monogenic secondary sequence exists for the group $G(13, 12)$.

§98: Obstructions to strong monogenicity:

Theorem VIII of the Introduction includes the hypothesis that $\widetilde{K}_0(\mathbb{Z}[C_q]) = 0$. The integers $q \geq 2$ with this property have been completely determined by Cassou-Noguès (cf [11]); thus we have:

(98.1) If q is prime then $\widetilde{K}_0(\mathbb{Z}[C_q]) = 0 \iff q \leq 19$.

(98.2) If q is composite then $\widetilde{K}_0(\mathbb{Z}[C_q]) = 0 \iff q = 4, 6, 8, 9, 10$ or 14.

Consider again the exact sequences of (74.10), namely

$$\mathcal{Z}(k) \qquad 0 \longrightarrow R(k+1) \longrightarrow P(i) \xrightarrow{\nearrow K(k) \searrow} \Lambda \longrightarrow R(k) \longrightarrow 0$$

in which $P(k)$ is projective of rank 1. The question arises:

(98.3) Can $\mathcal{Z}(k)$ be chosen so that $P(k) = \Lambda$?

To analyze (98.3) let $j^* : \mathcal{M}od_\Lambda \to \mathcal{M}od_{\mathbb{Z}[C_q]}$ be the restriction of scalars functor. This functor is exact so that applied to $\mathcal{Z}(k)$ it gives an exact sequence

(98.4) $0 \longrightarrow j^*(R(i+1)) \longrightarrow j^*(P(i)) \longrightarrow j^*(\Lambda) \longrightarrow j^*(R(i)) \longrightarrow 0.$

We observed in (68.14) that

(98.5) Each $j^*(R(i))$ is projective over $\mathbb{Z}[C_q]$.

Clearly $j^*(P(i))$ is projective over $\mathbb{Z}[C_q]$ whilst $j^*(\Lambda)$ is free over $\mathbb{Z}[C_q]$. Consequently the class of $j^*(P(i))$ in $\widetilde{K}_0(\mathbb{Z}[C_q])$ is given by

(98.6) $[j^*(P(i))] = [j^*(R(i+1))] - [j^*(R(i))].$

We saw in (69.18) that:

(98.7) $j^*(R(1))$ is free of rank $(p-1)/q$ over $\mathbb{Z}[C_q]$.

In particular,

(98.8) $[j^*(P(1))] = [j^*(R(2))] \in \widetilde{K}_0(\mathbb{Z}[C_q]).$

As $\mathbb{Z}[C_q]$ has no nontrivial stably free modules if $j^*(R(2))$ is not free over $\mathbb{Z}[C_q]$ then $P(1)$ cannot be chosen to be free over Λ; that is:

Theorem 98.9: *The class $[j^*(R(2))] \in \widetilde{K}_0(\mathbb{Z}[C_q])$ is the first obstruction to the existence of a strongly monogenic resolution of $R(1)$.*

The smallest value of q for which $\widetilde{K}_0(\mathbb{Z}[C_q]) \neq 0$ is $q = 12$. The first problematic group for our method is therefore $G(13, 12)$. We proceed to show the above obstruction is nonzero in this case.

§99: The unit group of $\mathbb{Z}[y]/(y^6 - 1)$:

The next two sections are preparatory to computing the projective modules over $\mathbb{Z}[C_{12}]$. In this section we compute the unit group of $\mathbb{Z}[C_6]$. As primitive third and sixth roots of unity we may take respectively

$$\zeta_3 = \frac{-1+\sqrt{-3}}{2}; \quad \zeta_6 = \frac{1+\sqrt{-3}}{2}.$$

As $\zeta_6 = 1 + \zeta_3$ then $\mathbb{Z}(\zeta_6) = \mathbb{Z}(\zeta_3)$. Putting $A = \mathbb{Z}(\zeta_3) = \mathbb{Z}(\zeta_6)$ then it is straightforward to observe that

(99.1) $$A^* \cong C_6 \text{ generated by } \zeta_6.$$

There is a ring isomorphism $\mathbb{Z}[y]/(y^2 + y + 1) \xrightarrow{\simeq} A$ induced from the correspondence $y \mapsto \zeta_3$ from which it follows that

(99.2) $$(\mathbb{Z}[y]/(y^2 + y + 1))^* \cong C_6 \text{ generated by } -y.$$

From the factorization $(y^3 - 1) = (y - 1)(y^2 + y + 1)$ we get a fibre product of ring homomorphisms as follows:

(99.3) $$\begin{array}{ccc} \mathbb{Z}[y]/(y^3 - 1) & \to & \mathbb{Z}[y]/(y^2 + y + 1) \\ \downarrow & & \downarrow \\ \mathbb{Z} & \to & \mathbb{F}_3 \end{array}$$

Clearly $\mathbb{Z}^* = \{\pm 1\} \cong C_2 \cong \mathbb{F}_3^*$ so (99.3) gives a fibre product of unit groups

$$\begin{array}{ccc} (\mathbb{Z}[y]/(y^3 - 1))^* & \to & C_6 \\ \downarrow & & \downarrow \\ C_2 & \xrightarrow{\simeq} & C_2 \end{array}$$

from which it follows easily that

(99.4) $$(\mathbb{Z}[y]/(y^3 - 1))^* \cong C_6 \text{ generated by } -y.$$

We also have a factorization $(y^3 + 1) = (y + 1)(y^2 - y + 1)$ which gives a corresponding fibre product of ring homomorphisms:

(99.5) $$\begin{array}{ccc} \mathbb{Z}[y]/(y^3+1) & \rightarrow & \mathbb{Z}[y]/(y^2 - y + 1) \\ \downarrow & & \downarrow \\ \mathbb{Z} & \rightarrow & \mathbb{F}_3 \end{array}$$

There is a ring isomorphism $\mathbb{Z}[y]/(y^2 - y + 1) \xrightarrow{\simeq} A$ induced from the correspondence $y \mapsto \zeta_6$ from which it follows that

(99.6) $$(\mathbb{Z}[y]/(y^2 - y + 1))^* \cong C_6 \text{ generated by } y.$$

From the fibre product (99.5) we now get a fibre product of groups

$$\begin{array}{ccc} (\mathbb{Z}[y]/(y^3+1))^* & \rightarrow & C_6 \\ \downarrow & & \downarrow \\ C_2 & \xrightarrow{\simeq} & C_2 \end{array}$$

from which we see that:

(99.7) $$(\mathbb{Z}[y]/(y^3 + 1))^* \cong C_6 \text{ generated by } y.$$

Finally from the factorization $(y^6 - 1) = (y^3 - 1)(y^3 + 1)$ we get a fibre product of ring homomorphisms

$$\begin{array}{ccc} \mathbb{Z}[y]/(y^6-1) & \rightarrow & \mathbb{Z}[y]/(y^3 + 1) \\ \downarrow & & \downarrow \\ \mathbb{Z}[y]/(y^3-1) & \rightarrow & \mathbb{F}_2[C_3]. \end{array}$$

As is well known, $\mathbb{F}_2[C_3]^* \cong C_3$ whilst we have shown that

$$(\mathbb{Z}[y]/(y^3 - 1))^* \cong (\mathbb{Z}[y]/(y^3 + 1))^* \cong C_6 \cong C_2 \times C_3.$$

Consequently we get a fibre product of unit groups

$$\begin{array}{ccc} (\mathbb{Z}[y]/(y^6-1))^* & \rightarrow & C_2 \times C_3 \\ \downarrow & & \downarrow \\ C_2 \times C_3 & \rightarrow & C_3 \end{array}$$

It now follows easily that:

(99.8) $$(\mathbb{Z}[y]/(y^6-1))^* \cong C_2 \times C_2 \times C_3 \cong C_2 \times C_6.$$

Moreover, one checks easily that y generates a copy of C_6 in $(\mathbb{Z}[y]/(y^6-1))^*$ whilst -1 generates the extra copy of C_2.

§100: The unit group of $\mathbb{F}_2[C_6]$:

Taking $C_6 = \langle y | y^6 = 1\rangle$ and $\mathbb{F}_2$ to be the field with two elements we represent the group ring $\mathbb{F}_2[C_6]$ as the quotient

$$\mathbb{F}_2[C_6] = \mathbb{F}_2[y]/(y^6-1).$$

As 3 is invertible in $\mathbb{F}_2$, the factorization $y^6 - 1 = (y^4 + y^2 + 1)(y^2 - 1)$ leads to direct product decomposition of rings

(100.1) $$\mathbb{F}_2[y]/(y^6-1) \cong \mathbb{F}_2[z]/(z^4+z^2+1) \times \mathbb{F}_2[w]/(w^2-1)$$

where, to avoid confusion, we use distinct variable letters in the two factors. Hence we also have direct product decomposition of unit groups

(100.2) $$\mathbb{F}_2[C_6]^* \cong (\mathbb{F}_2[z]/(z^4+z^2+1))^* \times (\mathbb{F}_2[w]/(w^2-1))^*.$$

To proceed we note that $\mathbb{F}_2[z]/(z^4+z^2+1) = \mathbb{F}_2[z]/(z^2+z+1)^2$ is a local ring with sixteen elements in which we note that

$$(z^2+z+1)^2 = (z^3+1)^2 = (z^3+z^2+z)^2 = 0.$$

Thus $\{0, z^2+z+1, z^3+1, z^3+z^2+z\}$ are all non-units and hence

$$|(\mathbb{F}_2[z]/(z^4+z^2+1))^*| \leq 12.$$

In fact, $(z+z^2)^2 = 1$ and it is straightforward now to check that

(100.3) $$(\mathbb{F}_2[z]/(z^4+z^2+1))^* \cong C_6 \times C_2 = \langle z, z+z^2\rangle.$$

Clearly we also have $(\mathbb{F}_2[w]/(w^2-1))^* = \{1, w\} \cong C_2$ and hence from (100.1) and (100.3) we see that $\mathbb{F}_2[C_6]^* \cong C_6 \times C_2 \times C_2$. We identify specific generators as follows: put

$$u = y + y^2 + y^3 + y^4 + y^5; \quad v = 1 + y + y^4;$$

then $u^2 = 1; v^2 = 1$ and as the ring $\mathbb{F}_2[C_6]$ is commutative it follows that $\langle u, v\rangle = \{1, u, v, uv\}$ is a subgroup of $\mathbb{F}[C_6]^*$ with $\langle u, v\rangle \cong C_2 \times C_2$. Also y generates a cyclic subgroup $\langle y\rangle = \{1, y, y^2, y^3, y^4, y^5\}$ of order six in $\mathbb{F}[C_6]^*$. As $uv = y^2 + y^3 + y^5$ one sees that $\langle y\rangle \cap \langle u, v\rangle = \{1\}$ so that we have a direct product:

(100.4) $$\mathbb{F}_2[C_6]^* \cong C_6 \times C_2 \times C_2 = \langle y, u, v\rangle.$$

Put $\Gamma = \langle y\rangle$ and $\Delta = \langle y, u\rangle$. It is useful to have an explicit listing of the elements of $\mathbb{F}_2[C_6]^*$ in the form $\mathbb{F}_2[C_6]^* = \Delta \cup \Delta \cdot v$

(100.5)

Δ	$\Delta \cdot v$
1	$1 + y + y^4$
y	$y + y^2 + y^5$
y^2	$1 + y^2 + y^3$
y^3	$y + y^3 + y^4$
y^4	$y^2 + y^4 + y^5$
y^5	$1 + y^3 + y^5$
$y + y^2 + y^3 + y^4 + y^5$	$y^2 + y^3 + y^5$
$1 + y^2 + y^3 + y^4 + y^5$	$1 + y^3 + y^4$
$1 + y + y^3 + y^4 + y^5$	$y + y^4 + y^5$
$1 + y + y^2 + y^4 + y^5$	$1 + y^2 + y^5$
$1 + y + y^2 + y^3 + y^5$	$1 + y + y^3$
$1 + y + y^2 + y^3 + y^4$	$y + y^2 + y^4$

§101: Projective modules over $\mathbb{Z}[C_{12}]$:

Let $n \geq 4$ be an even integer so that $\zeta_{2n} = \exp(\pi i/n)$ is a primitive $2n^{\text{th}}$ root of unity. Now $\mathbb{Z}(\zeta_{2n})$ is the ring of integers in the cyclotomic field $\mathbb{Q}(\zeta_{2n})$ and the unit group $\mathbb{Z}(\zeta_{2n})^*$ takes the following form (cf [25] p. 556):

(101.1) $$\mathbb{Z}(\zeta_{2n})^* \cong C_{2n} \times C_\infty^{r(n)}.$$

Here $C_\infty^{r(n)}$ is a free abelian group of rank $r(n) = \varphi(n) - 1$ where φ is Euler's totient function and C_{2n} is the cyclic group generated by ζ_{2n}. Here we are interested in the case where $n = 6$ when $\varphi(6) = 2$ and so

(101.2) $$\mathbb{Z}(\zeta_{12})^* \cong C_{12} \times C_\infty.$$

First consider the fibre product diagram

(101.3) $$\begin{array}{ccc} \mathbb{Z}[C_{12}] & \xrightarrow{\pi_+} & \mathbb{Z}[y]/(y^6+1) \\ \pi_- \downarrow & & \downarrow \psi_+ \\ \mathbb{Z}[y]/(y^6-1) & \xrightarrow{\psi_-} & \mathbb{F}_2[C_6] \end{array}$$

where π_- π_+ are the projections induced by the factorization

$$(y^{12} - 1) = (y^6 - 1)(y^6 + 1)$$

and ψ_-, ψ_+ are the reductions mod 2. From (101.3) we have a homomorphism $\psi_+ : (\mathbb{Z}[x]/(x^6 + 1))^* \to \mathbb{F}[C_6]^*$; again we have changed the variable in the domain to avoid confusion with the variable in the codomain. We claim:

Proposition 101.4: *ψ_+ is not surjective.*

Proof. Observe that the factorization $(x^6 + 1) = (x^2 + 1)(x^4 - x^2 + 1)$ gives rise to an injective ring homomorphism which imbeds $\mathbb{Z}[x]/(x^6 + 1)$ in $\mathbb{Z}[w]/(w^2 + 1) \times \mathbb{Z}[z]/(z^4 - z^2 + 1)$ as a subring of finite index thus:

$$\mathbb{Z}[x]/(x^6 + 1) \hookrightarrow \mathbb{Z}[w]/(w^2 + 1) \times \mathbb{Z}[z]/(z^4 - z^2 + 1).$$

Comparing discriminants, as in Chapter Seven, we see that the index is actually nine. Hence we have a corresponding imbedding of unit groups, again with finite index

$$(\mathbb{Z}[x]/(x^6 + 1))^* \hookrightarrow (\mathbb{Z}[w]/(w^2 + 1))^* \times (\mathbb{Z}[z]/(z^4 - z^2 + 1))^*.$$

Now $\mathbb{Z}[z]/(z^4 - z^2 + 1)$ is isomorphic to the ring $\mathbb{Z}(\zeta_{12})$ so that, by (101.2), we have $(\mathbb{Z}[z]/(z^4 - z^2 + 1))^* \cong C_{12} \times C_\infty$. Moreover it is straightforward to verify that $(\mathbb{Z}[w]/(w^2 + 1))^* = \{1, w, -1, -w\} \cong C_4$, giving a finite index imbedding $(\mathbb{Z}[x]/(x^6 + 1))^* \hookrightarrow C_4 \times C_{12} \times C_\infty$. It follows that

$$(\mathbb{Z}[x]/(x^6 + 1))^* \cong \Phi \times C_\infty$$

where the torsion subgroup Φ of $(\mathbb{Z}[x]/(x^6+1))^*$ is isomorphic to a subgroup of $C_4 \times C_{12}$. Writing w and ζ for the respective generators of C_4 and C_{12} we see that any element of Φ takes the form (w^a, ζ^b) for some a, b. Moreover, from the definition of ψ_+ it then follows that $\psi_+(w^a, \zeta^b) = y^{3a+b}$ so that $\psi_+(\Phi) \subset \Gamma$. As, x itself is a unit of of finite order in $(\mathbb{Z}[x]/(x^6+1))^*$ and $\psi_+(x) = y$ then we have:

$$\psi_+(\Phi) = \Gamma.$$

Now suppose that $\psi_+ : (\mathbb{Z}[x]/(x^6+1))^* \to \mathbb{F}_2[C_6]$ is surjective. Then there is an induced surjection $C_\infty \cong \Phi\backslash(\Phi \times C_\infty) \twoheadrightarrow \Gamma\backslash\mathbb{F}_2[C_6]^* \cong C_2 \times C_2$ which is a contradiction. Thus, as claimed, ψ_+ fails to be surjective. □

In fact, we have:

Proposition 101.5: $\mathrm{Im}(\psi_+) = \Delta$.

Proof. Put $\eta = x + x^2 + x^3 + x^4 + x^5 \in \mathbb{Z}[x]/(x^6+1)$. Then η is a unit with inverse $\eta^{-1} = -x + x^2 - x^3 + x^4 - x^5$. Moreover,

$$\psi_+(\eta) = y + y^2 + y^3 + y^4 + y^5 = u \in \Delta.$$

We observed above that $y \in \mathrm{Im}(\psi_+)$ and so $\Delta = \langle y, u\rangle \subset \mathrm{Im}(\psi_+)$. As Δ is a maximal proper subgroup of $\mathbb{F}_2[C_6]^*$ then $\mathrm{Im}(\psi_+) = \Delta$ as claimed. □

From the fibre product (101.3) we also have a homomorphism of unit groups

$$\psi_- : (\mathbb{Z}[y]/(y^6-1))^* \to \mathbb{F}[C_6]^*$$

and again, as $\mathbb{F}_2[C_6]$ is commutative, we have a group homomorphism

$$\psi : (\mathbb{Z}[y]/(y^6-1))^* \times (\mathbb{Z}[y]/(y^6+1))^* \to \mathbb{F}[C_6]^*; \psi(\xi, \eta) = \psi_-(\xi)\cdot\psi_+(\eta).$$

One checks easily that $\mathrm{Im}(\psi_-) = \Gamma \subset \mathrm{Im}(\psi_+)$. Consequently $\mathrm{Im}(\psi) = \Delta$ and so

$$\textbf{(101.6)} \qquad \mathrm{Coker}(\psi) = \Delta\backslash\mathbb{F}_2[C_6]^* \cong C_2.$$

There is a bijection between $\mathrm{Coker}(\psi)$ and the set of double cosets of (101.3) thus: $\mathrm{Coker}(\psi) \longleftrightarrow (\mathbb{Z}[y]/(y^6+1))^*\backslash\mathbb{F}_2[C_6]^*/(\mathbb{Z}[y]/(y^6-1))^*$. Hence we have a bijection

(101.7) $$(\mathbb{Z}[y]/(y^6+1))^*\backslash\mathbb{F}_2[C_6]^*/(\mathbb{Z}[y]/(y^6-1))^* \longleftrightarrow C_2.$$

If $\alpha \in \mathbb{F}_2[C_6]^*$ we denote by $[\alpha]$ the class of α in the double quotient

$$\mathbb{Z}[y]/(y^6+1)^*\backslash\mathbb{F}_2[C_6]^*/\mathbb{Z}[y]/(y^6-1)^*.$$

We note from the table of (100.5) that

(101.8) $v = 1 + y + y^4$ represents the nontrivial double coset in (101.7).

If $\alpha \in \mathbb{F}_2[C_6]^*$ we denote by $P(\alpha)$ the projective module over $\mathbb{Z}[C_{12}]$ obtained by glueing $\mathbb{Z}[y]/(y^6-1)$ to $\mathbb{Z}[y]/(y^6+1)$ via α. We recall

(101.9) $$P(\alpha) \text{ is free} \iff [\alpha] = [1].$$

In particular, it follows from (101.7) and (101.8) that $P(v) = P(1+y+y^4)$ is not free over $\mathbb{Z}[C_{12}]$. However $y + y^2 + y^5 = yv$ so that $[y+y^2+y^5] = [v]$ and so $P(y+y^2+y^5) \cong P(v)$. We note the following for future reference:

(101.10) $$P(y+y^2+y^5) \text{ is not free over } \mathbb{Z}[C_{12}].$$

Moreover, as $\mathbb{Z}[C_{12}]$ has no nontrivial stably free modules we see that:

(101.11) $$\widetilde{K}_0(\mathbb{Z}[C_{12}]) \neq 0.$$

§102: $R(1)$ is not strongly monogenic over $\mathbb{Z}[G(13,12)]$:

We proceed to show that the first obstruction to strong monogenicity of $R(1)$ is nonzero for the group $G(13,12)$. Thus take $\Lambda = \mathbb{Z}[G(13,12)]$; then $\mathbb{Z}[C_{13}]$ is a Λ-module under the Galois action of C_{12}. In fact, taking $C_{13} = \langle x | x^{13} = 1\rangle$ and $C_{12} = \langle y | y^{12} = 1\rangle$ then the isomorphism $C_{12} \xrightarrow{\simeq} \mathrm{Aut}(C_{13})$ is effected by the action $y(x) = x^2$. Let $I(C_{13})$ denote the augmentation ideal of $\mathbb{Z}[C_{13}]$; then $I(C_{13})$ is a Λ-submodule of $\mathbb{Z}[C_{13}]$ under this action. More generally, $(x-1)^r I(C_{13})$ is a Λ-submodule of $\mathbb{Z}[C_{13}]$ for each r in the range $1 \le r \le 12$. We proceed to give an explicit description of the $\mathbb{Z}[C_{12}]$-module $j^*((x-1)I(C_{13}))$ where $j : \mathbb{Z}[C_{12}] \to \Lambda$ is the inclusion.

The following collection $\{e_r\}_{1 \leq r \leq 12}$ forms a $\mathbb{Z}$-basis for $j^*((x-1)I(C_{13}))$.

$$\begin{aligned}
e_1 &= x^2 - 2x + 1 && (= (x-1)(x-1)) \\
e_2 &= x^3 - x^2 - x + 1 && (= (x^2-1)(x-1)) \\
e_3 &= x^4 - x^3 - x + 1 && (= (x^3-1)(x-1)) \\
e_4 &= x^5 - x^4 - x + 1 && (= (x^4-1)(x-1)) \\
e_5 &= x^6 - x^5 - x + 1 && (= (x^5-1)(x-1)) \\
e_6 &= x^7 - x^6 - x + 1 && (= (x^6-1)(x-1)) \\
e_7 &= x^8 - x^7 - x + 1 && (= (x^7-1)(x-1)) \\
e_8 &= x^9 - x^8 - x + 1 && (= (x^8-1)(x-1)) \\
e_9 &= x^{10} - x^9 - x + 1 && (= (x^9-1)(x-1)) \\
e_{10} &= x^{11} - x^{10} - x + 1 && (= (x^{10}-1)(x-1)) \\
e_{11} &= x^{12} - x^{11} - x + 1 && (= (x^{11}-1)(x-1)) \\
e_{12} &= -x^{12} - x + 2 && (= (x^{12}-1)(x-1)) .
\end{aligned}$$

Under the action of y the basis is transformed as follows:

$$\begin{aligned}
y(e_1) &= -e_1 + e_2 + e_3 \\
y(e_2) &= -e_1 + e_4 + e_5 \\
y(e_3) &= -e_1 + e_6 + e_7 \\
y(e_4) &= -e_1 + e_8 + e_9 \\
y(e_5) &= -e_1 + e_{10} + e_{11} \\
y(e_6) &= -e_1 + e_{12} \\
y(e_7) &= e_2 \\
y(e_8) &= -e_1 + e_3 + e_4 \\
y(e_9) &= -e_1 + e_5 + e_6 \\
y(e_{10}) &= -e_1 + e_7 + e_8 \\
y(e_{11}) &= -e_1 + e_9 + e_{10} \\
y(e_{12}) &= -e_1 + e_{11} + e_{12}
\end{aligned}$$

We denote this module by P. On general grounds we know that P is projective over $\mathbb{Z}[C_{12}]$. To determine whether P is free we put

$$P_+ = P/\mathrm{Ker}(y^6 + 1); \quad P_- = P/\mathrm{Ker}(y^6 - 1)$$

and proceed to describe these modules explicitly. Note that a sequence of elementary transformations gives the basis $\{E_r\}_{1\leq r\leq 6}\bigcup\{\varphi_s\}_{1\leq s\leq 6}$ for P:

$$\begin{cases} E_1 = e_1 & & & & & & & & & +e_{11} & - & e_{12} \\ E_2 = & e_2 & & & & & & & +e_{10} & & - & e_{12} \\ E_3 = & & e_3 & & & & & +e_9 & & & - & e_{12} \\ E_4 = & & & e_4 & & & +e_8 & & & & - & e_{12} \\ E_5 = & & & & e_5+e_7 & & & & & & - & e_{12} \\ E_6 = & & & & 2e_6 & & & & & & - & e_{12} \\ \varphi_1 = & & & & e_6 & & & & & & & \\ \varphi_2 = & & & & & e_7 & & & & & & \\ \varphi_3 = & & & & & & e_8 & & & & & \\ \varphi_4 = & & & & & & & e_9 & & & & \\ \varphi_5 = & & & & & & & & e_{10} & & & \\ \varphi_6 = & & & & & & & & & e_{11} & & \end{cases}$$

Put $K_+ = \mathrm{Span}\{E_r : \mid 1 \leq r \leq 6\}$. Then K_+ is a submodule of P on which $y^6 + \mathrm{Id}$ vanishes. Moreover, $y^6 - \mathrm{Id}$ vanishes on the quotient P/K_+ so that, as P/K_+ is torsion free it follows that $K_+ = \mathrm{Ker}(y^6 + \mathrm{Id})$. Hence $P_+ = P/K_+$. Let $\nu_+ : P \to P_+$ be the canonical map. Calculation of the y-action gives

$$\begin{cases} y(\varphi_1) = -E_1 & & & & & & & & & & & +\varphi_6 \\ y(\varphi_2) = & E_2 & & & & & +2\varphi_1 & & & & -\varphi_5 & \\ y(\varphi_3) = -E_1 & & +E_3 & +E_4 & & -E_6 & +2\varphi_1 & & -\varphi_3 & -\varphi_4 & & +\varphi_6 \\ y(\varphi_4) = -E_1 & & & & +E_5 & & +\varphi_1 & -\varphi_2 & & & & +\varphi_6 \\ y(\varphi_5) = -E_1 & & & & & +E_6 & -2\varphi_1 & +\varphi_2 & +\varphi_3 & & & +\varphi_6 \\ y(\varphi_6) = -E_1 & & & & & +E_6 & -2\varphi_1 & & & +\varphi_4 & +\varphi_5 & +\varphi_6 \end{cases}$$

Defining $\Phi_r = \nu_+(\varphi_r)$ we see that $\{\Phi_r\}_{1\leq r\leq 6}$ is a basis for P_+ with respect to which the action of y is given by the following representation ρ_+:

$$\rho_+(y) = \begin{pmatrix} 0 & 2 & 2 & 1 & -2 & -2 \\ 0 & 0 & 0 & -1 & 1 & 0 \\ 0 & 0 & -1 & 0 & 1 & 0 \\ 0 & 0 & -1 & 0 & 0 & 1 \\ 0 & -1 & 0 & 0 & 0 & 1 \\ 1 & 0 & 1 & 1 & 1 & 1 \end{pmatrix}.$$

The free module of rank 1 over $\mathbb{Z}[C_6]$ is described by the matrix

$$\lambda_+(y) = \begin{pmatrix} 0 & 0 & 0 & 0 & 0 & 1 \\ 1 & 0 & 0 & 0 & 0 & 0 \\ 0 & 1 & 0 & 0 & 0 & 0 \\ 0 & 0 & 1 & 0 & 0 & 0 \\ 0 & 0 & 0 & 1 & 0 & 0 \\ 0 & 0 & 0 & 0 & 1 & 0 \end{pmatrix}.$$

Putting
$$A+ = \begin{pmatrix} 1 & 0 & -2 & -3 & -1 & -4 \\ 0 & 0 & 0 & 0 & 0 & 1 \\ 0 & 0 & 0 & 1 & 0 & 1 \\ 0 & 0 & 1 & 1 & 0 & 1 \\ 0 & 0 & 1 & 1 & 1 & 1 \\ 0 & 1 & 1 & 1 & 1 & 1 \end{pmatrix}$$

one checks easily that $\rho_+(y)A_+ = A_+\lambda_+(y)$. As $\det(A_+) = 1$ then A_+ defines an isomorphism $\mathbb{Z}[C_6]$-modules $A_+ : \mathbb{Z}[C_6] \to P_+$; that is:

(102.1) P_+ is a free module of rank 1 over $\mathbb{Z}[C_6]$.

Next we consider P_-. Here we note that another sequence of elementary transformations gives the following basis $\{F_r\}_{1\leq r\leq 6} \bigcup \{\psi_s\}_{1\leq s\leq 6}$ for P:

$$\begin{cases} F_1 = e_1 & & & & & & & & & & -e_{11} & \\ F_2 = & e_2 & & & & & & & & -e_{10} & & \\ F_3 = & & e_3 & & & & & & -e_9 & & & \\ F_4 = & & & e_4 & & & & -e_8 & & & & \\ F_5 = & & & & e_5 - e_7 & & & & & & & \\ F_6 = & & & & & & & & & & & e_{12} \\ \psi_1 = & & & & & e_6 & & & & & & \\ \psi_2 = & & & & & & e_7 & & & & & \\ \psi_3 = & & & & & & & e_8 & & & & \\ \psi_4 = & & & & & & & & e_9 & & & \\ \psi_5 = & & & & & & & & & e_{10} & & \\ \psi_6 = & & & & & & & & & & e_{11} & \end{cases}$$

Put $K_- = \text{Span}\{F_r : \mid 1 \leq r \leq 6\}$. Then K_- is a submodule of P on which $y^6 - \text{Id}$ vanishes. Moreover, $y^6 + \text{Id}$ vanishes on the quotient P/K_-

so that, as P/K_- is torsion free it follows that $K_- = \mathrm{Ker}(y^6 - \mathrm{Id})$. Hence $P_- = P/K_-$. Let $\nu_- : P \to P_-$ be the canonical map. Calculation of the y-action gives

$$\begin{cases} y(\psi_1) = -F_1 \qquad\qquad\qquad +F_6 \qquad\qquad\qquad\qquad -\psi_6 \\ y(\psi_2) = \quad F_2 \qquad\qquad\qquad\qquad\qquad\qquad +\psi_5 \\ y(\psi_3) = -F_1 \quad +F_3 +F_4 \qquad\qquad +\psi_3 +\psi_4 \qquad -\psi_6 \\ y(\psi_4) = -F_1 \qquad\qquad +F_5 \quad +\psi_1 +\psi_2 \qquad\qquad -\psi_6 \\ y(\psi_5) = -F_1 \qquad\qquad\qquad\qquad +\psi_2 +\psi_3 \qquad -\psi_6 \\ \\ y(\psi_6) = -F_1 \qquad\qquad\qquad\qquad\qquad +\psi_4 +\psi_5 -\psi_6 \end{cases}$$

Putting $\Psi_r = \nu_-(\psi_r)$ we see that $\{\Psi_r\}_{1\le r\le 6}$ is a basis for P_- with respect to which the action of y is given by the following representation ρ_-:

$$\rho_-(y) = \begin{pmatrix} 0 & 0 & 0 & 1 & 0 & 0 \\ 0 & 0 & 0 & 1 & 1 & 0 \\ 0 & 0 & 1 & 0 & 1 & 0 \\ 0 & 0 & 1 & 0 & 0 & 1 \\ 0 & 1 & 0 & 0 & 0 & 1 \\ -1 & 0 & -1 & -1 & -1 & -1 \end{pmatrix}.$$

The free module of rank 1 over $\mathbb{Z}[y]/(y^6+1)$ is described by the matrix

$$\lambda_+(y) = \begin{pmatrix} 0 & 0 & 0 & 0 & 0 & -1 \\ 1 & 0 & 0 & 0 & 0 & 0 \\ 0 & 1 & 0 & 0 & 0 & 0 \\ 0 & 0 & 1 & 0 & 0 & 0 \\ 0 & 0 & 0 & 1 & 0 & 0 \\ 0 & 0 & 0 & 0 & 1 & 0 \end{pmatrix}.$$

Putting
$$A_- = \begin{pmatrix} 0 & 0 & 0 & 0 & 0 & 1 \\ 1 & 0 & 1 & 0 & 0 & 1 \\ 0 & 0 & 1 & 1 & 1 & 1 \\ 0 & 0 & 0 & 0 & 1 & 0 \\ 0 & 1 & 0 & 0 & 0 & -1 \\ 0 & 0 & -1 & 0 & -1 & -1 \end{pmatrix}$$

then $\rho_-(y)A_- = A_-\lambda_-(y)$ and $\det(A_-) = 1$. Thus A_- defines an isomorphism of $\mathbb{Z}[y]/(y^6+1)$-modules $A_- : \mathbb{Z}[y]/(y^6+1) \to P_-$. In particular,

(102.2) P_- is a free module of rank 1 over $\mathbb{Z}[y]/(y^6+1)$.

Write $\Lambda_+ = \mathbb{Z}[C_6]$ and $\Lambda_- = \mathbb{Z}[y]/(y^6+1)$ and observe also that:

(102.3) $$\rho_+(y) \equiv \rho_-(y) \bmod 2.$$

(102.4) $$\lambda_+(y) \equiv \lambda_-(y) \bmod 2.$$

Replacing A_+ and A_- by their mod 2 reductions gives an automorphism

$$A_-^{-1} \circ A_+ : \mathbb{F}_2[C_6] \xrightarrow{\simeq} \mathbb{F}_2[C_6].$$

Hence P is isomorphic to the projective module $\widehat{P}$ defined by the fibre product diagram

(102.5) $$\begin{array}{ccc} \widehat{P} & \to & \Lambda_+ \\ \downarrow & & \downarrow \\ \Lambda_- & \to & \mathbb{F}_2[C_6] \end{array}$$

given by the glueing automorphism $A_-^{-1} \circ A_+$. Calculation shows that

$$A_-^{-1} \circ A_+ = \begin{pmatrix} 0 & 1 & 0 & 0 & 1 & 1 \\ 1 & 0 & 1 & 0 & 0 & 1 \\ 1 & 1 & 0 & 1 & 0 & 0 \\ 0 & 1 & 1 & 0 & 1 & 0 \\ 0 & 0 & 1 & 1 & 0 & 1 \\ 1 & 0 & 0 & 1 & 1 & 0 \end{pmatrix} \quad (\text{mod } 2).$$

Under the identification $\mathrm{End}(\mathbb{F}_2[C_6]) \cong \mathbb{F}_2[C_6]$ we have a correspondence

$$A_-^{-1} \circ A_+ \longleftrightarrow y + y^2 + y^5.$$

In the notation of §101, $P \cong P(y+y^2+y^5)$. It follows from (101.10) that;

(102.6) P is not free over $\mathbb{Z}[C_{12}]$.

In consequence we have:

(102.7) $R(1)$ is not strongly monogenic over $\mathbb{Z}[G(13,12)]$.

In fact, we can make a stronger statement, namely that over $\Lambda = \mathbb{Z}[G(13,12)]$.

Theorem 102.8: $R(2) \oplus [y-1) \notin \Omega_3(\mathbb{Z})$.

Proof. If $R(2) \oplus [y-1) \in \Omega_3(\mathbb{Z})$ then there exists an exact sequence of Λ-modules

$$0 \to R(2) \oplus [y-1) \to \Lambda^{(b)} \to \Lambda^{(a)} \to \Lambda \to \mathbb{Z} \to 0$$

and hence an exact sequence of $\mathbb{Z}[C_q]$-modules

$$0 \to j^*(R(2)) \oplus j^*([y-1)) \to \mathbb{Z}[C_q]^{(pb)} \to \mathbb{Z}[C_q]^{(pa)} \to \mathbb{Z}[C_q]^{(p)} \to \mathbb{Z} \to 0.$$

Comparing this with the standard exact sequence

$$0 \to (y-1) \to \mathbb{Z}[C_q] \to \mathbb{Z}[C_q] \to \mathbb{Z}[C_q] \to \mathbb{Z} \to 0$$

we see that, for some positive integers m, n

$$\mathbb{Z}[C_q]^{(m)} \oplus j^*(R(2)) \oplus j^*([y-1)) \cong \mathbb{Z}[C_q]^{(n)} \oplus (y-1).$$

However, $j^*([y-1)) \cong \mathbb{Z}[C_q]^{(e)} \oplus (y-1)$ where $e = \frac{(p-1)(q-1)}{q}$ and so

$$\mathbb{Z}[C_q]^{(m+e)} \oplus j^*(R(2)) \oplus (y-1) \cong \mathbb{Z}[C_q]^{(n)} \oplus (y-1).$$

As $j^*(R(2))$ is projective, it follows from (46.13) that

$$\mathbb{Z}[C_q]^{(m+e)} \oplus j^*(R(2)) \cong \mathbb{Z}[C_q]^{(n)};$$

that is, $j^*(R(2))$ is stably free over $\mathbb{Z}[C_q]$. However, $\mathbb{Z}[C_q]$ satisfies the Eichler condition and so every stably free module is free. In particular, $j^*(R(2))$ is free in contradiction to (102.6). Hence $R(2) \oplus [y-1) \notin \Omega_3(\mathbb{Z})$. □

Chapter Fourteen

The D(2) property

Of the Theorems stated in the Introduction, Theorem I and Corollary II were proved in Chapter Eleven and Theorem III was proved in Chapter Twelve. In this chapter we will prove Theorems IV–VIII. We finish by giving a brief survey of the status of the $D(2)$-property for the groups $G(p, q)$ in the light of Theorem VIII.

§103: Proof of Theorem IV:

In Chapter Nine we showed the existence of a projective resolution of $R(1)$ with period $2q$ having the following form

$$
\begin{array}{ccccccccccccccccc}
 & & & & & K(q) & & & & K(q-1) & & K(2) & & & & K(1) & & & \\
 & & & & \nearrow & & \searrow & & \nearrow & & \searrow & \nearrow & \searrow & & \nearrow & & \searrow & & \\
(**) & 0 & \to & R(1) & \to & \Lambda & \longrightarrow & \Lambda & \longrightarrow & P(q-1) & \longrightarrow \cdots \longrightarrow & \Lambda & \longrightarrow & P(1) & \longrightarrow & \Lambda & \longrightarrow & R(1) & \to & 0 \\
 & & & & & & & \searrow & \nearrow & & & \searrow & \nearrow & & & & & & \\
 & & & & & & & R(q) & & & & R(2) & & & & & & &
\end{array}
$$

in which each $P(k)$ is projective of rank 1 over Λ and $\bigoplus_{k=1}^{q-1} P(k) \cong \Lambda^{q-1}$. Hence we see that $R(k+1)$ is a *generalized syzygy* of $R(1)$; in fact, $R(k+1) \in D_{2k}(R(1))$. We wish to improve this to the stronger statement $R(k+1) \in \Omega_{2k}(R(1))$. To this end, consider the following condition $\mathcal{P}(k)$ on Λ:

$\mathcal{P}(k)$: *There exists an exact sequence Λ-modules*

$$0 \to R(k+1) \to P(k) \to \Lambda \to R(k) \to 0$$

where $P(k)$ is a projective module such that $[P(k)] \in \mathcal{LF}(\Lambda)$.

We first prove:

Proposition 103.1: *If $\mathcal{P}(k)$ is satisfied then $R(k+1) \in \Omega_2(R(k))$.*

Proof. Suppose given an exact sequence $0 \to R(k+1) \to P(k) \to \Lambda \to R(k) \to 0$ of Λ-modules where $P(k)$ is a projective module such that

$[P(k)] \in \mathcal{LF}(\Lambda)$ and split the sequence as follows:

(I) $$0 \to R(k+1) \to P(k) \to K \to 0;$$

(II) $$0 \to K \to \Lambda \to R(k) \to 0.$$

By (41.8) and (41.9) we see that $\mathrm{Im}(S^K) = \mathrm{Im}(S_{R(k)})$. Now by (93.11) we have $\mathrm{Im}(S_{R(k)}) = \mathcal{LF}(\Lambda)$. Thus $[P(k)] \in \mathrm{Im}(S^K)$ and so, by (43.1), there exists an exact sequence

(I)′ $$0 \to R(k+1) \to \Lambda \to K \to 0.$$

Splicing **(I)′** and **(II)** gives $0 \to R(k+1) \to \Lambda \to \Lambda \to R(k) \to 0$ from which we see that, as claimed, $R(k+1) \in \Omega_2(R(k))$. □

As a corollary we get immediately:

(103.2) If $\mathcal{P}(k)$ is satisfied then $R(k+1) \in \Omega_{2k}(R(1))$ for $1 \le k \le q-1$.

From the exact sequence $0 \to [y-1) \to \Lambda \to \Lambda \to [y-1) \to 0$ it follows that $[y-1) \in \Omega_2([y-1))$ and hence $[y-1) \in \Omega_{2k}([y-1))$. Now assume that each $\mathcal{P}(k)$ is satisfied. Then it follows from (103.2) that $R(k+1) \oplus [y-1) \in \Omega_{2k}(R(1) \oplus [y-1))$. However, by (71.6), $R(1) \oplus [y-1) \cong I_G \in \Omega_1(\mathbb{Z})$ so that we have established the following which is Theorem IV of the Introduction:

(103.3) If each $\mathcal{P}(k)$ is satisfied then $R(k+1) \oplus [y-1) \in \Omega_{2k+1}(\mathbb{Z})$ for $1 \le k \le q-1$.

§104: Proof of Theorem V:

In the proof of Theorem V it is necessary to impose the two conditions mentioned in the Introduction namely that

(*) $$\widetilde{K}_0(\mathbf{Z}[C_q]) = 0$$

and

(**) $$Inj(p,q) \text{ holds.}$$

As in Chapter Seven, fix an odd prime p, put $\zeta_p = \exp(2\pi i/p)$, let $q \geq 2$ be an integer which divides $p-1$ and define A to be the fixed point ring

$$A = \mathbb{Z}(\zeta_p)^{C_q}$$

under the natural action of C_q on $\mathbb{Z}(\zeta_p)$. The inclusion $A \hookrightarrow \mathbb{Z}(\zeta_p)$ induces a homomorphism of ideal class groups,

$$\nu : Cl(A) \to Cl(\mathbb{Z}(\zeta_p))$$
$$\nu([P]) = [P \cdot \mathbb{Z}(\zeta_p)].$$

We first consider the condition

Inj(p, q) : $\qquad \nu : Cl(A) \to Cl(\mathbb{Z}(\zeta_p))$ *is injective.*

Put $\mathcal{C}_q = \mathcal{C}_q(\mathbb{Z}(\zeta_p), \theta)$ where $\theta \in \mathrm{Aut}(C_p)$ is our chosen element of order q and let $\iota : \mathbb{Z}(\zeta_p) \hookrightarrow \mathcal{C}_q$ be the canonical inclusion. Then ι gives rise to a *restriction of scalars* functor $\iota^* : \mathcal{M}od_{\mathcal{C}_q} \to \mathcal{M}od_{\mathbb{Z}(\zeta_p)}$ and hence to a homomorphism of reduced class groups

$$\iota^* : \widetilde{K}_0(\mathcal{AF}(\mathcal{C}_q)) \to \widetilde{K}_0(\mathbb{Z}(\zeta_p)) = Cl(\mathbb{Z}(\zeta_p)).$$

We shall consider also the following condition:

Inj$(p, q)^*$: $\iota^* : \widetilde{K}_0(\mathcal{AF}(\mathcal{C}_q)) \to \widetilde{K}_0(\mathbb{Z}(\zeta_p)) = Cl(\mathbb{Z}(\zeta_p))$ *is injective.*

Consider the homomorphism $\iota^* : \widetilde{K}_0(\mathcal{AF}(\mathcal{C}_q)) \to Cl(\mathbb{Z}(\zeta_p))$ induced from the inclusion $\iota : \mathbb{Z}(\zeta_p) \hookrightarrow \mathcal{C}_q$ by restricting scalars. With respect to the isomorphism $\widetilde{K}_0(\mathcal{AF}(\mathcal{C}_q)) \cong Cl(A)$ it is straightforward to see that ι^* corresponds to the homomorphism $\nu : Cl(A) \to Cl(\mathbb{Z}(\zeta_p))$. Hence:

(104.1) Inj(p, q) and Inj$(p, q)^*$ are equivalent conditions.

Now consider again the homomorphism $(\pi_1, \pi_2) : \widetilde{K}_0(\Lambda) \to \widetilde{K}_0(\mathbb{Z}[C_q]) \times \widetilde{K}_0(\mathcal{AF}(\mathcal{C}_q))$ and let $i^* : \mathcal{M}od_\Lambda \to \mathcal{M}od_{\mathbb{Z}[C_p]}$ be the restriction of scalars functor; then:

Theorem 104.2: *Suppose that Inj(p,q) holds and let P be a projective module over Λ; if $i^*(P)$ is free over $\mathbb{Z}[C_p]$ then $\pi_2([P]) = 0 \in \widetilde{K}_0(\mathcal{AF}(\mathcal{T}_q(A, \pi))$.*

Proof. We first note that there are two 'restriction of scalars' functors in operation; $i^* : \mathcal{M}od_\Lambda \to \mathcal{M}od_{\mathbb{Z}[C_p]}$ is induced from the inclusion of group rings

$$i : \mathbb{Z}[C_p] \hookrightarrow \Lambda$$

whereas $\iota^* : \mathcal{M}od_{\mathcal{C}_q} \to \mathcal{M}od_{\mathbb{Z}(\zeta_p)}$ is induced from the inclusion

$$\iota : \mathbb{Z}(\zeta_p) \hookrightarrow \mathcal{C}_q$$

of $\mathbb{Z}(\zeta_p)$ into the cyclic algebra $\mathcal{C}_q(\mathbb{Z}(\zeta_p), \theta)$. Furthermore, noting, as in (59.7), that $\mathcal{C}_q(\mathbb{Z}(\zeta_p), \theta) \cong \mathcal{T}_q(A, \pi)$ we shall find it more convenient to deal with $\mathcal{T}_q(A, \pi)$ rather than $\mathcal{C}_q(\mathbb{Z}(\zeta_p), \theta)$. Finally we describe $\mathbb{Z}(\zeta_p)$ in the guise of the dual augmentation ideal I_C^* so that under the condition $Inj(p, q)$ we now assume the injectivity of

$$\iota^* : \widetilde{K}_0(\mathcal{AF}(\mathcal{T}_q(A, \pi))) \to Cl(I_C^*).$$

We shall prove the contrapositive, namely that if $\pi_2([P]) \neq 0 \in \widetilde{K}_0(\mathcal{AF}(\mathcal{T}_q(A, \pi)))$ then the projective module $i^*(P)$ is not free over $\mathbb{Z}[C_p]$. First take the case where P has rank 1 over Λ and, by (51.9), represent $\pi_2(P)$ as

$$\pi_2(P) = \left(\bigoplus_{k=1}^{q-1} R(k)\right) \oplus (R(q) \otimes \mathbf{a})$$

where $\mathbf{a} \neq 0 \in Cl(A)$. Now consider the homomorphism of unreduced groups $(\pi_1, \pi_2) : K_0(\Lambda) \to K_0(\mathbb{Z}[C_q]) \times K_0(\mathcal{AF})$. As $\iota^*(R(k)) \cong I_C^*$, it follows that $\iota^*(\pi_2(P)) \cong {I_C^*}^{(q-1)} \oplus \mathbf{b}$ for some ideal $\mathbf{b} \in Cl(I_C^*)$.

As $\mathbf{a} \neq 0$ and ι^* is injective it follows that $\mathbf{b} \neq 0 \in Cl(I_C^*)$. After consideration of the fibre square,

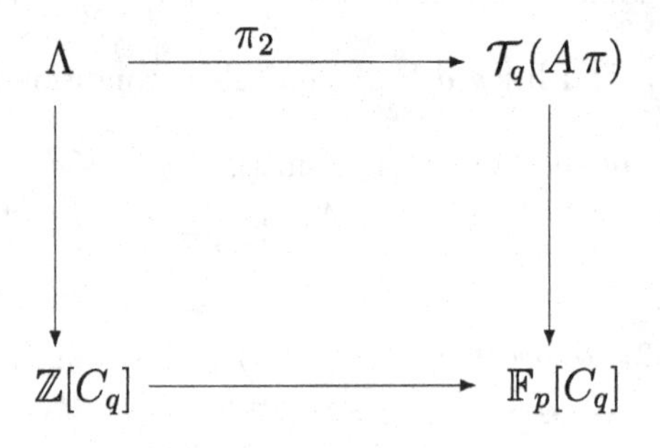

we describe P as a pullback module

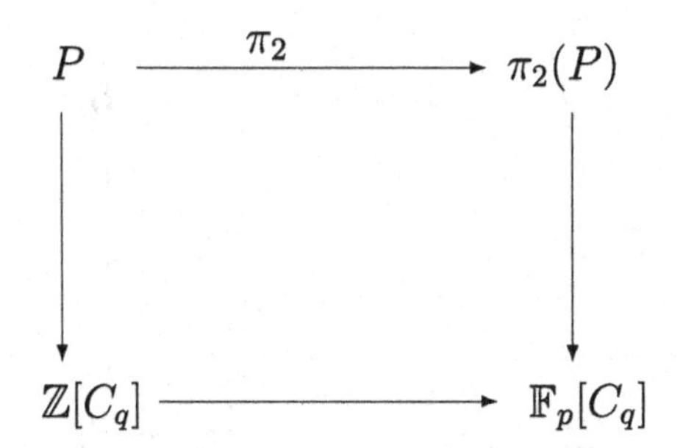

where, as above, $\pi_2(P) = \left(\bigoplus_{k=1}^{q-1} R(k)\right) \oplus (R(q) \otimes \mathbf{a})$. On considering the homomorphism of fibre squares

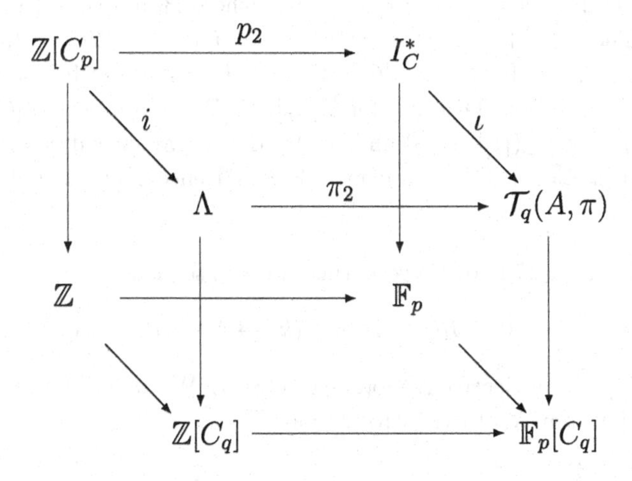

and applying the restriction of scalars functors, we see that $i^*(P)$ is described as a pullback module

$$\begin{array}{ccc} i^*(P) & \longrightarrow & I_C^{*(q-1)} \oplus \mathbf{b} \\ \downarrow & & \downarrow \\ \mathbb{Z}^{(q)} & \longrightarrow & \mathbb{F}_p^{(q)} \end{array}$$

By Rim's Theorem, $i^*(P) \cong \mathbb{Z}[C_p]^{(q-1)} \oplus Q$ where Q is described as a pullback module

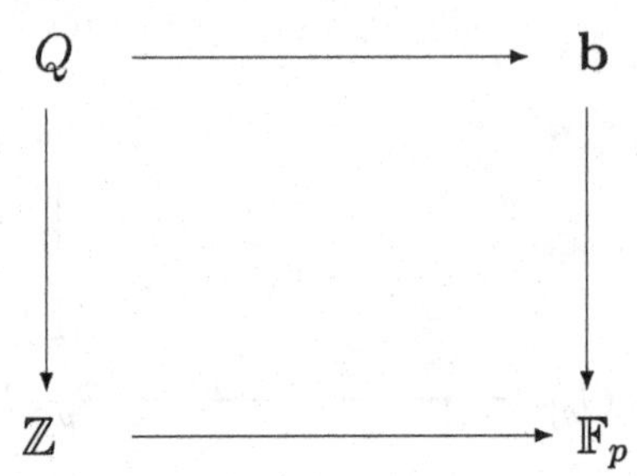

As $\mathbf{b} \neq 0 \in Cl(I_C^*)$, it follows, again from Rim's Theorem, that $Q \neq 0 \in \widetilde{K}_0(\mathbb{Z}[C_p])$. Hence $i^*(P)$ is not free over $\mathbb{Z}[C_p]$.

In general we can write $P = P_0 \oplus \Lambda^{(e)}$ where P_0 has rank 1 over Λ and $e \geq 1$. Then $i^*(P) \cong i^*(P_0) \oplus i^*(\Lambda)^{(e)} \cong i^*(P_0) \oplus \mathbb{Z}[C_p]^{(qe)}$. Suppose now that $i^*(P)$ is free over $\mathbb{Z}[C_p]$; then $i^*(P_0)$ is stably free so that, by the Swan-Jacobinski Theorem for $\mathbb{Z}[C_p]$, $i^*(P_0)$ is free over $\mathbb{Z}[C_p]$. However $\pi_2([P_0]) = \pi_2([P]) \neq 0$ and so, by the above special case, $i^*(P_0)$ is not free over $\mathbb{Z}[C_p]$. This is a contradiction. Hence $i^*(P)$ is not free over $\mathbb{Z}[C_p]$. □

We recall that in (74.10) we constructed exact sequences

$$\mathcal{Z}(k) \qquad 0 \to R(k+1) \to P(k) \to \Lambda \to R(k) \to 0$$

where $P(k)$ is a projective Λ-module of rank 1. We arrive at the following, which is Theorem V of the Introduction.

Theorem 104.3: *If* $\widetilde{K}_0(\mathbf{Z}[C_q]) = 0$ *and* $Inj(p,q)$ *holds then* $[P(k)] \in \mathcal{LF}(\Lambda)$ *for each* $1 \leq k \leq q-1$.

Proof. Applying the restriction of scalars functor $i^* : \mathcal{M}od_\Lambda \to \mathcal{M}od_{\mathbb{Z}[C_p]}$ to $\mathcal{Z}(k)$ and observing that $i^*(R(k)) \cong i^*(R(k+1)) \cong I(C_p)$ and $i^*(\Lambda) \cong \mathbb{Z}[C_p]^{(q)}$ it follows that $i^*(\mathcal{Z}(k))$ takes the form

$$i^*(\mathcal{Z}(k)) \qquad 0 \to I(C_p) \to i^*(P(k)) \to \mathbb{Z}[C_p]^{(q)} \to I(C_p) \to 0.$$

It follows from (46.8) that $i^*(P(k))$ is free. As we are assuming that $Inj(p,q)$ holds then by (104.2) we see that $\pi_2[P(k)] = 0$. As $\widetilde{K}_0(\mathbf{Z}[C_q]) = 0$ then we also have $\pi_1[P(k)] = 0$. Hence $[P(k)] \in \mathcal{LF}(\Lambda) = \mathrm{Ker}(\pi_1, \pi_2)$. □

§105: Proof of Theorems VI, VII and VIII:

In its naive definition, $G(p,q)$ has two generators and three relations thus:

$$G(p,q) = \langle x, y | yxy^{-1} = x^a, x^p = 1, y^q = 1\rangle$$

where $a \in \{2, \ldots, p-1\}$ is an integer whose residue class mod p has order q in $\mathbb{F}_p^*$. We wish to describe $G(p,q)$ using only two relations. To this end we define

$$\mathcal{G}(p,q;m) = \langle X, Y | YX^mY^{-1} = X^{m+1}, X^p = Y^q\rangle.$$

Theorem 105.1: *$G(p,q) \cong \mathcal{G}(p,q;m)$ if $(a-1)\cdot m \equiv 1 (\mathrm{mod}\ p)$.*

Proof. Suppose $(a-1)\cdot m \equiv 1(\mathrm{mod} p)$ and write $(a-1)\cdot m = 1 + kp$. Then in $\mathcal{G}(p,q;m)$ we see that

$$YX^{mp}Y^{-1} = (YX^mY^{-1})^p = X^{(m+1)p}$$

As $X^{mp} = Y^{mq}$ then Y commutes with X^{mp} so that $X^{mp} = X^{mp}X^p$ and hence $X^p = 1$. Hence also $Y^q = 1$. However $(YX^mY^{-1})^{a-1} = X^{(a-1)(m+1)}$ and so

$$\begin{aligned} YX^{1+kp}Y^{-1} &= YX^{(a-1)m}Y^{-1} \\ &= X^{(a-1)(m+1)} \\ &= X^{(a-1)}X^{(a-1)m} \\ &= X^{(a-1)} \cdot X \cdot X^{kp}. \end{aligned}$$

However Y commutes with $X^{kp} = (X^p)^k$ and so $YXY^{-1} = X^a$. As the generators of $\mathcal{G}(p,q;m)$ satisfy the relations for $G(p,q)$ then the correspondence $x \mapsto X; y \mapsto Y$ induces a surjective homomorphism $G(p,q) \twoheadrightarrow \mathcal{G}(p,q;m)$. To show this is an isomorphism we must show conversely that the relations for $\mathcal{G}(p,q;m)$ are satisfied by the generators of $G(p,q)$. From the relation $yxy^{-1} = x^a$ we have

$$yx^my^{-1} = x^{am} = x^{1+m+kp}.$$

As $x^{kp} = 1$ then $yx^my^{-1} = x^{m+1}$. The conclusion follows as $x^p = y^q = 1$. □

If $a \in \{2, \ldots, p-1\}$ we may choose an integer $m \in \{2, \ldots, p-1\}$ such that $(a-1)\cdot m \equiv 1$ (mod p) giving the following theorem of Wamsley [67]:

(105.2) If $(a-1)\cdot m \equiv 1(\mathrm{mod} p)$ then $\mathcal{G}(p,q;m)$ is a balanced presentation for $G(p,q)$.

Now suppose that $R(2) \oplus [y-1)$ is an element of $\Omega_3(\mathbb{Z})$. However, a straightforward rank calculation shows that $\mathrm{rk}_{\mathbb{Z}}(R(k+1) \oplus [y-1)) = pq - 1 < |\Lambda|$ so that:

(105.3) $R(2) \oplus [y-1)$ is a minimal element of $\Omega_3(\mathbb{Z})$.

However, $R(2) \oplus [y-1)$ is straight, by Corollary II so that

(105.4) $R(2) \oplus [y-1)$ is the *unique* minimal element of $\Omega_3(\mathbb{Z})$.

The syzygy $\Omega_3(\mathbb{Z})$ is the critical one as regards the $D(2)$-property as witness the following (cf. [33], Theorem III, p. 216).

(105.5) Let G be a finite group and suppose that each minimal module $J \in \Omega_3(\mathbb{Z})$ is both realizable and full; then the $D(2)$-property holds for G.

For the sake of completeness we give a proof of (105.5) in Appendix C.

From (105.4) and the balanced presentation of (105.2) we see that $R(2) \oplus [y-1)$ is realized as the second homotopy group of a geometric 2-complex. As $R(2) \oplus [y-1)$ is also full, (105.5) now implies the following which is Theorem VI of the Introduction:

(105.6) If $R(2) \oplus [y-1) \in \Omega_3(\mathbb{Z})$ then $G(p,q)$ has the $D(2)$-property.

By a subproof of (103.3) it follows that if condition $\mathcal{P}(1)$ holds then $R(2) \oplus [y-1)$ is an element of $\Omega_3(\mathbb{Z})$; hence we have:

(105.7) The $D(2)$-property holds for $G(p,q)$ provided that condition $\mathcal{P}(1)$ holds.

Thus we have proved also Theorem VII of the Introduction. If we now drop the assumption that $\mathcal{P}(1)$ holds but assume instead that $\widetilde{K}_0(\mathbb{Z}[C_q]) = 0$ and that *Inj*(p,q) holds then, by (104.3), $\mathcal{P}(1)$ follows as a consequence. We finally arrive at the following which is Theorem VIII of the Introduction.

(105.8) The $D(2)$-property holds for $G(p,q)$ if $\widetilde{K}_0(\mathbb{Z}[C_q]) = 0$ and $\mathrm{Inj}(p,q)$ holds.

§106: The hypotheses $\widetilde{K}_0(\mathbb{Z}[C_q]) = 0$ and *Inj*(p,q):

In the proof of Theorem VIII we imposed two hypotheses, namely $\widetilde{K}_0(\mathbb{Z}[C_q]) = 0$ and *Inj*(p,q). The first of these is both straightforward and stringent. The values of q for which $\widetilde{K}_0(\mathbb{Z}[C_q]) = 0$ have been completely determined by Cassou-Noguès [11] who shows:

(106.1) $\widetilde{K}_0(\mathbb{Z}[C_q]) = 0 \iff 2 \le q \le 11$ or $q = 13, 14, 17, 19$.

In practice we ignore the case $q = 2$ since, as we have shown in [33], $G(p, 2) = D_{2p}$ always has the $D(2)$ property. We say that q is *atypical* when $\widetilde{K}_0(\mathbb{Z}[C_q]) = 0$; otherwise we say that q is *typical*. In particular, the smallest integer q for which $\widetilde{K}_0(\mathbb{Z}[C_q]) \neq 0$ is $q = 12$, when $\widetilde{K}_0(\mathbb{Z}[C_{12}]) \cong \mathbb{Z}/2$. The smallest metacyclic group $G(p, q)$ which fails this hypotheses is $G(13, 12)$ and in Chapter Thirteen we observed the effect that failing this hypothesis has on our method. However, whilst our method fails for $G(13, 12)$, this does not, as yet, provide a counterexample to the $D(2)$ property although it certainly makes the task of deciding the question more difficult in this case. It also prompts the following:

Question: Is it true that $R(2) \oplus [y - 1) \notin \Omega_3(\mathbb{Z})$

(i) for all groups $G(p, 12)$?
(ii) for all groups $G(p, p-1)$ where $p \ge 13$?
(iii) for all groups $G(p, q)$ where q is typical and $p \ge 13$?

The second condition *Inj*(p, q) is more obscure and its practical effect more difficult to estimate. We review it briefly below and in more detail in Appendix B. Thus put

$$\zeta_p = \exp(2\pi i/p) \quad \text{and} \quad \mu_p = \zeta_p + \zeta_p^{-1}.$$

It is known that the norm homomorphism $N : Cl(\mathbb{Z}(\zeta_p)) \to Cl(\mathbb{Z}(\mu_p))$ is surjective. Thus we see that

(106.2) $$h(p) = h^-(p) \cdot h^+(p).$$

where $h^+(p) = |Cl(\mathbb{Z}(\mu_p))|$, $h_-(p) = |\mathrm{Ker}(N : Cl(\mathbb{Z}(\zeta_p)) \to Cl(\mathbb{Z}(\mu_p)))|$. Using a theorem of Kida [40], Richard Hill [27] has shown:

(106.3) If q is coprime to $h_+(p)$ then Inj(p, q) holds.

We give an account of Hill's theorem in Appendix B. This in turn implies the following consequence of our Main Theorem.

(106.4) The $D(2)$-property holds for $G(p, q)$ provided q is an atypical divisor of $p - 1$ and is coprime to $h_+(p)$.

In particular:

(106.5) The $D(2)$-property holds for $G(p, q)$ provided q is an atypical divisor of $p-1$ and $h_+(p) = 1$.

It is clear that there are many such pairs (p, q) satisfying the hypotheses of (106.4). Indeed, it tempting to conjecture that there are infinitely many. However, at this point we encounter the fact that $h_+(p)$ is far more difficult to calculate than $h_-(p)$. In this connection, R. Schoof has done extensive calculations of a number $\widetilde{h}(p)$ which is conjecturally the same as $h_+(p)$. For primes $p < 10,000$ the values of $\widetilde{h}(p)$ are tabulated in an appendix to [68]. Assuming the validity of the identity $\widetilde{h}(p) = h_+(p)$ and the accuracy of this table, then the first three primes p with $\widetilde{h}(p) \neq 1$ are $p = 163, \widetilde{h} = 4$; $p = 191, \widetilde{h} = 11$ and $p = 229, \widetilde{h} = 3$. In a private communication Richard Hill has indicated a proof that in the first two cases $\mathrm{Inj}(p, q)$ continues to hold for all divisors q of $p-1$, thereby proving:

(106.6) *Inj*(p, q) holds for all divisors q of $p-1$ provided $p \leq 227$.

However, Hill points out the following, regardless of whether or not $\widetilde{h}(p) = h_+(p)$:

(106.7) *Inj*(p, q) fails when $p = 229$ and $q = 3$.

We note also that for small primes it is more likely than not that $h^+(p) = 1$. Thus in Schoof's table, 925 of the 1228 odd primes p in the range $p < 10000$ take the value $\widetilde{h}(p) = 1$ and, again assuming the validity of the conjectured identity, for these primes the $D(2)$-property holds for $G(p, q)$ whenever q is an atypical divisor of $p-1$. It would be tiresome to list them all. However, as a sample, below we list all groups $G(p, q)$ where $p < 100$ is an odd prime, $q \neq 2$ is an atypical divisor of $p-1$ and for which, therefore, the $D(2)$-property holds by the above criterion.

$G(5,4)$; $G(7,3)$, $G(7,6)$; $G(11,5)$, $G(11,10)$; $G(13,3)$, $G(13,4)$, $G(13,6)$;
$G(17,4)$, $G(17,8)$; $G(19,3)$, $G(19,6)$, $G(19,9)$; $G(23,11)$;
$G(29,4)$, $G(29,7)$, $G(29,14)$; $G(31,3)$, $G(31,5)$, $G(31,6)$, $G(31,10)$;
$G(37,3)$, $G(37,4)$, $G(37,6)$, $G(37,9)$; $G(41,4)$, $G(41,5)$, $G(41,8)$, $G(41,10)$
$G(43,3)$, $G(43,6)$, $G(43,7)$, $G(43,14)$; $G(53,4)$, $G(53,13)$;
$G(61,3)$, $G(61,4)$, $G(61,5)$, $G(61,6)$, $G(61,10)$; $G(67,3)$, $G(67,6)$,
$G(67,11)$; $G(71,5)$, $G(71,7)$, $G(71,10)$, $G(71,14)$
$G(73,3)$, $G(73,4)$, $G(73,6)$, $G(73,8)$, $G(73,9)$;

$G(79,3)$, $G(79,6)$, $G(79,13)$; $G(89,4)$, $G(89,8)$, $G(89,11)$;
$G(97,3)$, $G(97,4)$, $G(97,6)$, $G(97,8)$.

Of these examples, the D(2)-property for $G(7,3)$ was previously shown in the thesis of Jonathan Remez [53]. More recently, the D(2)-property for $G(5,4)$ and $G(7,6)$ was established in the thesis of Jason Vittis [65] by means of direct calculation.

Appendix A

Examples

§A1: The dihedral groups $D_{2p} = G(p, 2)$:

In this section we take Λ to be the integral group ring $\mathbb{Z}[G(p,2)]$; an alternative description is $\mathbb{Z}[D_{2p}]$. In this case the basic sequence

$$\mathcal{Z}_+ \qquad \begin{array}{ccccccccc} & & & & K & & & & \\ & & & \nearrow & & \searrow & & & \\ 0 & \longrightarrow & R(1) \longrightarrow \Lambda & & \longrightarrow & & \Lambda \twoheadrightarrow R(2) & \longrightarrow & 0 \end{array}$$

is enough to construct a strongly monogenic resolution for $R(1)$. To see this, let $\mathbb{Z}_-$ be the Λ-lattice whose underlying abelian group is $\mathbb{Z}$ on which x acts trivially and y acts as multiplication by -1. Apply $-\otimes\mathbb{Z}_-$ and write

$$\mathcal{Z}_- \qquad \begin{array}{ccccccccc} & & & & K\otimes\mathbb{Z}_- & & & & \\ & & & \nearrow & & \searrow & & & \\ 0 & \longrightarrow & R(1)\otimes\mathbb{Z}_- \longrightarrow \Lambda\otimes\mathbb{Z}_- & & \longrightarrow & & \Lambda\otimes\mathbb{Z}_- \twoheadrightarrow R(2)\otimes\mathbb{Z}_- & \longrightarrow & 0. \end{array}$$

Then $R(1)\otimes\mathbb{Z}_- \cong R(2)$, $\Lambda\otimes\mathbb{Z}_- \cong \Lambda$ and $R(2)\otimes\mathbb{Z}_- \cong R(1)$. Writing $L = K\otimes\mathbb{Z}_-$ then $\mathcal{Z}_-$ has the form

$$\mathcal{Z}_- \qquad \begin{array}{ccccccccc} & & & & L & & & & \\ & & & \nearrow & & \searrow & & & \\ 0 & \longrightarrow & R(2) \longrightarrow \Lambda & & \longrightarrow & & \Lambda \longrightarrow R(1) & \longrightarrow & 0. \end{array}$$

We obtain a strongly monogenic resolution on taking the Yoneda product $\mathcal{Z}_+ \circ \mathcal{Z}_-$:

$$\begin{array}{ccccccccccccc} & & & K & & & & & & L & & & \\ & & \nearrow & & \searrow & & & & \nearrow & & \searrow & & \\ 0 \longrightarrow & R(1) & & \longrightarrow & & \Lambda & \longrightarrow & \Lambda & & \longrightarrow & & \Lambda \longrightarrow R(1) \longrightarrow & 0. \\ & & & & & & \searrow \quad \nearrow & & & & & & \\ & & & & & & R(2) & & & & & & \end{array}$$

Writing $P = R(1)$, $R = R(2)$ and splicing together multiple copies of $\mathcal{Z}_+ \circ \mathcal{Z}_-$ we obtain the following exact sequence which is a complete minimal resolution, in the sense of Tate cohomology (c.f. [2]) of each of P, R, K and L:

$$
\begin{array}{ccccccccccccccccc}
L & & & & K & & & & L & & & & K & & & & \\
& \searrow^{\partial_2^+} & & & \nearrow \searrow^{\partial_0^+} & & & & \nearrow \searrow^{\partial_2^+} & & & & \nearrow \searrow^{\partial_0^+} & & & & \\
\xrightarrow{\partial_2^+} & \Lambda & \xrightarrow{\partial_1^+} & \Lambda & \xrightarrow{\partial_0^+} & \Lambda & \xrightarrow{\partial_3^+} & \Lambda & \xrightarrow{\partial_2^+} & \Lambda & \xrightarrow{\partial_1^+} & \Lambda & \xrightarrow{\partial_0^+} & \Lambda & \xrightarrow{\partial_3^+} & & \\
& & \searrow \nearrow & & & & \searrow \nearrow & & & & \searrow \nearrow & & & & \searrow & & \\
& & P & & & & R & & & & P & & & & R & &
\end{array}
$$

Finding expressions for the differentials is more delicate. However, on writing

$$p = 2n + 1; \qquad \theta = \sum_{r=0}^{n-1} x^r; \qquad \Sigma_x = \sum_{r=0}^{2n} x^r$$

then the following maps ∂_i^+ give a complete resolution as above:

$$
\textbf{(A1.1)} \qquad
\begin{cases}
\partial_0^+ = (1 - y)\theta + \Sigma_x \cdot y; \\[2ex]
\partial_1^+ = (x^{n+1} - x)(y - 1); \\[2ex]
\partial_2^+ = (1 + y)\theta - \Sigma_x \cdot y; \\[2ex]
\partial_3^+ = (x^{n+1} - x)(y + 1).
\end{cases}
$$

Indeed, with these choices, the above sequence is exact even when $2n + 1$ is not prime. In that case the quasi-triangular representation of Chapter Seven is not available so that the modules P, R are no longer defined as row modules. However, it remains true that $[(x^{n+1} - x)(y - 1)] \cong \overline{I_C}$ and $[(x^{n+1} - x)(y - 1)] \cong \overline{I_C^*}$ and for any $n \geq 1$, the above sequence gives a complete resolution of $\overline{I_C}$ over the dihedral group D_{4n+2}. The complete details in this more general case can be found in [36].

As the dihedral groups $G = D_{4n+2}$ have free period four then for any lattice M over $\mathbb{Z}[D_{4n+2}]$, the syzygies $\Omega_r(M)$ satisfy $\Omega_r(M) = \Omega_{r+4}(M)$ and we obtain the following duality statements:

$$\textbf{(A1.2)} \qquad \Omega_r^*(M) = \Omega_{4-r}(M).$$

In fact the corresponding relations already hold at the level of module isomorphism; thus writing

(A1.3)
$$\begin{cases} K &= [(1-y)\theta + \Sigma_x \cdot y); \\ P &= [(x^{n+1} - x)(y-1)) \\ L &= [(1+y)\theta - \Sigma_x \cdot y); \\ R &= [(x^{n+1} - x)(y+1)) \end{cases}$$

then the following duality isomorphism hold:

(A1.4)
$$\begin{cases} K^* &\cong K \\ P^* &\cong R; \\ L^* &\cong L \\ R^* &\cong P. \end{cases}$$

One should perhaps stress that, as D_{4n+2} has cohomological period four, no two of K, P, L, R are stably isomorphic. Indeed, under tensor product the stable modules $[K]$, $[P]$, $[L]$, $[R]$ form the cyclic group of order 4, generated either by $[P]$ or $[R]$, with $[K]$ as identity. John Evans, [18], has shown that the precise relations between the modules K, P, L, R with respect to tensor product are as follows:

(A1.5)

$\otimes$	K	P	L	R
K	$K \oplus \Lambda^{n+1}$	$P \oplus \Lambda^n$	$L \oplus \Lambda^{n+1}$	$R \oplus \Lambda^n$
P	$P \oplus \Lambda^n$	$L \oplus \Lambda^{n-1}$	$R \oplus \Lambda^n$	$K \oplus \Lambda^{n-1}$
L	$L \oplus \Lambda^{n+1}$	$R \oplus \Lambda^n$	$K \oplus \Lambda^{n+1}$	$P \oplus \Lambda^n$
R	$R \oplus \Lambda^n$	$K \oplus \Lambda^{n-1}$	$P \oplus \Lambda^n$	$L \oplus \Lambda^{n-1}$

As in Chapter Nine, the above complete resolution forms the positive strand of a diagonal resolution of $\mathbb{Z}$. Taken together with the negative strand

$$\cdots \to \Lambda \stackrel{y-1}{\to} \Lambda \stackrel{y+1}{\to} \Lambda \stackrel{y-1}{\to} \Lambda \stackrel{y+1}{\to} \Lambda \stackrel{y-1}{\to} \Lambda \stackrel{y+1}{\to} \Lambda \stackrel{y-1}{\to} \Lambda \stackrel{y+1}{\to}$$

one can read off the syzygies $\Omega_r(\mathbb{Z})$ directly as follows:

$$\textbf{(A1.6)} \qquad \Omega_r(\mathbb{Z}) \sim \begin{cases} [K] \oplus [y+1] & r \equiv 0 \mod 4 \\ [P] \oplus [y-1] & r \equiv 1 \mod 4 \\ [L] \oplus [y+1] & r \equiv 2 \mod 4 \\ [R] \oplus [y-1] & r \equiv 3 \mod 4. \end{cases}$$

As it is clear *a priori* that $\Omega_0(\mathbb{Z}) = [\mathbb{Z}]$ we are left with the following identity between stable modules

$$\textbf{(A1.7)} \qquad [\mathbb{Z}] = [K] \oplus [y+1].$$

This identity has a more precise form which holds true over any metacyclic group $G(p, q)$. To explain, take the basic sequence of (72.7)

$$\begin{array}{ccccccccccc} & & & & & K(q) & & & & & \\ & & & & \nearrow & & \searrow & & & & \\ 0 & \longrightarrow & \overline{I_C} & \longrightarrow & \Lambda & \longrightarrow & \Lambda & \twoheadrightarrow & \overline{I_C^*} & \longrightarrow & 0 \end{array}$$

and extract from it the short exact sequence $0 \to \overline{I_C} \to \Lambda \to K \to 0$ where $K = K(q)$. We also have a exact sequence $0 \to [y-1) \to \Lambda \to [y+1) \to 0$. Taking direct sums gives an exact sequence

$$0 \to \overline{I_C} \oplus [y-1) \to \Lambda \oplus \Lambda \to K \oplus [y+1) \to 0.$$

However, by (71.5) $\overline{I_C} \oplus [y-1)$ is isomorphic to the augmentation ideal of Λ, giving an alternative exact sequence

$$0 \to \overline{I_C} \oplus [y-1) \to \Lambda \to \mathbb{Z} \to 0.$$

Comparing the two using the dual to Schanuel's Lemma (23.5) we see that

$$\textbf{(A1.8)} \qquad K \oplus [y+1) \oplus \Lambda \cong \mathbb{Z} \oplus \Lambda \oplus \Lambda.$$

Jacobinski's Cancellation Theorem now gives the following precise form of (A1.7):

$$\textbf{(A1.9)} \qquad K \oplus [y+1) \cong \mathbb{Z} \oplus \Lambda.$$

This illustrates a somewhat paradoxical aspect of the theory of stable modules, namely that whilst a module (in this case the trivial module $\mathbb{Z}$) may be indecomposable, its stable class may decompose non-trivially. This phenomenon was pointed out, though without an explicit example, in the paper of Gruenberg and Roggenkamp ([24], Proposition 1). They attribute the original observation to E.C. Dade.

§A2: The groups $G(p, 3)$:

Here we assume that p is a prime of the form $p = 3d + 1$. We note that the duality relation (53.3) here assumes the form

$$\textbf{(A2.1)} \qquad \begin{cases} R(1)^* \cong R(3) \\ R(2)^* \cong R(2) \\ R(3)^* \cong R(1). \end{cases}$$

In addition to the basic sequence

$$\mathcal{Z}(3) \qquad 0 \longrightarrow R(1) \longrightarrow \Lambda \xrightarrow{\nearrow K(3) \searrow} \Lambda \longrightarrow\!\!\!\!\to R(3) \longrightarrow 0$$

we shall suppose we can construct a secondary sequence

$$\mathcal{Z}(1) \qquad 0 \longrightarrow R(2) \longrightarrow \Lambda \xrightarrow{\nearrow K(1) \searrow} \Lambda \longrightarrow\!\!\!\!\to R(1) \longrightarrow 0$$

As $\widetilde{K}_0(\mathbb{Z}[C_3]) = 0$, then by the conditions given in Chapter Fourteen, a sufficient condition for the existence of such a secondary sequence is that $\mathit{Inj}(p, 3)$ should hold. However, it is not clear whether this condition is actually necessary. The first problematic case is the group $G(229, 3)$, where, as we have indicated, condition $\mathit{Inj}(229, 3)$ fails. In this case, the question of whether such a secondary sequence $\mathcal{Z}(1)$ exists resides, at present, in the realms of heroic computation.

However, under the assumption that such a secondary sequence exists, we may take the dual of $\mathcal{Z}(1)$ to get

$$\mathcal{Z}(1)^* \qquad 0 \longrightarrow R(1)^* \longrightarrow \Lambda \xrightarrow{\nearrow K(1)^* \searrow} \Lambda \longrightarrow\!\!\!\!\to R(2)^* \longrightarrow 0.$$

As $R(2)^* \cong R(2)$ and $R(1)^* \cong R(3)$ the Yoneda product $\mathcal{Z}(3) \circ \mathcal{Z}(1)^* \circ \mathcal{Z}(1)$ is then a strongly monogenic resolution

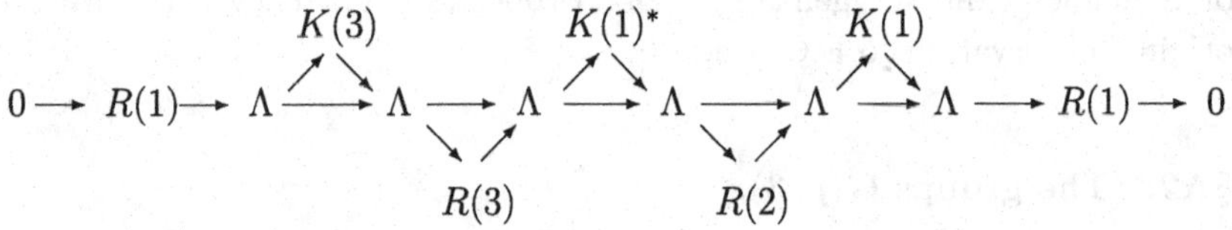

$$
\begin{array}{ccccccccccccccccc}
 & & & K(3) & & & & K(1)^* & & & & K(1) & & & & & \\
 & & \nearrow & & \searrow & & \nearrow & & \searrow & & \nearrow & & \searrow & & & & \\
0 \to R(1) \to & \Lambda & & \longrightarrow & & \Lambda \to \Lambda & & \longrightarrow & & \Lambda \to \Lambda & & \longrightarrow & & \Lambda \to R(1) \to 0 & & & \\
 & & & & & \searrow \;\; \nearrow & & & & \searrow \;\; \nearrow & & & & & & & \\
 & & & & & R(3) & & & & R(2) & & & & & & &
\end{array}
$$

Taking iterated extensions and putting $K(2) = K(1)^*$ we again obtain a complete minimal resolution in the sense of Tate cohomology of each $R(i)$ and each $K(j)$:

$$
(\mathfrak{T}) \qquad
\begin{array}{cccccccccccccccc}
K(1) & & & & K(3) & & & & K(2) & & & & K(1) & & & K(3) \\
\searrow & & & \nearrow & & \searrow & & \nearrow & & \searrow & & \nearrow & & \searrow & & \nearrow \\
\to \Lambda & \xrightarrow{\partial_1^+} & \Lambda & & \xrightarrow[\partial_0^+]{} & & \Lambda \xrightarrow{\partial_5^+} \Lambda & & \xrightarrow[\partial_4^+]{} & & \Lambda \xrightarrow{\partial_3^+} \Lambda & & \xrightarrow[\partial_2^+]{} & & \Lambda \xrightarrow{\partial_1^+} \Lambda & \to \\
\searrow & \nearrow & & & & & \searrow \;\; \nearrow & & & & \searrow \;\; \nearrow & & & & \searrow \;\; \nearrow & \\
R(1) & & & & & & R(3) & & & & R(2) & & & & R(1) &
\end{array}
$$

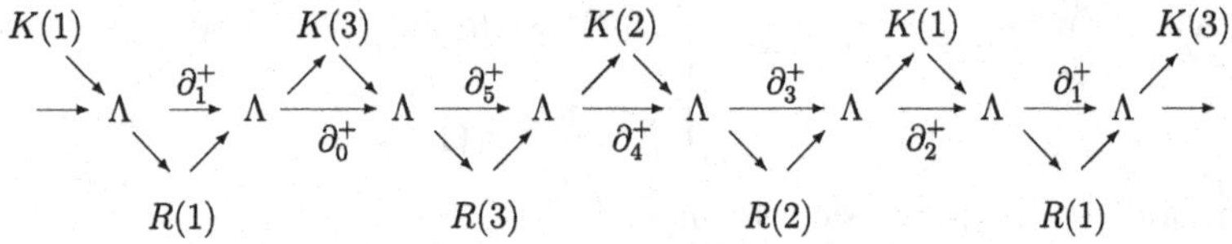

The precise form of the differentials becomes more complicated as p gets larger. In the simplest case, $p = 7$, Remez ([53], [54]) gives the following exact sequence which, after taking iterated Yoneda products with itself, gives a complete resolution, in the sense of Tate cohomology.

$$
0 \longrightarrow \mathbb{Z} \xrightarrow{\epsilon^*} \Lambda \xrightarrow{\partial_5} \Lambda^{(2)} \xrightarrow{\partial_4} \Lambda^{(2)} \xrightarrow{\partial_3} \Lambda^{(2)} \xrightarrow{\partial_2} \Lambda^{(2)} \xrightarrow{\partial_1} \Lambda \xrightarrow{\epsilon} \mathbb{Z} \longrightarrow 0
$$

Here ϵ is the augmentation map, $\epsilon^* = (\sum_{r=0}^{p-1} x^r)(1 + y + y^2)$ is the dual of ϵ and the differentials ∂_r are as follows:

$$
\partial_5 = \begin{pmatrix} (1 + x - x^4 - x^5) + (1 + x^2 - x^4 - x^6)y + (1 - x)y^2 \\ y^2 - y \end{pmatrix}
$$

$$
\partial_4 = \begin{pmatrix} 1 + x^4 - y & 0 \\ 0 & y^2 + y + 1 \end{pmatrix}
$$

$$
\partial_3 = \begin{pmatrix} (1 - x^4) + (1 - x + x^2 - x^3)y + (1 - x)y^2 & 0 \\ 0 & y^2 - y \end{pmatrix}
$$

$$
\partial_2 = \begin{pmatrix} 1 + x - y^2 & 0 \\ 0 & y^2 + y + 1 \end{pmatrix}
$$

$$
\partial_1 = ((-1 + x - x^2 + x^6) + (-x^3 + x^6)y + (x - x^3 - x^5 + x^6)y^2, y^2 - y)
$$

§A3: The groups $G(p, 4)$:

Here we assume that p is a prime of the form $p = 4d + 1$. To show the existence of a complete strongly monogenic resolution of $R(1)$ it suffices that, in addition to the basic sequence

$$\mathcal{Z}(4) \qquad 0 \longrightarrow R(1) \longrightarrow \Lambda \xrightarrow{\nearrow K(4) \searrow} \Lambda \twoheadrightarrow R(4) \longrightarrow 0$$

one can construct a secondary sequence of the form

$$\mathcal{Z}(1) \qquad 0 \longrightarrow R(2) \longrightarrow \Lambda \xrightarrow{\nearrow K(1) \searrow} \Lambda \twoheadrightarrow R(1) \longrightarrow 0.$$

In this case there is again a Λ-lattice $\mathbb{Z}_-$ on which x acts trivially and y acts as multiplication by -1. Moreover, we then have

(A3.1) $$R(1) \otimes \mathbb{Z}_- \cong R(3);$$

(A3.2) $$R(2) \otimes \mathbb{Z}_- \cong R(4).$$

Thus defining $\mathcal{Z}(2) = \mathcal{Z}(4) \otimes \mathbb{Z}_-$ and $\mathcal{Z}(3) = \mathcal{Z}(1) \otimes \mathbb{Z}_-$ we have

$$\mathcal{Z}(2) \qquad 0 \longrightarrow R(3) \longrightarrow \Lambda \xrightarrow{\nearrow K(2) \searrow} \Lambda \twoheadrightarrow R(2) \longrightarrow 0$$

$$\mathcal{Z}(3) \qquad 0 \longrightarrow R(4) \longrightarrow \Lambda \xrightarrow{\nearrow K(3) \searrow} \Lambda \twoheadrightarrow R(3) \longrightarrow 0$$

where $K(2) = K(4) \otimes \mathbb{Z}_-$ and $K(3) = K(1) \otimes \mathbb{Z}_-$. The Yoneda product $\mathcal{Z}(4) \cdot \mathcal{Z}(3) \cdot \mathcal{Z}(2) \cdot \mathcal{Z}(1)$ is then a strongly monogenic resolution of $R(1)$

$$0 \to R(1) \to \Lambda \xrightarrow{\nearrow K(4) \searrow} \Lambda \underset{\searrow R(4) \nearrow}{\longrightarrow} \Lambda \xrightarrow{\nearrow K(3) \searrow} \Lambda \underset{\searrow R(3) \nearrow}{\longrightarrow} \Lambda \xrightarrow{\nearrow K(2) \searrow} \Lambda \underset{\searrow R(2) \nearrow}{\longrightarrow} \Lambda \xrightarrow{\nearrow K(1) \searrow} \Lambda \to R(1) \to 0.$$

Moreover, taking iterated extensions we obtain an eightfold periodic complete resolution in the sense of Tate thus:

$$\begin{array}{ccccccccccccccccccccc}
K(1) & & & & & K(4) & & & & K(3) & & & & K(2) & & & & K(1) & & & & K(4) \\
\longrightarrow & \Lambda & \longrightarrow & \Lambda & \longrightarrow & & \Lambda & \longrightarrow & \Lambda & \longrightarrow & \Lambda & \longrightarrow & \Lambda & \longrightarrow & \Lambda & \longrightarrow & \Lambda & \longrightarrow & \Lambda & \longrightarrow & \Lambda & \longrightarrow \\
& & R(1) & & & & & R(4) & & & & R(3) & & & & R(2) & & & & R(1) & &
\end{array}$$

In the case $p = 5$ this was shown explicitly in the thesis of Nadim [48]. As we have shown, the existence of $\mathcal{Z}(1)$ is guaranteed provided that $\mathrm{Inj}(p, 4)$ holds.

§A4: The groups $G(p, 6)$:

Here we consider primes p of the form $p = 6d + 1$; again in addition to the primary sequence

$$\mathcal{Z}(6) \qquad 0 \longrightarrow R(1) \longrightarrow \Lambda \xrightarrow{\nearrow K(6) \searrow} \Lambda \longrightarrow\!\!\!\!\rightarrow R(6) \longrightarrow 0$$

it is sufficient to construct a secondary sequence

$$\mathcal{Z}(1) \qquad 0 \longrightarrow R(2) \longrightarrow \Lambda \xrightarrow{\nearrow K(1) \searrow} \Lambda \longrightarrow\!\!\!\!\rightarrow R(1) \longrightarrow 0.$$

As before, $\mathcal{Z}(1)$ exists provided $\mathrm{Inj}(p, 6)$ holds. In this case, there is again a Λ-lattice $\mathbb{Z}_-$ on which x acts trivially and y acts as multiplication by -1; we have:

$$\begin{cases} R(1) \cong R(4) \otimes \mathbb{Z}_-; \\ R(2) \cong R(5) \otimes \mathbb{Z}_- \\ R(3) \cong R(6) \otimes \mathbb{Z}_-. \end{cases}$$

Now define
$$\begin{cases} \mathcal{Z}(3) = \mathcal{Z}(6) \otimes \mathbb{Z}_- \\ \mathcal{Z}(4) = \mathcal{Z}(1) \otimes \mathbb{Z}_- \end{cases}$$

and, taking duals, put
$$\begin{cases} \mathcal{Z}(2) = \mathcal{Z}(4)^\bullet \\ \mathcal{Z}(5) = \mathcal{Z}(1)^\bullet. \end{cases}$$

For $2 \leq i \leq 5$, $\mathcal{Z}(i)$ takes the form

$$\mathcal{Z}(i) \qquad 0 \longrightarrow R(i+1) \longrightarrow \Lambda \xrightarrow{\nearrow K(i) \searrow} \Lambda \twoheadrightarrow R(i) \longrightarrow 0$$

where $K(2) = K(1)^{\bullet} \otimes \mathbb{Z}_{-}$, $K(3) = K(6) \otimes \mathbb{Z}_{-}$, $K(4) = K(1) \otimes \mathbb{Z}_{-}$ and $K(5) = K(1)^{\bullet}$. Again, taking iterated Yoneda products we obtain a complete resolution in the sense of Tate:

$$\begin{array}{ccccccccccccccccccccccc}
 & & & K(5) & & & & K(4) & & & & K(3) & & & & K(2) & & & & K(1) & & & & K(0) & & \\
\Lambda & \to & \Lambda & \nearrow\searrow & \Lambda & \to & \Lambda & \nearrow\searrow & \Lambda & \to & \Lambda & \nearrow\searrow & \Lambda & \to & \Lambda & \nearrow\searrow & \Lambda & \to & \Lambda & \nearrow\searrow & \Lambda & \to & \Lambda & \nearrow\searrow & \Lambda & \to \\
 & R(6) & & & & R(5) & & & & R(4) & & & & R(3) & & & & R(2) & & & & R(1) & & & & R(6)
\end{array}$$

Appendix B

Class field theory and condition $\mathrm{Inj}(p, q)$

In Chapter Fourteen we referred to the theorem of R.M. Hill that $\mathrm{Inj}(p, q)$ holds provided q is coprime to $h_+(p)$. In this Appendix we give a proof of Hill's theorem.

§B1: Functorial decomposition of class groups:

We first recall the following standard theorem from Class Field Theory (cf [68], p. 185).

Theorem B1.1: *Let $K' \hookrightarrow K$ be a finite abelian extension of algebraic number fields and suppose that any field F with $K' \subsetneq F \subseteq K$ is a ramified extension of K'; let A (resp. A') be the ring of integers in K (resp K'); then the norm homomorphism on ideal class groups $N : Cl(A) \to Cl(A')$ is surjective.*

Let $\mathfrak{A}$ be a finite abelian group; then $\mathfrak{A}$ decomposes as a direct sum

$$\mathfrak{A} = \bigoplus_{\pi} \mathfrak{A}_{(\pi)}$$

where π ranges over the rational primes and $\mathfrak{A}_{(\pi)}$ denotes the Sylow-π subgroup of $\mathfrak{A}$. This decomposition is functorial; any group homomorphism $f : \mathfrak{A} \to \mathfrak{B}$ restricts to homomorphisms $f_\pi : \mathfrak{A}_{(\pi)} \to \mathfrak{B}_{(\pi)}$ from which f can be entirely re-constituted thus:

$$f = \bigoplus_{\pi} f_\pi.$$

In what follows we shall also consider the coarser decomposition

$$\mathfrak{A} = \mathfrak{A}_{(2)} \bigoplus \mathfrak{A}_{odd}$$

where $\mathfrak{A}_{(2)}$ is the 2-Sylow subgroup and $\mathfrak{A}_{odd}$ is the odd order subgroup. Clearly

$$\mathfrak{A}_{odd} = \bigoplus_{\pi \neq 2} \mathfrak{A}_{(\pi)}.$$

Now let $\mathbb{K}$ be an algebraic number field; as the classgroup $Cl(\mathbb{K})$ is a finite abelian group we may apply the above decompositions to obtain

$$Cl(\mathbb{K}) = \bigoplus_{\pi} Cl(\mathbb{K})_{(\pi)} = Cl(\mathbb{K})_{(2)} \bigoplus Cl(\mathbb{K})_{odd}.$$

If $\mathbb{F} \subset \mathbb{K}$ is a subfield of K, the inclusion $i : \mathbb{F} \hookrightarrow \mathbb{K}$ induces a homomorphism $i_* : Cl(\mathbb{F}) \to Cl(\mathbb{K})$ which decomposes functorially to give homomorphisms

$$i_* : Cl(\mathbb{F})_{(\pi)} \to Cl(\mathbb{K})_{(\pi)}.$$

Likewise the norm homomorphism $N : Cl(\mathbb{K}) \to Cl(\mathbb{F})$ decomposes functorially to give homomorphisms

$$N : Cl(\mathbb{K})_{(\pi)} \to Cl(\mathbb{F})_{(\pi)}.$$

In what follows, q will denote the degree of an extension of algebraic number fields $i : \mathbb{F} \hookrightarrow \mathbb{K}$. Then $N \circ i_* : Cl(\mathbb{F}) \to Cl(\mathbb{F})$ is multiplication by q. If q is coprime to $|Cl(\mathbb{F})|$ then $N \circ i_*$ is injective. Hence i_* is also injective and we have:

(B1.2) If q is coprime to $|Cl(\mathbb{F})|$ then $i_* : Cl(\mathbb{F}) \to Cl(\mathbb{K})$ is injective.

More generally, $N \circ i_* : Cl(\mathbb{F})_{(\pi)} \to Cl(\mathbb{F})_{(\pi)}$ is also multiplication by q and hence is an isomorphism on $Cl(\mathbb{F})_{(\pi)}$; thus

(B1.3) If q is coprime to π then $i_* : Cl(\mathbb{F})_{(\pi)} \to Cl(\mathbb{K})_{(\pi)}$ is injective.

In particular

(B1.4) If q is odd then $i_* : Cl(\mathbb{F})_{(2)} \to Cl(\mathbb{K})_{(2)}$ is injective.

§B2: CM fields:

We now turn our attention to CM fields. When $(\mathbb{K}, \tau)$ is a CM-field, the complex structure τ induces an automorphism $\tau_* : Cl(\mathbb{K}) \to Cl(\mathbb{K})$ satisfying $\tau_*^2 = \mathrm{Id}$.

We define $$\begin{cases} Cl(\mathbb{K})^-_{odd} = \{P \in Cl(\mathbb{K})_{odd} \mid \sigma(P) = -P\} \\ Cl(\mathbb{K})^+_{odd} = \{P \in Cl(\mathbb{K})_{odd} \mid \sigma(P) = P\}. \end{cases}$$

As multiplication by 2 is an isomorphism on finite abelian groups of odd order we have a decomposition

(B2.1) $$Cl(\mathbb{K})_{odd} = Cl(\mathbb{K})^-_{odd} \oplus Cl(\mathbb{K})^+_{odd}.$$

Likewise, for each odd prime π we have

(B2.2) $$Cl(\mathbb{K})_{(\pi)} = Cl(\mathbb{K})^-_{(\pi)} \oplus Cl(\mathbb{K})^+_{(\pi)}$$

where $$\begin{cases} Cl(\mathbb{K})^-_{(\pi)} = \{P \in Cl(\mathbb{K})_{(\pi)} \mid \sigma(P) = -P\} \\ Cl(\mathbb{K})^+_{(\pi)} = \{P \in Cl(\mathbb{K})_{(\pi)} \mid \sigma(P) = P\}. \end{cases}$$

Now suppose that $\mathbb{K}$ is a subfield of $\mathbb{Q}(\zeta_p)$ and put $\mathbb{K}_\mathbb{R} = \mathbb{K}^\tau$, the fixed field of $\mathbb{K}$ under τ. As p is totally ramified in $\mathbb{Q}(\zeta)$ then the extension $\mathbb{K}_\mathbb{R} \subset \mathbb{K}$ satisfies the hypotheses of (B1.1) so that that the norm homomorphism $N : Cl(\mathbb{K}) \to Cl(\mathbb{K}_\mathbb{R})$ is surjective. Hence $N : Cl(\mathbb{K})_{odd} \to Cl(\mathbb{K}_\mathbb{R})_{odd}$ is also surjective. We observe, for any $P \in Cl(\mathbb{K})$ that $N(\tau(P)) = N(P)$. Let $P \in Cl(\mathbb{K})^-_{odd}$ so that $\tau(P) = -P$. As N is a homomorphism, we have $N(\tau(P)) = -N(P)$. However, by the above remark, $N(\tau(P)) = N(P)$. Hence $N(P) = -N(P)$ and so $2N(P) = 0$. As $N(P) \in Cl(\mathbb{K}_\mathbb{R})_{odd}$ then $N(P) = 0$ and so $Cl(\mathbb{K})^-_{odd} \subset \mathrm{Ker}(N)$. Hence we see that $N : Cl(\mathbb{K})^+_{odd} \to Cl(\mathbb{K}_\mathbb{R})_{odd}$ is surjective. We claim that the restriction of N to $Cl(\mathbb{K})^+_{odd}$ is injective. Thus suppose that $P \in Cl(\mathbb{K})^+_{odd}$ satisfies $N(P) = 0$. If $j : \mathbb{K}_\mathbb{R} \hookrightarrow \mathbb{K}$ denotes the inclusion it follows that $j_*N(P) = 0$. However $j_*N(P) = P + \tau(P)$. As $P \in Cl(\mathbb{K})^+_{odd}$ then $\tau(P) = P$ and so $2P = 0$. As P has odd order then $P = 0$ and the restriction of N to $Cl(\mathbb{K})^+_{odd}$ is injective as claimed; thus we have:

Proposition B2.3: $N : Cl(\mathbb{K})^+_{odd} \xrightarrow{\simeq} Cl(\mathbb{K}_\mathbb{R})_{odd}$ *is an isomorphism.*

§B3: Kida's Lemma and its consequences:

The following observation is due to Kida [40].

Lemma B3.1: *Let q be an odd prime and let $j : (\mathbb{F}, \tau) \hookrightarrow (\mathbb{K}, \tau)$ be an extension of CM-fields of degree q. If $\mathbb{F}$ contains no units of order q^a where $a \geq 1$ then*

$$j_* : Cl(\mathbb{F})^-_{(q)} \to Cl(\mathbb{K})^-_{(q)}$$

is injective.

With the hypothesis of (B3.1), then writing $Cl(\mathbb{F})^-_{odd} = Cl(\mathbb{F})^-_{(q)} \oplus \left(\bigoplus_{\pi \neq q} Cl(\mathbb{F})^-_{(\pi)}\right)$ it follows that $j_* : Cl(\mathbb{F})^-_{(q)} \to Cl(\mathbb{K})^-_{(q)}$ is injective by (B3.1) whilst

$$j_* : \bigoplus_{\pi \neq q} Cl(\mathbb{F})^-_{(\pi)} \longrightarrow \bigoplus_{\pi \neq q} Cl(\mathbb{K})^-_{(\pi)}$$

is injective by (B1.3). As a consequence we have:

(B3.2) Let q be an odd prime and let $j : (\mathbb{F}, \tau) \hookrightarrow (\mathbb{K}, \tau)$ be an extension of CM-fields of degree q. If $\mathbb{F}$ contains no units of order q^a where $a \geq 1$ then

$$j_* : Cl(\mathbb{F})^-_{odd} \to Cl(\mathbb{K})^-_{odd}$$

is injective.

As a generalization we have:

Proposition B3.3: *Let $q = q_1 \dots q_N$ be a product of odd primes $(q_r)_{1 \leq r \leq N}$, possibly with repetitions, and suppose that*

$$(\mathbb{K}_N, \tau_N) \xrightarrow{j_N} (\mathbb{K}_{N-1}, \tau_{N-1}) \xrightarrow{j_N} \cdots \xrightarrow{j_1} (\mathbb{K}_1, \tau_1) \xrightarrow{j_1} (\mathbb{K}_0, \tau_0) = (\mathbb{K}, \tau)$$

is a sequence of CM-fields such that $j_r : \mathbb{K}_r \to \mathbb{K}_{r-1}$ has degree q_r. Denote by j the composite $j = j_1 \circ \cdots \circ j_N$. If $\mathbb{K}$ contains no units of any order $q_1^{e_1} \dots q_N^{e_N}$ with $e_1 + \cdots + e_N \geq 1$ then

$$j_* : Cl(\mathbb{K}_N)_{(2)} \oplus Cl(\mathbb{K}_N)^-_{odd} \longrightarrow Cl(\mathbb{K})_{(2)} \oplus Cl(\mathbb{K})^-_{odd}$$

is injective.

Proof. By (B3.2), each $(j_r)_* : Cl(\mathbb{K}_r)^-_{odd} \longrightarrow Cl(\mathbb{K}_{r-1})^-_{odd}$ is injective so that $j_* : Cl(\mathbb{K}_N)^-_{odd} \longrightarrow Cl(\mathbb{K})^-_{odd}$ is injective by composition. However, as the degree q of $j : (\mathbb{K}_N, \tau_N) \to (\mathbb{K}, \tau)$ is odd then by (B1.4), $j_* : Cl(\mathbb{K}_N)_{(2)} \to Cl(\mathbb{K})_{(2)}$ is also injective. □

Theorem B3.4: *Let p be an odd prime, and let $j : \mathbb{Q}(\zeta_p)^{C_q} \to \mathbb{Q}(\zeta_p)$ be the inclusion of the fixed field $\mathbb{Q}(\zeta_p)^{C_q}$ under the Galois action of C_q where q is an odd divisor of $p-1$; then*

$$j_* : Cl\left(\mathbb{Q}(\zeta_p)^{C_q}\right)_{(2)} \oplus Cl\left(\mathbb{Q}(\zeta_p)^{C_q}\right)^-_{odd} \longrightarrow Cl(\mathbb{Q}(\zeta_p))_{(2)} \oplus Cl(\mathbb{Q}(\zeta_p))^-_{odd}$$

is injective.

Proof. Put $\mathbb{K} = \mathbb{Q}(\zeta_p)$. Then the units of finite order in $\mathbb{K}$ have order dividing $2p$ (cf [25], p. 536; also [68], p. 144). Now write q as a product $q = q_1 \dots q_N$ odd primes possibly with repetitions. As q divides $p-1$ it follows that $\mathbb{K}$ has no nontrivial units of any order $q_1^{e_1} \dots q_N^{e_N}$. For $1 \leq r \leq N$ put $\Gamma(r) = C_{q_1 \dots q_r} \subset C_q$ and put $\mathbb{K}_r = \mathbb{Q}(\zeta_p)^{\Gamma(r)}$ so that we have a filtration of CM-fields

$$\mathbb{Q}(\zeta_p)^{C_q} = \mathbb{K}_N \subset \mathbb{K}_{N-1} \subset \cdots \subset \mathbb{K}_1 \subset \mathbb{K}_0 = \mathbb{Q}(\zeta_p)$$

where $\mathbb{K}_r \subset \mathbb{K}_{r-1}$ has degree q_r. The conclusion now follows from (B3.3). □

We now have the theorem of Richard Hill [27].

Theorem B3.5: *Let p be an odd prime, let q be a divisor of $p-1$ and let $j : \mathbb{Q}(\zeta_p)^{C_q} \to \mathbb{Q}(\zeta_p)$ be the inclusion of the fixed field under the Galois action of C_q. If q is coprime to $h^+(p)$ then $j_* : Cl\left(\mathbb{Q}(\zeta_p)^{C_q}\right) \longrightarrow Cl(\mathbb{Q}(\zeta_p))$ is injective.*

Proof. Put $\mathbb{K} = \mathbb{Q}(\zeta_p)$ and $\mathbb{F} = \mathbb{Q}(\zeta_p)^{C_q}$ and consider first the special case where q is odd. By (B3.4), $j_* : Cl(\mathbb{F})_{(2)} \oplus Cl(\mathbb{F})^-_{odd} \longrightarrow Cl(\mathbb{K})_{(2)} \oplus Cl(\mathbb{K})^-_{odd}$ is injective so it suffices to show that $j_* : Cl(\mathbb{F})^+_{odd} \longrightarrow Cl(\mathbb{K})^+_{odd}$ is injective.

By (B2.3), $|Cl(\mathbb{K})^+_{odd}| = |Cl(\mathbb{K}_{\mathbb{R}})_{odd}| = |Cl(\mathbb{Q}(\zeta_p + \zeta_p^{-1})_{odd}|$ and this last divides $|Cl(\mathbb{Q}(\zeta_p + \zeta_p^{-1})| = h^+(p)$. By (B1.1) the norm mapping $N : Cl(\mathbb{K})^+_{odd} \to Cl(\mathbb{F})^+_{odd}$ is surjective so that $|Cl(\mathbb{F})^+_{odd}|$ divides $|Cl(\mathbb{K})^+_{odd}|$ and hence $|Cl(\mathbb{F})^+_{odd}|$ also divides $h^+(p)$. Now $N \circ j_* : Cl(\mathbb{F})^+_{odd} \to Cl(\mathbb{F})^+_{odd}$ is multiplication by q which, being coprime to $h^+(p)$ is, by the above, also coprime to $|Cl(\mathbb{F})^+_{odd}|$. It now follows from (B1.2) that $j_* : Cl(\mathbb{F})^+_{odd} \longrightarrow Cl(\mathbb{K})^+_{odd}$ is injective as required.

In the general case, write $q = 2^a t$ where $a \geq 1$ and t is odd. Put $\mathbb{E} = \mathbb{Q}(\zeta_p)^{C_t}$. Taking $j_1 : \mathbb{F} \hookrightarrow \mathbb{E}$ and $j_2 : \mathbb{E} \hookrightarrow \mathbb{K}$ to be the inclusions we have $j_* = (j_2)_* \circ (j_1)_*$. We note that, by the above special case, $(j_2)_* : Cl(\mathbb{E}) \to Cl(\mathbb{K})$ is injective. We note also that, as q is coprime to $h^+(p)$ and q is even, then $h^+(p)$ is odd. Now $\mathbb{F} \subset \mathbb{K}_{\mathbb{R}} = \mathbb{Q}(\mu_p)$ and as the norm homomorphism $N : Cl(\mathbb{Q}(\mu_p)) \to Cl(\mathbb{F})$ is surjective it follows that $|Cl(\mathbb{F})|$ divides $h^+(p)$. Hence $|Cl(\mathbb{F})|$ is odd. Now consider the norm mapping $N : Cl(\mathbb{E}) \to Cl(\mathbb{F})$. Then $N \circ (j_1)_* : Cl(\mathbb{F}) \to Cl(\mathbb{F})$ is multiplication by 2^a. As $|Cl(\mathbb{F})|$ is odd then $N \circ (j_1)_*$ is injective and so $(j_1)_* : Cl(\mathbb{F}) \to Cl(\mathbb{E})$ is injective by (B1.2). However, as we have already observed, $(j_2)_* : Cl(\mathbb{E}) \to Cl(\mathbb{K})$ is injective. Hence $j_* = (j_2)_* \circ (j_1)_*$ is injective as required. □

Appendix C

A sufficient condition for the D(2) property

In this Appendix, Λ will denote the integral group ring $\Lambda = \mathbb{Z}[G]$ of a finite group G. We first recall (cf [31], [33]) that an algebraic 2-complex is an exact sequence of Λ-modules of the form

$$0 \to J \to C_2 \to C_1 \to C_0 \to \mathbb{Z} \to 0$$

where each C_i is finitely generated and stably free over Λ. By a *strict algebraic 2-complex* we mean such a sequence in which each C_i is finitely generated and free. It is straightforward to see that, by internal stabilization, any algebraic 2-complex is chain homotopy equivalent to a strict algebraic 2-complex. An algebraic 2-complex is said to be *geometrically realizable* when it is chain homotopy equivalent over Λ to the cellular chain complex of a finite 2-complex with fundamental group G.

§C1: A sufficient condition:

$\Omega_3^{min}(\mathbb{Z})$ will denote the set of isomorphism classes at the minimal level of $\Omega_3(\mathbb{Z})$. We will prove:

Sufficient condition: If each $J \in \Omega_3^{min}(\mathbb{Z})$ is realizable and full then the $D(2)$-property holds for G.

We shall break down the proof into a number of steps arranged around the following statements :

R(2) : Every algebraic 2-complex over Λ is geometrically realizable.

D(2) : The $D(2)$-property holds for G.

Next consider the following statements where $J \in \Omega_3(\mathbb{Z})$:

RF(J) : J is realizable and full;

R$(2, J)$: Every algebraic 2-complex $0 \to J \to F_2 \to F_1 \to F_0 \to \mathbb{Z} \to 0$ is geometrically realizable.

§C2: $R(2) \Longrightarrow D(2)$:

In this section we will prove:

Theorem C2.1: *If every algebraic 2-complex over Λ is geometrically realizable then G has the $D(2)$-property.*

We note that the converse is also true, the two parts constituting the Realization Theorem of [31], [33].

In what follows, G will denote a finite group with integral group ring $\Lambda = \mathbb{Z}[G]$. We recall the following:

(C2.2) If $0 \to F \to A \to B \to 0$ is an exact sequence of Λ-modules in which F is free and B is finitely generated and torsion free over $\mathbb{Z}$ then $A \cong F \oplus B$.

Until further notice X will denote a finite 3-complex with $\pi_1(X) \cong G$ and universal cover $\widetilde{X}$ such that $H^3(X, \mathcal{B}) = H_3(\widetilde{X}, \mathbb{Z}) = 0$ for all local coefficient systems $\mathcal{B}$ on X. Putting $K = X^{(2)}$ then we have a pair of exact sequences

$$0 \to Z_3(\tilde{X}) \to C_3(\tilde{X}) \stackrel{\partial_3}{\to} \mathrm{Im}(\partial_3) \to 0;$$

$$0 \to \mathrm{Im}(\partial_3) \to Z_2(\tilde{X}) \to H_2(\tilde{X}; \mathbb{Z}) \to 0$$

which, spliced together give an exact sequence

$$0 \to Z_3(\tilde{X}) \to C_3(\tilde{X}) \stackrel{\partial_3}{\to} Z_2(\tilde{X}) \to H_2(\tilde{X}; \mathbb{Z}) \to 0.$$

As $K = X^{(2)}$ we have $Z_2(\tilde{X}) = H_2(\tilde{K})$, whilst $Z_3(\tilde{X}) = H_3(\tilde{X}; \mathbb{Z}) = 0$, so that the above exact sequence reduces to $0 \to C_3(\tilde{X}) \stackrel{\partial_3}{\to} H_2(\tilde{K}; \mathbb{Z}) \to H_2(\tilde{X}; \mathbb{Z}) \to 0$. By the Hurewicz Theorem, $\pi_2(K) \cong H_2(\tilde{K}; \mathbb{Z})$ and $\pi_2(X) \cong H_2(\tilde{X}; \mathbb{Z})$; thus:

(C2.3) There is an exact sequence $0 \to C_3(\tilde{X}) \stackrel{\partial_3}{\to} \pi_2(K) \to \pi_2(X) \to 0$. Consequently

(C2.4) $C_3(\tilde{X}) \cong \mathrm{Im}(\partial_3)$.

Now $\mathrm{Tor}(\pi_2(X)) = \mathrm{Tor}(H_2(\tilde{X};\mathbb{Z})) = \mathrm{Tor}(H^3(\tilde{X};\mathbb{Z}))$ so that, by the cohomology assumption on X, $\pi_2(X)$ is torsion free. We have a short exact sequence of finitely generated Λ-modules $0 \to C_3(\tilde{X}) \xrightarrow{\partial_3} \pi_2(K) \to \pi_2(X) \to 0$ in which $C_3(\tilde{X})$ is free, of rank n say. It follows from (C2.2) that:

(C2.5) $$\pi_2(K) \cong \pi_2(X) \oplus \mathrm{Im}(\partial_3) \cong \pi_2(X) \oplus \Lambda^n.$$

In particular, we see that

(C2.6) $\pi_2(X) \cong \pi_2(K)/\mathrm{Im}(\partial_3)$.

Moreover, as $\pi_2(K) \in \Omega_3(\mathbb{Z})$ then we also have:

(C2.7) $\pi_2(X) \in \Omega_3(\mathbb{Z})$.

We may refine the above analysis slightly. Firstly, since $H_3(\tilde{X};\mathbb{Z}) = 0$ the boundary map $\partial_3 : C_3(\tilde{X}) \to C_2(\tilde{X})$ is injective. Moreover, from the exactness of the following sequence of $\mathbb{Z}[G]$-modules $0 \to \pi_2(K)/\mathrm{Im}(\partial_3) \to C_2(\tilde{X})/\mathrm{Im}(\partial_3) \xrightarrow{\partial_2} C_1(\tilde{X})$. we have:

(C2.8) $\pi_2(X) \cong \mathrm{Ker}(\partial_2 : C_2(\tilde{X})/\mathrm{Im}(\partial_3) \to C_1(\tilde{X}))$.

We now have an exact sequence $0 \to \pi_2(X) \to C_2(\tilde{X})/\mathrm{Im}(\partial_3) \xrightarrow{\partial_2} C_1(\tilde{X})$ in which $\pi_2(X)$, being a representative of $\Omega_3(\mathbb{Z})$, is torsion free over $\mathbb{Z}$. However, $C_1(\tilde{X})$ is free over $\mathbb{Z}[G]$ and hence free over $\mathbb{Z}$. Thus $C_2(\tilde{X})/\mathrm{Im}(\partial_3)$ is also torsion free over $\mathbb{Z}$. From (C2.2) and the exact sequence

$$0 \to \mathrm{Im}(\partial_3) \to C_2(\tilde{X}) \to C_2(\tilde{X})/\mathrm{Im}(\partial_3) \to 0$$

we see that:

(C2.9) $C_2(\tilde{X})/\mathrm{Im}(\partial_3)$ is stably free over Λ.

It now follows that we have an algebraic 2-complex;

(C2.10) $\langle X\rangle = (0 \to \pi_2(X) \xrightarrow{j} C_2(\tilde{X})/\mathrm{Im}(\partial_3) \to C_1(\tilde{X}) \to C_0(\tilde{X}) \to \mathbb{Z} \to 0)$.
We can now prove:

Theorem C2.11: *If every algebraic 2-complex over $\Lambda = \mathbb{Z}[G]$ is geometrically realizable then G has the $D(2)$-property:*

Proof. We assume that every algebraic 2-complex over $\Lambda = \mathbb{Z}[G]$ is geometrically realizable. Suppose X is a finite 3-complex with $\pi_1(X) \cong G$ and universal cover $\widetilde{X}$ such that for all local coefficient systems $\mathcal{B}$ on X, $H^3(X, \mathcal{B}) = H_3(\widetilde{X}, \mathbb{Z}) = 0$. We must show that there is a finite 2-complex L which is homotopy equivalent to X. Let $\langle X \rangle$ denote the algebraic 2-complex constructed in (C2.10). By hypothesis, $\langle X \rangle$ is geometrically realized. This means there is a finite 2-complex L with $\pi_1(L) = G$ and a chain homotopy equivalence $f : \langle X \rangle \to C_*(L)$. As usual, let K denote the 2-skeleton of X; without loss of generality, we may suppose that

(i) $L^{(1)} = K^{(1)}$, and that
(ii) $f_r = \mathrm{Id} : C_r(\widetilde{K}) \equiv C_r(\widetilde{L})$ for $r \leq 1$.

Furthermore, assuming X has N cells of dimension three, we may write:

$$X = K \cup_{\alpha_1} E_1^{(3)} \cup_{\alpha_2} E_2^{(3)} \cdots\cdots \cup_{\alpha_N} E_N^{(3)}.$$

Observe that f induces an isomorphism $f_* : \pi_2(\langle X \rangle) = \pi_2(X) \stackrel{\simeq}{\to} \pi_2(L)$. The chain map $f \circ \nu : C_*(K) \to C_*(L)$ has the property that $(f \circ \nu)_r = \mathrm{Id} : C_r(\tilde{K}) \equiv C_r(\tilde{L})$ for $r \leq 1$. On may now construct a cellular map $g : K \to L$ such that

(i) $f_* = Id : \pi_1(K) \to \pi_1(L)$, and
(ii) $g_* = f_* \circ \nu_* : \pi_2(K) \to \pi_2(L)$.

In particular, since f_* is an isomorphism, we obtain.

$$\mathrm{Ker}(g_* : \pi_2(K) \to \pi_2(L)) = \mathrm{Ker}(\nu_*) = \mathrm{Im}(\partial_3 : C_3(\tilde{X}) \to \pi_2(K)) \cong \mathbb{Z}[G]^N.$$

Each homotopy class $[\alpha_j]$ belongs to $\mathrm{Im}(\partial_3) = \mathrm{Ker}(gi_* : \pi_2(K) \to \pi_2(L))$, so that $g : K \to L$ extends over the 3-cells of X to a map $h : X \to L$ such that

(iii) $h_* = Id : \pi_1(X) \to \pi_1(L)$ and
(iv) $h_* = f_* : \pi_2(X) \to \pi_2(L)$.

Since $H_r(\tilde{X}; \mathbb{Z}) = H_r(\tilde{L}; \mathbb{Z}) = 0$ for $2 < r$, we see by Whitehead's Theorem that h is a homotopy equivalence as required. □

§C3: RF(J) $\Longrightarrow$ R($2, J$).

In this section we will prove:

Theorem C3.1: *If $J \in \Omega_3(\mathbb{Z})$ is realizable and full then every algebraic 2-complex of the form $0 \to J \to F_2 \to F_1 \to F_0 \to \mathbb{Z} \to 0$ is geometrically realizable.*

Proof. Suppose $\mathcal{E} = (0 \to J \to E_2 \to E_1 \to E_0 \to \mathbb{Z} \to 0)$ is geometric realization of J and that $\mathcal{F} = (0 \to J \to F_2 \to F_1 \to F_0 \to \mathbb{Z} \to 0)$ is an algebraic 2-complex. Decompose $\mathcal{E}$ and $\mathcal{F}$ as Yoneda products

$$\mathcal{E} = \mathcal{E}_+ \circ \mathcal{E}_-, \mathcal{F} = \mathcal{F}_+ \circ \mathcal{F}_-$$

where

$$\mathcal{E}_+ = (0 \to J \xrightarrow{j} E_2 \to K \to 0) \quad ; \quad \mathcal{E}_- = (0 \to K \to E_1 \to E_0 \to \mathbb{Z} \to 0);$$

$$\mathcal{F}_+ = (0 \to J \to F_2 \to K' \to 0) \quad ; \quad \mathcal{F}_- = (0 \to K' \to F_1 \to F_0 \to \mathbb{Z} \to 0)$$

and first consider the special case where $K \cong K'$. Let $c(\mathcal{F}) \in \mathrm{Ext}^3(\mathbb{Z}, J)$ be the congruence class of $\mathcal{F}$ under the Yoneda interpretation of congruence for $\mathrm{Ext}^3(-,-)$ and let $c' \in \mathrm{Ext}^1(K, J)$ be the class corresponding to $c(\mathcal{F})$ under the dimension shifting isomorphism $\mathrm{Ext}^1(K, J) \cong \mathrm{Ext}^3(\mathbb{Z}, J)$ and let

$$\mathcal{P} = (0 \to J \to P \to K \to 0)$$

be the extension classified by c'; then there is a congruence in the Yoneda sense

$$\mathcal{F} \equiv \mathcal{P} \circ \mathcal{E}_-$$

so that, in particular, there is a chain homotopy equivalence $\mathcal{F} \simeq_{Id_{\mathbb{Z}}} \mathcal{P} \circ \mathcal{E}_-$. Now $\mathrm{Ext}^1(K, J) \cong \mathrm{Ext}^3(\mathbb{Z}, J) \cong \mathbb{Z}/|G|$. As each F_i is free then it follows from Swan's projectivity criterion that $c(\mathcal{F})$ corresponds to an element of $(\mathbb{Z}/|G|)^*$. Hence c' also corresponds to an element of $(\mathbb{Z}/|G|)^*$ and so again, by Swan's projectivity criterion, P is projective and $P \cong \underrightarrow{\lim}(\alpha, j)$ where $\alpha : J \to J$ is a Λ homomorphism defining an element of $\mathrm{Aut}_{\mathcal{D}\mathrm{er}}(J)$ and where, as above, $\mathcal{E}_+ = (0 \to J \xrightarrow{j} E_2 \to K \to 0)$. In particular, $S_J(\alpha) = [P] \in \widetilde{K}_0(\Lambda)$. However, the modules E_i, F_j are free so that $[P] = 0$ and hence $\alpha \in \mathrm{Ker}(S_J)$. By hypothesis, J is full and so there exists $\widetilde{\alpha} \in \mathrm{Aut}_\Lambda(J)$ such that $\nu(\widetilde{\alpha}) = \alpha$ where $\nu : \mathrm{Aut}_\Lambda(J) \to \mathrm{Aut}_{\mathcal{D}\mathrm{er}}(J)$ is the canonical homomorphism. We now have an isomorphism of exact sequences over Id_K as follows:

$$\begin{array}{ccccccccc} 0 \to & J & \xrightarrow{i} & E_2 & \to & K & \to 0 \\ & \downarrow \widetilde{\alpha} & & \downarrow \widehat{\alpha} & & \downarrow \mathrm{Id}_K & \\ 0 \to & J & \xrightarrow{\bar{i}} & P & \to & K & \to 0. \end{array}$$

Hence $\mathcal{E}_+ \cong_{\mathrm{Id}_K} \mathcal{P}$ and so $\mathcal{E} = \mathcal{E}_+ \circ \mathcal{E}_- \simeq \mathcal{P} \circ \mathcal{E}_- \simeq \mathcal{F}$. In particular, $\mathcal{F}$ is realized up to homotopy equivalence by $\mathcal{E}$. This completes the proof in the special case where $K \cong K'$.

If $K \not\cong K'$ then by iterating Schanuel's Lemma we have $K \oplus L \cong K' \oplus L'$ where $L \cong F_1 \oplus E_0$ and $L' \cong F_0 \oplus E_1$. Let $\mathcal{E}'$, $\mathcal{F}'$ be the complexes obtained from $\mathcal{E}$, $\mathcal{F}$ by internal stabilization thus:

$$\begin{aligned} \mathcal{E}' &= (0 \to J \to E_2 \oplus L \to E_1 \oplus L \to E_0 \to \mathbb{Z} \to 0) \\ \mathcal{F}' &= (0 \to J \to F_2 \oplus L' \to F_1 \oplus L' \to F_0 \to \mathbb{Z} \to 0). \end{aligned}$$

There are now Yoneda congruences $\mathcal{E} \equiv \mathcal{E}'$ and $\mathcal{F}' \equiv \mathcal{F}$. By the above special case, $\mathcal{E}' \simeq \mathcal{F}'$. Hence $\mathcal{E} \simeq \mathcal{F}$ and again $\mathcal{F}$ is geometrically realized. □

§C4: Proof of the Sufficient Condition:

We begin by proving:

(C4.1) If each $J \in \Omega_3^{\min}(\mathbb{Z})$ is realizable and full then each $J \in \Omega_3(\mathbb{Z})$ is realizable and full.

Proof. We first observe that, with regard to the tree structure of $\Omega_3(\mathbb{Z})$ described in Chapter Three, $\Omega_3(\mathbb{Z})$ is a fork with finitely many prongs as in **I** below with the degenerate possibility that $\Omega_3(\mathbb{Z})$ is straight as in **II** below:

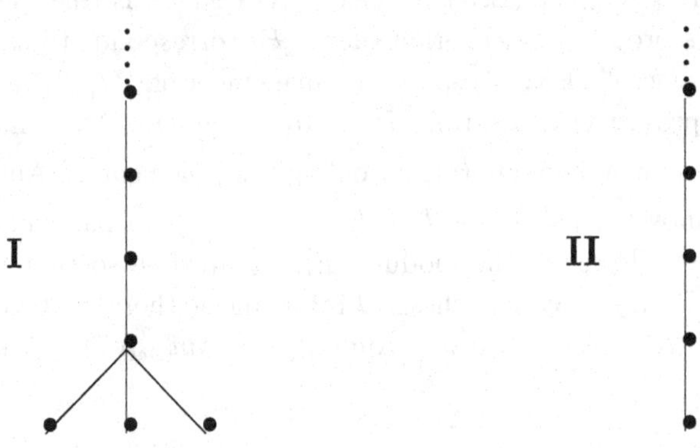

In either case there is no branching above level 1. Hence each $J \in \Omega_3(\mathbb{Z})$ has the form $J \cong J' \oplus \Lambda^m$ for some $J' \in \Omega_3(\mathbb{Z})$ and some $m \geq 0$. However, if $J \in \Omega_3(\mathbb{Z})$ is full then $J \oplus \Lambda^m$ is trivially full whilst if $J \in \Omega_3(\mathbb{Z})$ is realizable, then, by adding m trivial relations, $J \oplus \Lambda^m$ is also realizable. The conclusion now follows. □

From (C3.1) we now have:

(C4.2) If each $J \in \Omega_3^{\min}(\mathbb{Z})$ is realizable and full then each algebraic 2-complex over Λ is geometrically realizable.

From (C2.11) we now obtain the following which is the stated Sufficient Condition.

(C4.3) If each $J \in \Omega_3^{\min}(\mathbb{Z})$ is realizable and full then the $D(2)$ property holds for G.

References

[1] A.A. Albert; Structure of algebras; A.M.S. Colloquium Publications vol XXIV, American Mathematical Society, 1961.

[2] M.F. Atiyah and C.T.C. Wall; Cohomology of groups: Chapter IV of 'Algebraic Number Theory' (J.W.S. Cassels and A. Fröhlich eds), Academic Press (1967).

[3] M. Auslander and D.S. Rim; Ramification index and multiplicity; Illinois *J. Math.* **7** (1963) 566–581.

[4] R. Baer; Erweiterung von Gruppen und ihren Isomorphismen: *Math. Zeit.* **38** (1934) 375–416.

[5] S. Bentzen and I. Madsen; On the Swan subgroup of certain periodic groups: *Math. Ann.* **264** (1983) 447–474.

[6] B.J. Birch; Cyclotomic fields and Kummer extensions: Chapter III of 'Algebraic Number Theory' (J.W.S. Cassels and A. Fröhlich eds), Academic Press (1967).

[7] N. Bourbaki; Algèbre (livre 2, chapitre 8). Hermann (Paris) 1958.

[8] N. Bourbaki; Commutative Algebra. Hermann-Addison Wesley (1972).

[9] J.F. Carlson; Modules and Group Algebras: ETH Lecture Notes in Mathematics, Birkhäuser Verlag (1996).

[10] H. Cartan and S. Eilenberg; Homological Algebra: Princeton University Press (1956).

[11] P. Cassou-Noguès; Classes d'ideaux de l'algebre d'un groupe abelian. *Comptes Rendus Acad Sci Paris* **276** (1973) A973–A975.

[12] P. M. Cohn; Some remarks on the invariant basis property: Topology 5 (1966) 215–228.

[13] P.M. Cohn; On the structure of the GL_2 of a ring: Publications Mathematiques IHES **30** (1966) 5–53.

[14] C.W. Curtis and I. Reiner; Methods of Representation Theory; Wiley-Interscience, Volume I. 1981. Volume II. 1987.

[15] M.N. Dyer and A.J. Sieradski; Trees of homotopy types of two-dimensional CW complexes. *Comment. Math. Helv.* **48** (1973) 31–44.

[16] G.Eastlund; Projective modules over finite groups. Dissertation, University College London (2016).

[17] M. Eichler; Über die Idealklassenzahl total definiter Quaternionalgebren: *Math. Zeit.* **43** (1938) 102–109.

[18] J.D.P. Evans; Group algebras of metacyclic type: Ph.D Thesis. University College London. 2017.

[19] K.L. Fields; On the Brauer-Speiser Theorem. *Bull. A.M.S.* **77** (1971) 223.

[20] R.H. Fox; Free differential calculus V. *Ann. of Math.* **71** (1960) 408–422.

[21] A. Fröhlich and M.J. Taylor; Algebraic theory of numbers: Cambridge University Press (1991).

[22] S. Galovich, I. Reiner and S. Ullom; Class groups for integral representions of metacyclic groups. *Mathematika* **19** (1972) 105–111.

[23] L. Guthrie; The smallest example of a nontrivial projective Swan module. Dissertation, University College London (2014).

[24] K.W. Gruenberg and K.W. Roggenkamp; Decomposition of the augmentation ideal and of the relation modules of a finite group. *Proc. Lond. Math. Soc.* **31** (1975) 149–166.

[25] H. Hasse; Number Theory. Springer-Verlag (1962).

[26] A. Heller; Indecomposable modules and the loop space operation: *Proc. Amer. Math. Soc.* **12** (1961) 640–643.

[27] R.M. Hill; Private communication (August 2019).

[28] H. Jacobinski; Genera and decompositions of lattices over orders. *Acta. Math.* **121** (1968) 1–29.

[29] F.E.A. Johnson; Flat algebraic manifolds: (in "Geometry of low-dimensional manifolds: I" Proceedings of the 1989 Durham Conference, Conference, L.M.S. Lecture Notes vol. 150, pp. 73–91. C.U.P. 1991).

[30] F.E.A. Johnson; Explicit homotopy equivalences in dimension two. *Math. Proc. Camb. Phil. Soc.* **133** (2002) 411–430.

[31] F.E.A. Johnson; Stable modules and Wall's D(2) problem. I: *Comment. Math. Helv.* **78** (2003) 18–44.

[32] F.E.A. Johnson; Minimal 2-complexes and the D(2)-problem. *Proc. A.M.S.* **132** (2003), 579–586.

[33] F.E.A. Johnson; Stable modules and the D(2)-problem. LMS Lecture Notes In Mathematics, vol. 301. CUP 2003.

[34] F.E.A. Johnson; Homotopy classification and the generalized Swan homomorphism: *Journal of K-Theory*, **4** (2009) 491–536.

[35] F.E.A. Johnson; Syzygies and homotopy theory. Springer-Verlag 2011.

[36] F.E.A. Johnson; Syzygies and diagonal resolutions for dihedral groups. *Communications in Algebra* **44** (2015) 2034–2047.

[37] F.E.A. Johnson; A cancellation theorem for generalized Swan modules. *Illinois Journal of Mathematics.* **63** (2019) 103–125.

[38] F.E.A. Johnson and J.J. Remez: Diagonal resolutions for metacyclic groups. *J. Algebra.* **474** (2017) 329–360.

[39] I. Kaplansky; Modules over Dedekind rings and valuation rings. *Trans. A.M.S.* **72** (1952) 327–340.

[40] Y. Kida; l-extensions of CM-fields and cyclotomic invariants. *Journal of Number Theory* **12** (1980) 519–528.

[41] L. Klingler; Modules over the integral group ring of a non-abelian group of order pq. Memoirs A.M.S, no. 341 (1986).

[42] S. MacLane; Homology: Springer-Verlag, (1963).

[43] B. Magurn; An algebraic introduction to K-theory, Cambridge University Press (2002).

[44] J. Milnor; Groups which act on S^n without fixed points: *Amer. J. Math.* **79** (1957) 623–630.

[45] J. Milnor; Whitehead torsion: Bull. A.M.S. (72) 1966, 358–426.

[46] J. Milnor; Introduction to Algebraic K-Theory: Annals of Mathematics Studies vol 72, Princeton University Press 1971.

[47] R.J. Milgram; Evaluating the Swan finiteness obstruction for periodic groups: preprint, Stanford University 1979.

[48] A.J. Nadim. A periodic monogenic resolution. PhD Thesis, University College London 2015.

[49] T. Petrie; Free metacyclic group actions on homotopy spheres. *Ann. of Math.* **94** (1971) 108–124.

[50] R.S. Pierce; Associative Algebras: Graduate Texts in Mathematics no. 88 Springer-Verlag (1982).

[51] L.C. Pu. Integral representations of non-abelian groups of order pq. *Michigan Journal of Mathematics* **12** (1965) 231–246.

[52] I. Reiner; Maximal Orders. Academic Press, 1975.

[53] J.J. Remez; Diagonal resolutions for the metacyclic groups $G(pq)$. PhD Thesis, University College London 2012.

[54] J.J. Remez; An explicit diagonal resolution for a non-abelian metacyclic group. *Mathematika* **63** (2017) 499–517.

[55] D.S. Rim; Modules over finite groups. *Ann. of Math.* **69** (1959) 700–712.

[56] M. Rosen; Representations of twisted group rings. PhD Thesis, Princeton University 1963.

[57] H.J.S. Smith; On systems of linear indeterminate equations and congruences. *Phil. Trans.* **151** (1861) 293–326.

[58] E. Steinitz; Rechteckige Systeme und Moduln in algebraischen Zahlkörpen. *I: Math. Ann.* **71** (1911) 328–354; *II: Math. Ann.* **72** (1912) 297–345.

[59] R.G. Swan; Induced representations and projective modules: *Ann. of Math.* **71** (1960) 552–578.

[60] R.G. Swan; Periodic resolutions for finite groups: *Ann. of Math.* **72** (1960) 267–291.

[61] R.G. Swan; K-Theory of finite groups and orders. (Notes by E.G. Evans) Lecture Notes in Mathematics 149. Springer-Verlag 1970.

[62] R.G. Swan; Projective modules over binary polyhedral groups. *Journal für die Reine und Angewandte Mathematik.* **342** (1983) 66–172.

[63] R.G. Swan; Torsion-free cancellation over orders. *Illinois Journal of Mathematics* **32** (1988), 329–360.

[64] J. Vittis; The D(2)-problem for some metacyclic groups. PhD Thesis, University College London 2019.

[65] C.T.C. Wall; Finiteness conditions for CW Complexes: *Ann. of Math.* **81** (1965) 56–69.

[66] J.W. Wamsley; The deficiency of metacyclic groups: *Proc. A.M.S.* **24** (1970) 724–726.

[67] L.C. Washington; Introduction to cyclotomic fields. Graduate Texts in Mathematics vol 83, 2nd edition. Springer-Verlag, 1997.

[68] N. Yoneda; On the homology theory of modules: *J. Fac. Sci. Tokyo, Sec. I* **7** (1954) 193–227.

Index

Printed in the United States
by Baker & Taylor Publisher Services

Printed in the United States
by Baker & Taylor Publisher Services